THE BIOLOGY OF ANIMAL STRESS
Basic Principles and Implications for Animal Welfare

THE BIOLOGY OF ANIMAL STRESS
Basic Principles and Implications for Animal Welfare

Edited by

G.P. Moberg

and

J.A. Mench
Department of Animal Science
University of California
Davis
USA

CABI *Publishing*

CABI Publishing is a division of CAB International

CABI Publishing
CAB International
Wallingford
Oxon OX10 8DE
UK

Tel: +44 (0)1491 832111
Fax: +44 (0)1491 833508
Email: cabi@cabi.org
Web site: www.cabi-publishing.org

CABI Publishing
875 Massachusetts Avenue
7th Floor
Cambridge, MA 02139
USA

Tel: +1 617 395 4056
Fax: +1 617 354 6875
Email: cabi-nao@cabi.org

A catalogue record for this book is available from the British Library, London, UK.

Library of Congress Cataloging-in-Publication Data
The biology of animal stress: basic principles and implications for animal welfare / edited by G.P. Moberg and J.A. Mench.
p. cm.
Includes bibliographical references (p.).
ISBN 0-85199-359-1 (alk. paper)
1. Stress (Physiology) 2. Animal welfare I. Moberg, Gary P. II. Mench, Joy A.

QP82.2.S8 B55 2000
571.9'51--dc

ISBN 0 85199 359 1

First printed 2000
Reprinted 2001, 2005

Typeset by AMA DataSet Ltd, UK
Printed and bound in the UK by Biddles Ltd, King's Lynn

Contents

Contributors

J.L. Barnett, *Agriculture Victoria, Victorian Institute of Animal Science, Werribee, Victoria 3030, Australia*

F. Blecha, *Department of Anatomy and Physiology, College of Veterinary Medicine, 228 Veterinary Medical Sciences, 1600 Denison Avenue, Manhattan, KS 66506-5602, USA*

K. Carlstead, *National Zoological Park, Smithsonian Institution, Washington, DC 20008, USA* (present address: Honolulu Zoo, 151 Kapahulu Avenue, Honolulu, HI 96815)

C.J. Cook, *Technology Development Group, Horticultural and Food Research Institute of New Zealand, Private Bag 3123, Hamilton, New Zealand*

J.P. Capitanio, *Department of Psychology and California Regional Primate Research Center, University of California, One Shields Avenue, Davis, CA 95616, USA*

J.A. Carroll, *Animal Physiology Research Unit, Agricultural Research Service, United States Department of Agriculture, Room S-107 Animal Sciences Research Center, University of Missouri, Columbia, MO 65211, USA*

C.J. Dyer, *Animal Physiology Research Unit, Agricultural Research Service, United States Department of Agriculture, Room S-107 Animal Sciences Research Center, University of Missouri, Columbia, MO 65211, USA*

T.H. Elsasser, *USDA, Agricultural Research Service, Growth Biology Lab, Building 200, Room 120, BARC-East, Beltsville, MD 20705, USA*

N. Filipov, *Department of Physiology, School of Veterinary Medicine, University of Georgia, Athens, GA 30602, USA* (present address:

Department of Pharmacology, Albany Medical Center, Albany, New York)

P.J. Harris, *Technology Development Group, Horticultural and Food Research Institute of New Zealand, Private Bag 3123, Hamilton, New Zealand*

P.H. Hemsworth, *Animal Welfare Centre, University of Melbourne and Agriculture Victoria, Victorian Institute of Animal Science, Werribee, Victoria 3030, Australia*

J.R. Ingram, *Technology Development Group, Horticultural and Food Research Institute of New Zealand, Private Bag 3123, Hamilton, New Zealand*

K.C. Klasing, *Department of Animal Science, University of California, One Shields Avenue, Davis, CA 95616, USA*

J. Ladewig, *Department of Animal Science and Animal Health, Division of Ethology and Health, The Royal Veterinary and Agricultural University, Bülowsvej 13, DK-1870, Fredericksberg C, Denmark*

D.C. Lay Jr, *Department of Animal Science, Iowa State University, Ames, IA 50011, USA*

W.A. Mason, *California Regional Primate Research Center, University of California, One Shields Avenue, Davis, CA 95616, USA*

R.L. Matteri, *Animal Physiology Research Unit, Agricultural Research Service, United States Department of Agriculture, Room S-107 Animal Sciences Research Center, University of Missouri, Columbia, MO 65211, USA*

L.R. Matthews, *Animal Behaviour and Welfare Research Center, AgResearch, Ruakura Research Center, Private Bag 3123, Hamilton, New Zealand*

D.J. Mellor, *Animal Welfare Science and Bioethics Center, Institute of Food, Nutrition and Human Health, Massey University, Palmerston North, Private Bag 11222, New Zealand*

S.P. Mendoza, *Department of Psychology, University of California, One Shields Avenue, Davis, CA 95616, USA*

J.A. Mench, *Department of Animal Science, University of California, One Shields Avenue, Davis, CA 95616, USA*

G.P. Moberg, *Stress Research Unit, Department of Animal Science, One Shields Avenue, University of California, Davis, CA 95616, USA (deceased)*

T.G. Pottinger, *NERC Institute of Freshwater Ecology, Windermere Laboratory, Far Sawrey, Ambleside, Cumbria LA22 0LP, UK*

J. Rushen, *Dairy and Swine Research and Development Center, Agriculture and Agri-Food Canada, PO Box 90, Lennoxville, Quebec, Canada J1M 1Z3*

C.B. Schreck, *Oregon Cooperative Fish and Wildlife Research Unit, Biological Resources Division, USGS, Oregon State University, Corvallis, OR 97331-3803, USA*

D. Shepherdson, *Oregon Zoo, 4001 S.W. Canyon Rd, Portland, OR 97221, USA*

K.J. Stafford, *Animal Welfare Science and Bioethics Center, Institute of Veterinary Animal and Biomedical Science, Massey University, Palmerston North, Private Bag 11222, New Zealand*

F. Toates, *Department of Biology, The Open University, Walton Hall, Milton Keynes MK7 6AA, UK*

F. Thompson, *Department of Physiology, School of Veterinary Medicine, University of Georgia, Athens, GA 30602, USA*

T.L. Wolfle, *215 Severn Avenue, Annapolis, MD 21403, USA* (former Director of the Institute for Laboratory Animal Research, National Research Council, National Academy of Sciences)

Preface

The study of animal welfare continues to struggle with two persistent, interrelated problems: how to define animal welfare, and how to determine which measures should be used to evaluate welfare. One potential indicator of an animal's welfare is the presence or absence of stress. Considerable effort has been spent to determine if various management practises or conditions induce stress, resulting in the commonly held belief that any situation that results in stress should be avoided or prohibited. But because animals have evolved sophisticated behavioural and physiological mechanisms to deal with stress, stress jeopardizes the animal's welfare only if the stress results in some significant biological change that places that animal's well-being at risk. Unfortunately, many of the classical behavioural and physiological measures used to evaluate stress do not tell us if such meaningful biological changes have occurred, and thus the links between animal welfare and stress are still unclear.

In 1983, a major conference on animal stress was held at the University of California, Davis. By 1998 it seemed timely to get people together to readdress this topic. Although much had been learned in the interim, there were still many controversial and difficult areas, as well as new considerations and approaches to be discussed. The current volume is an outgrowth of that 1998 conference at Davis intended to explore the biology of animal stress and its implications for animal welfare.

Our goals in selecting speakers were to draw people from different disciplines working with a variety of vertebrate species in contexts ranging from farms to laboratories to zoos. The chapters in this volume similarly reflect this diversity. The first five chapters present summaries of the physiological, immunological, and behavioural responses to stress in animals. The next chapter provides information about some innovative, and non-invasive, methods for measuring stress. One of the most challenging problems in animal welfare is assessing the effects of long-term or intermittent physical or social stressors on animals, topics that are addressed in Chapters 7, 8 and 11. Our appreciation for the role that the animal's perception of the stressor plays in the stress response continues to grow, and this is addressed with respect to pain and behavioural responses in Chapters 8 and 9. Developmental events are also critical in later stress responding, and pre- and postnatal influences on the stress response are reviewed in chapters 12 and 13. The final chapters in the book tackle ways to ameliorate stress and improve welfare using genetic selection, environmental enrichment, and improved human–animal interactions.

Many individuals contributed to the success of the conference and the completion of this volume. The members of the international and local conference organizing committees, Ria deGrassi, Glenn Grey, Lynette Hart, Paul Hemsworth, Jan Ladewig, Sally Mendoza, Julie Morrow-Tesch, Edward Price, Dee Ann Reeder, Jeff Rushen, Carl Schreck, Carolyn Stull, and Philip Tillman, gave generously of their time and expertise in recommending speakers, helping with local arrangements, and reviewing manuscripts. Thanks are also due to Kathryn Bayne, Joseph Cech, Kannan Govindarajan, Joseph Garner, William Hershberger, Pamela Hullinger, Hank Kattesh, Donald Lay, Ruth Newberry, Catherine Rivier, Margaret Shea-Moore, Janice Swanson, and Tina Widowski, for their help in reviewing manuscripts for this book.

The conference could never have taken place without the able administrative support provided by Nicole Gibson and Jan Campbell of the Center for Animal Welfare and the organizational skills of Susan Kancir of the Dean's Office in the College of Agricultural and Environmental Sciences. Nor would it have been possible without the support of our conference sponsors and endorsers, the Center for Animal Welfare, the College of Agricultural and Environmental Sciences at the University of California, Davis, the USDA-CSREES National Research Initiative Competitive Grants Program, and the USDA W-173 Regional Research Committee on Animal Stress.

Special thanks are due to my Associate Director, Sue Heekin, who patiently edited and re-edited papers in this volume, tracked down errant references, checked proofs, and gently reminded authors about due dates. Without her help, preparing this book would have been a much more onerous task. On the other side of the pond, Tim Hardwick and Zoe Gipson from CAB *International* similarly waited patiently, and issued equally gentle reminders, while papers were edited and re-edited, references were tracked down, and galleys were checked, and we thank them for their encouragement and assistance.

My conference co-chair, co-editor, and friend Gary Moberg died unexpectedly before this volume could be published. Gary's influence on the fields of stress physiology and animal welfare cannot be overstated. His model of the biological costs of stress, and the risk-assessment approach that he developed, provided critical underpinnings to our understanding of the relationship between stress and animal welfare. His final, and most comprehensive, paper on these topics is the lead chapter in this volume. In addition to his writings, Gary had also a lasting personal influence on many colleagues throughout the world, and especially on the students that he taught and mentored. He was a warm, generous, humorous, enthusiastic, visionary and larger-than-life person who always challenged and stimulated. I miss Gary's friendship more than I can say. I hope that this volume stands as a small tribute to his many contributions.

Joy Mench
Professor of Animal Science and
Director of the Center for Animal Welfare
University of California, Davis

Biological Response to Stress: Implications for Animal Welfare

1

G.P. Moberg*

Stress Research Unit, Department of Animal Science, University of California, Davis, California, USA

Stress is a part of life and is not inherently bad. All life forms have evolved mechanisms to cope with the stresses of their lives. In fact we frequently seek stress, and we relish its biological effects as being exhilarating, even psychologically rewarding. This is why we ski, ride roller coasters and climb mountains. Yet no one denies that stress can have a damaging effect on the individual. We are only too aware of the human diseases associated with a stress-filled life, and we seem preoccupied about the toll that stress takes on us. Gradually we have come to accept that animals also suffer from the burden of stress, and that when suffering from stress they develop very similar pathologies. Like humans, while experiencing severe stress, animals can succumb to disease or fail to reproduce or develop properly (Moberg, 1985). It is recognition of these harmful effects of stress that has sensitized us to the importance of stress to an animal's welfare, or well-being. Our challenge is to differentiate between the little non-threatening stresses of life and those stresses that adversely affect an animal's welfare.

Because the term 'stress' has been used so broadly in biology, no clear definition of stress has emerged. Unlike most diseases, stress has no defined aetiology or prognosis. For this reason, our intuitive feelings about stress often guide our use of the term. For this discussion, I will define stress as the biological response elicited when an individual perceives a threat to its homeostasis. The threat is the 'stressor'. When the stress response truly threatens the animal's well-being, then the animal experiences 'distress'.

* Deceased

Previously, I have shied away from using the term distress (Moberg, 1987b); however, I now believe that the term distress helps us to differentiate between a non-threatening stress response (often referred to as 'good stress') and a biological state where the stress response has a deleterious effect on the individual's welfare (or 'bad stress'). Simply stated, our challenge is to determine when stress becomes distress.

Our second challenge is to determine how to measure stress and distress in animals. We have relied on a variety of endocrine, behavioural, autonomic nervous system and, more recently, immunological end-points to measure stress. Unfortunately, none of these measures has proved to be a litmus test for stress. One reason for their failure is that we loosely apply the term stress to a vast array of situations which often have little commonality. It is unreasonable to expect that a single indicator of stress will be appropriate for all types of stressors. Further complicating our use of these biological responses as measures of stress is that these systems frequently have comparable responses to both threatening and innocuous stimuli. For example, we have long associated the increased secretion of the adrenal glucocorticosteroid, cortisol, with stress, and investigators frequently cite an increase in circulating cortisol as a proof of stress occurring. Indeed, under carefully controlled experimental conditions cortisol can be a reliable indicator of stress. However, Colborn *et al.* (1991) found that stallions secreted similar amounts of cortisol whether the stallions were restrained, exercised or permitted to mate with a mare. With respect to the stallions' welfare, it is difficult to equate the effects of being restrained with mating. These data clearly illustrate the difficulty in simply using the secretion of cortisol or any other hormone to differentiate between non-threatening stress and distress.

Monitoring plasma cortisol or other physiological responses to diagnose stress becomes even more problematic when we attempt to evaluate stress outside the laboratory setting. The very act of monitoring these systems is stressful, confounding our interpretation of the results (Cook *et al.*, Chapter 6, this volume). Further complicating the measurement of stress is interanimal variability in the stress response (Moberg, 1985). As I will subsequently discuss, there are a variety of events that mould the stress response. The fact that different disease patterns emerge in animals experiencing the same stress has been attributed to differences in stress responses (Engle, 1967; Weiss, 1972; Henry, 1976, 1992).

When one considers all of the problems encountered in our attempt to measure stress, including what to measure, comparable biological responses to good and bad stress, difficulties in monitoring biological responses and interanimal variability, it is tempting to conclude that measuring stress is too daunting a problem. However, the impact of stress on animal welfare is too important to ignore. While controversy may swirl around how to define animal welfare, no one disagrees with the argument that suffering from stress threatens an animal's welfare. Therefore, in spite of the problems associated with measuring stress and defining distress, we must strive to develop an understanding of the biological response to stress.

Before we can address the challenges of how to differentiate biologically between stress and distress and how to diagnose stress, we must first develop a framework for organizing our current understanding of animal stress into a unifying and testable concept (for further discussion, see Moberg, 1985). Because there are many studies on stress involving a variety of species and numerous end-points, such a framework could assist us in focusing on the aspects of stress that are most relevant to the animal's welfare. To provide this framework, I will propose a model of animal stress to serve as the theoretical basis for my discussion on defining and recognizing distress in animals. I believe this model is applicable to understanding stress in all animal species, whether in humans, domestic and laboratory animals, wild animals or invertebrates. This model also provides an overview of the organization of subsequent chapters of this book.

Model of animal stress

The model of animal stress depicted in Fig. 1.1 evolved from an earlier model (Moberg, 1985) that divided the stress response into three general stages: the recognition of a stressor, the biological defence against the stressor and the consequences of the stress response. It is this last stage of the stress response that will determine whether an animal is suffering from distress or merely experiencing a brief episode in its life that will have no significant impact on its welfare.

A stress response begins with the central nervous system perceiving a potential threat to homeostasis. Whether or not the stimulus is actually a threat is not important; it is only the perception of a threat that is critical. That is why psychological stressors can be so devastating (McEwen and Stellar, 1993). They cause significant biological changes in the animal which may serve no value to the animal in alleviating the perceived stressor. Once the central nervous system perceives a threat, it develops a biological response or defence that consists of some combination of the four general biological defence responses: the behavioural response, the autonomic nervous system response, the neuroendocrine response or the immune response (Fig. 1.2).

In the case of many stressors, the first and undoubtedly most biologically economical response is a behavioural one. The animal may be successful in avoiding the stressor by simply removing itself from the threat. Thus an enemy may be avoided by escaping, or an animal may seek shade if its body temperature becomes elevated. Obviously, behavioural responses are not appropriate for all stressors, and animals may also find themselves in situations where their behavioural options are limited or thwarted. This is especially a concern when an animal's behavioural options are limited by

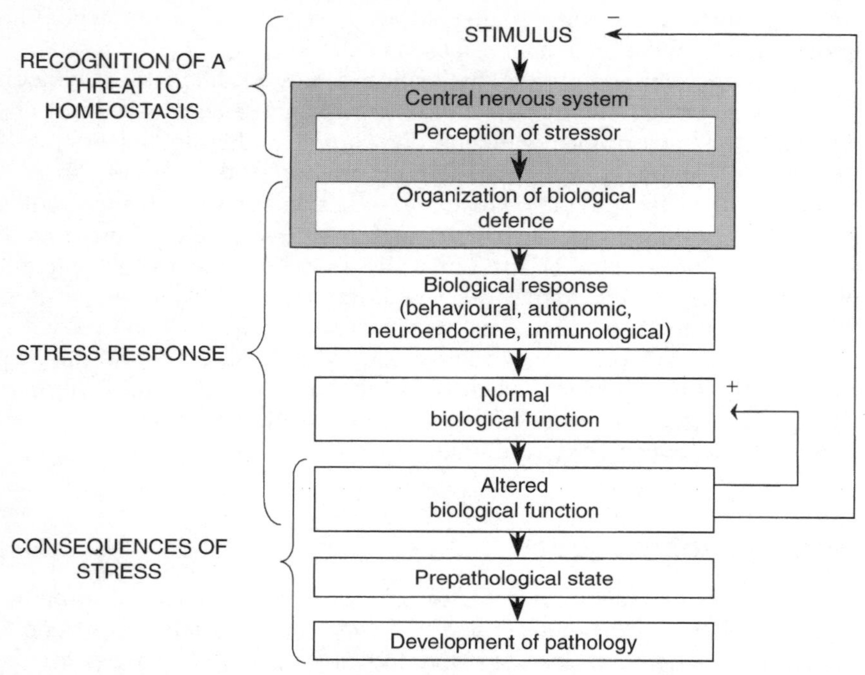

Fig. 1.1. A model of the biological response of animals to stress (reprinted with permission from Moberg, 1999).

confinement (Ladewig, Chapter 8, this volume). Even in those situations where a behavioural response will not alleviate the stressor, some component of behaviour may still be a part of every stress response. In those situations, behaviour may provide potential clues to distress (Bohus *et al.*, 1987; Rushen, Chapter 2, this volume). Unfortunately, our current lack of understanding of the behaviour of animals during stress limits the value of using behaviour as a means for predicting distress (Rushen, Chapter 2, this volume).

An animal's second line of defence during stress is the autonomic nervous system. This system was the basis of Cannon's proposed 'flight or fight' response during stress (Cannon, 1929). During stress, the autonomic nervous system affects a diverse number of biological systems, including the cardiovascular system, the gastrointestinal system, the exocrine glands and the adrenal medulla. The results are changes in heart rate, blood pressure and gastrointestinal activity, many of the physical signs we personally associate with stress. However, because the autonomic responses affect very specific biological systems and the biological effects are of relatively short duration, it might be argued that stress activation of the autonomic nervous system does not have a significant impact on an animal's long-term welfare. Furthermore, the value of monitoring the autonomic nervous system's

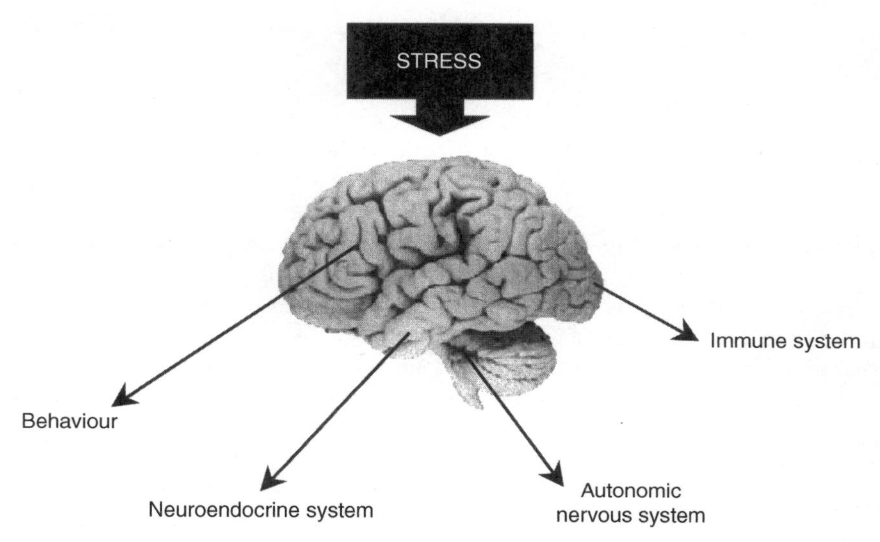

Fig. 1.2. The general types of biological responses available to the animal for coping with stress.

activity as a measure of stress is severely limited by the difficulty of measuring the system's response, especially in studies conducted outside the controlled confines of the laboratory. For this reason, the autonomic nervous system response during stress has been of only limited interest in the study of animal welfare.

In contrast to the effects of the autonomic nervous system, the hormones secreted from the hypothalamic–pituitary neuroendocrine system have a broad, long-lasting effect on the body. Virtually all of the biological functions that are affected by stress, including immune competence, reproduction, metabolism and behaviour, are regulated by these pituitary hormones. We know that the secretion of these pituitary hormones is altered either directly or indirectly during stress (Matteri *et al.*, Chapter 3, this volume). Stress-induced changes in the secretion of pituitary hormones have been implicated in failed reproduction (Moberg, 1987a; Rivier, 1995) and altered metabolism (Elsasser *et al.*, Chapter 4, this volume), immune competence (Blecha, Chapter 5, this volume) and behaviour. In the majority of stress studies, the hypothalamic–pituitary–adrenal (HPA) axis has been the primary neuroendocrine axis monitored. Increases in the circulation of the adrenal glucocorticosteroids (cortisol and corticosterone) have long been equated with stress. However, the secretion of prolactin and somatotropin (or growth hormone) has proven to be equally sensitive to stress. Likewise, thyroid-stimulating hormone and the gonadotropins (luteinizing hormone and follicle-stimulating hormone) are either directly or indirectly modulated by stress. There can be little question that the neuroendocrine system is one of the keys to our understanding of how stress alters biological function, resulting in distress.

We have long attributed the increased incidence of disease in animals suffering from stress to the suppression of immune competence during stress. Our early view was that, during stress, the immune system was modulated by other stress-responsive systems, especially the HPA. However we have recently come to appreciate the direct role that the central nervous system plays in regulating the immune system during stress, and also that the immune system in its own right is one of the major defence systems responding to a stressor (Dunn, 1988). Measurements of immune competence offer us a potentially powerful tool for evaluating the disease components of distress, but we await further developments in our understanding of the immune system in order to comprehend fully how to evaluate the system's response to stress.

As stated earlier, one of the challenges in stress biology is to develop clinical measures of stress and distress. Based on the importance of these four defence systems during stress, it seems logical to conclude that the solution to this problem would be to monitor these stress defence systems. However, as discussed, the technical challenges of measuring these systems without stressing the animal have hampered this approach. There are two additional problems complicating the use of such measurements. First, although all four biological defence systems are available to the animal to respond to a stressor, not all four are necessarily utilized by the animal to defend its homeostasis against a stressor. Contrary to Selye's prediction (1950), there is no non-specific stress response that applies to all stressors. As Mason (1968, 1975) so elegantly demonstrated, different stressors elicit very different types of biological responses. Therefore, if we are going to monitor these systems to diagnose stress we will need specific measures for each stressor.

Perhaps the greatest problem in measuring stress is interanimal variations in the stress response. Even when faced with the same stressor, each animal's central nervous system will use a combination of stress responses to cope with the stressor (Moberg, 1985). We now recognize that which of the four defence systems is used in response to a stressor is influenced by a variety of factors. These factors, or modifiers (Fig. 1.3), may influence how an animal perceives a stimulus and whether or not it views the stimulus as a threat to its homeostasis. In addition, these modifiers shape the animal's organization of its biological defences. Early experience (Moberg, 1985; Mason *et al.*, 1991; Lay, Chapter 12, this volume; Mason, Chapter 13, this volume), genetics (Marple *et al.*, 1972), age (Blecha *et al.*, 1983), social relationships (Henry, 1992) and human–animal interactions (Hemsworth *et al.*, 1981; Hemsworth and Barnett, Chapter 15, this volume) have all been found to modify the nature of the stress response. Under laboratory conditions, we may be able to identify these modifiers and account for their influence on the stress response. In fact, we sometimes capitalize on the differences in

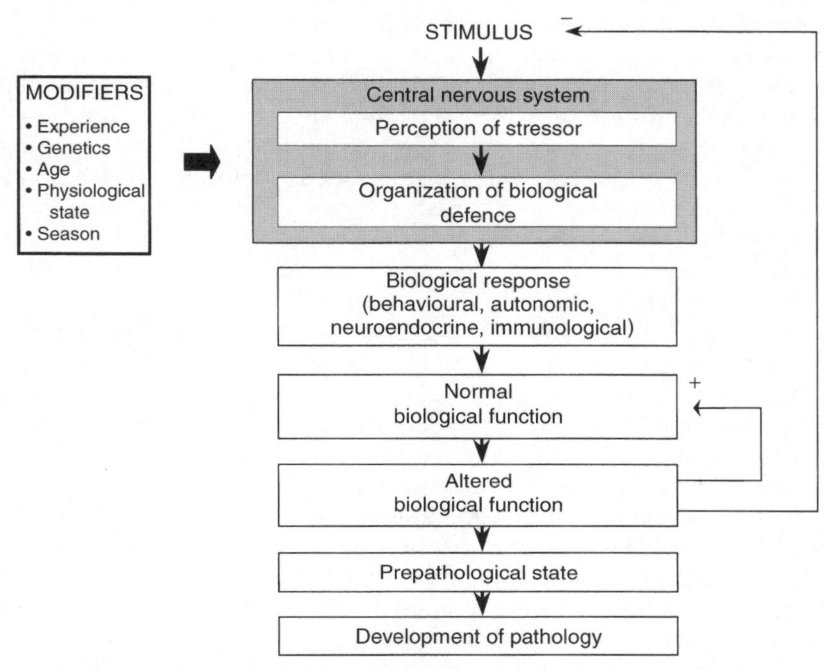

Fig. 1.3. Modifiers of the central nervous system's perception of a stressor and its organization of a biological defence.

biological responses to a stressor to develop biomedical models for studying depression or psychological development. However, outside the laboratory, it becomes almost impossible to account for these interanimal differences in the biological responses to stress. If we were asked to evaluate stress in a herd of cattle, or a population of wild animals, for example, there is no way that we can account for each animal's prior experiences, its social relationship with its peers or its genetic predisposition to have a heightened stress response. This example further underscores the problems we face in measuring the defence systems during stress. However, we must remember that it is not the type of biological defence an animal utilizes that is ultimately important to its welfare, but the resulting change in biological function that determines if there is a threat to an animal's welfare.

All of the biological defenses used to cope with a stressor alter biological function (Fig. 1.1). The autonomic nervous system may increase heart rate. Secreted glucocorticoids such as cortisol will alter glucose metabolism, increasing the blood concentration of glucose. A behavioural response alters ongoing behavioural patterns. It is these stress-induced changes in biological functions that will directly affect the animal's welfare. Whether or not the stress-altered functions are beneficial in helping the animal to cope is not important; the changes in biological function during stress result in a shift of biological resources away from biological activities occurring before the stressor. For example, energy originally utilized for growth or reproduction might be needed by the animal to cope with the stress. This change in biological function during stress is the 'biological cost of stress'. Hippocrates refers to the *ponos* of disease as the work the body exerts to cope with disease (Dubos, 1959). The *ponos* of stress, the work of stress, is the shift in biological functions away from non-stress functions or activities in order for the animal to cope with the stressor.

For most stressors the biological cost is negligible because the stressors are short-lived. During prolonged stress or when stress is severe, the biological cost is significant and the work of stress becomes a significant burden to the body. It is during such stress that the animal enters the next stages of stress: prepathology and pathology (Fig. 1.1). The prepathological state occurs when the stress response alters biological function sufficiently to place the animal at risk of developing pathologies. The most obvious example is infectious disease. The change in biological function occurring during a stress response may suppress immune competence, rendering the animal susceptible to pathogens that may be present in the environment. If the animal succumbs to these pathogens and becomes ill, it enters a pathological state. For example, the increased incidence of respiratory disease observed in cattle being transported is attributed to a suppression of the immune system caused by transportation stress (Blecha, Chapter 5, this volume). The longer an animal is stressed the longer the animal is in a prepathological state and the greater is the opportunity for a pathology to develop. However, when considering stress, we must not restrict our view of pathology to disease but use the term in the broadest sense. Disease is only one type of pathology that occurs during prolonged stress. When metabolism is shifted during stress, growing animals no longer grow normally. Stress can suppress reproduction or can result in deleterious behaviours such as tail biting in swine or self-mutilation in monkeys (Moberg, 1985). All are the pathologies of stress.

From our model of stress (Fig. 1.1), it is evident that the stress response is a cascade of biological events, the nature of which may vary between animals. Relying on one, or even two, of these biological defence responses to measure stress is not a reliable approach to evaluating animal welfare. I believe that we should not focus on the nature of the biological defences used during stress, but on their impact on the animal. It is the change in biological function that is important to welfare, not the mechanism that induces the change. By this argument, we could use the development of the pathological state to indicate a threat to welfare. However, depending on the development of pathologies to measure animal welfare is both impractical and inhumane. Therefore, I believe we must focus on the stress-induced changes in biological function as a measure of welfare because it is this biological cost of stress that is the key to understanding when stress becomes distress, placing an animal's welfare at risk.

When stress becomes distress

The key to differentiating distress from a non-threatening stress is the biological cost of the stress. Animals have evolved mechanisms to cope with the short-term stressors of their lives. An animal escapes from a predator or it dies. Survival is what is important, and the cost of stress is not important to the animal. However, this stress response is of little consequence to the animal because the stress is brief. The biological cost is minimal and non-threatening because sufficient reserves of biological resources exist to cope with the stressor and meet the biological cost of the stress. This concept is illustrated in Fig. 1.4, where the biological cost is met by reserves with no impact on other biological functions. An example is the body's utilization of glycogen stores during stress. The catecholamines secreted during stress convert the glycogen to readily utilizable glucose or other metabolic products required for gluconeogenesis. Once the stressor is alleviated, the glycogen stores are quickly replenished by gluconeogenesis to pre-stress levels. The overall biological cost is inconsequential to the animal's welfare.

When there are insufficient biological reserves to satisfy the biological cost of the stress response, then resources must be shifted away from other

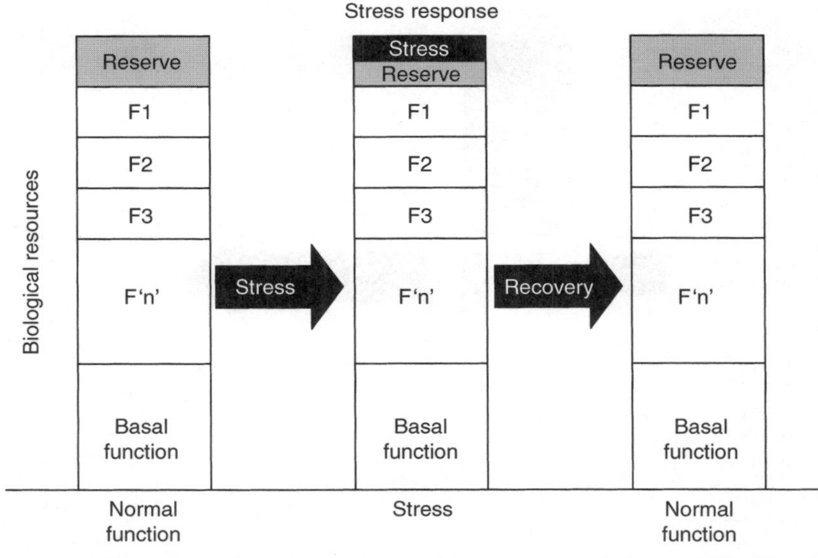

Fig. 1.4. A hypothetical scheme of how mild stress diverts biological resources. In this scheme, biological resources are arbitrarily assigned to various biological functions (F1–F′n′). During mild stress, only reserve resources are used to cope with the stressor. The total stress response extends from the time biological resources are diverted until the reserves have been replenished (reprinted with permission from Moberg, 1999).

biological functions. As shown in Fig. 1.5, when sufficient resources are diverted from these functions, these biological functions are impaired. For example, when stress shifts metabolism away from growth, the young animal no longer thrives and growth is stunted. When energy is shifted from supporting reproduction, reproductive success is diminished. In these cases, animals have entered the prepathological–pathological states and are experiencing distress (Fig. 1.1). This period of distress will last until the animal replenishes its biological resources sufficiently to restore normal functions (Fig. 1.5).

Distress can result from both acute and chronic stress. The physiological mechanisms utilized by these two types of stressors are similar, but we differentiate between acute and chronic stress based on our interpretation of the duration of the stressor. Acute stress is usually considered to be a relatively brief exposure to a single stressor. Although brief in nature, the biological cost of the stressor may be sufficient to alter biological functions and induce distress. Acute stress disrupts biological function by two fundamentally different mechanisms of action: by disrupting critical biological events, or by diverting biological resources away from other biological functions.

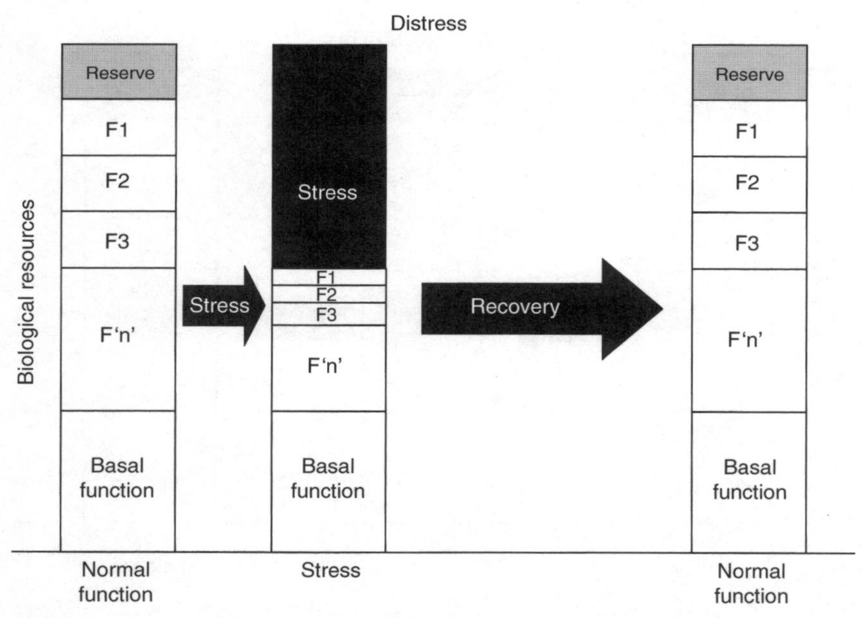

Fig. 1.5. A hypothetical scheme of how the diversion of biological resources necessary to cope with severe stress significantly impairs other biological functions, leading to distress. As compared with mild stress (Fig. 1.4), the biological cost of distress requires a much longer recovery period (reprinted with permission from Moberg, 1999).

In the body, there are several biological events that rely on critical timing for success. If this critical timing is disturbed, normal function is lost. Ovulation is perhaps one of the best examples. Successful ovulation depends on an exquisite timing between the preovulatory release of the luteinizing hormone (LH) to induce ovulation and the expression of oestrus behaviour. If this timing is disrupted, the opportunity to reproduce is lost (Moberg, 1987a). It is for this reason that ovulation is especially sensitive to stress and provides an excellent example of how an acute stressor can cause distress by blocking a critical biological event.

Acute stressors such as restraint, transportation or isolation have been found to disrupt the events surrounding ovulation (Moberg, 1987a). One possible physiological mechanism by which stress prevents the normal preovulatory secretion of LH, an essential step leading to ovulation, is activation of the HPA (Moberg, 1987a; Nangalama and Moberg, 1991). There are several possible mechanisms by which stress activation of the HPA may block successful ovulation. Stress activation of the HPA begins with the secretion of corticotropin-releasing hormone (CRH) from the neurosecretory cells in the hypothalamus. Recent evidence indicates that in rats and primates (Rivier, 1995) CRH acts at the hypothalamic level of the central nervous system to inhibit the secretion of gonadotropin-releasing hormone (GnRH) which stimulates LH secretion. However, it is possible that this central effect of CRH on GnRH secretion may not occur in all species (Naylor *et al.*, 1990). Another mechanism arises from the primary function of CRH, which is to stimulate the pituitary secretion of adrenocorticotropic hormone (ACTH). This pituitary hormone in turn induces the synthesis and secretion of the adrenal glucocorticosteroids such as cortisol. Increases in circulating cortisol during stress have been linked to reduced secretion of LH (Moberg, 1987a) and offer the second plausible way in which acute stress may block the preovulation secretion of LH (Nangalama and Moberg, 1991). Either of these two mechanisms could determine how a stressor could disrupt the critical events needed for successful ovulation. If the same acute stressor occurred at another time in the reproductive cycle, it would probably have no effect on reproduction and might not even be a significant biological cost to the animal. However, because of unfortunate timing, a stress response of the HPA (whose primary role is to shift carbohydrate metabolism to maintain energy during the stress) has the secondary, unrelated effect of disrupting critical events that are essential for successful reproduction. In this manner, an acute stressor causes distress by forcing the animal into a prepathological state (failure to secrete LH) that results in pathology (failure to ovulate and the loss of an opportunity to reproduce).

The second way in which acute stress leads to distress is by shifting biological resources away from other critical functions. Although the acute stress may be relatively brief, the biological cost of the stress may be of sufficient magnitude that there are inadequate biological resources in reserve to meet this challenge. Resources must be diverted away from other biological functions in order to cope (Fig. 1.5). The growing animal provides an example of how acute stress could cause distress by such a significant shift in resources. During its rapid growth phase, the young animal must devote considerable biological resources to support its growth. A stress-induced shift in metabolic resources away from supporting growth might impair the animal's development. To test this hypothesis, we measured the metabolic consequences of a single 4-h period of restraint in 31-day-old mice. Twenty-four hours after exposure to this acute stressor, there was a significant reduction in the mice's growth rate and a marked loss in both lean energy and fat energy reserves as compared with non-stressed controls. Furthermore, it took an additional 24 h before these metabolic parameters returned to control levels (Laugero and Moberg, unpublished). These findings demonstrate that exposure to a relatively brief acute stressor can have an extended effect on biological function because of the stress-induced shift in biological resources away from growth. In this example, the acute stress caused distress by shifting biological resources, which resulted in inadequate metabolic resources (prepathological state) and thus retarded normal growth (pathology).

We usually assume that an animal undergoing chronic stress is experiencing a long-term, continuous stress. In fact, it is difficult to envisage a single chronic stressor in non-experimental animals, with the possible exception of unrelenting pain or prolonged exposure to an extreme environmental condition such as severe cold. For most animals, as for most people, chronic stress undoubtedly results from experiencing a series of acute stressors whose accumulative biological cost forces the animal into a prepathological state, and possibly leads to a pathological condition. Such an accumulation of biological costs might result from repeated exposure to the same acute stressor, or might be the consequences of the summation of several active stressors' biological costs (Moberg, 1985).

Figure 1.6 suggests how an animal experiencing repeated exposure to the same stressor, without any opportunity to replenish its biological resources, could accumulate sufficient biological cost to affect other biological functions. The result of this accumulated biological cost would be a state of distress. To test this scheme, we again utilized the growing mouse as a model. Four hours of restraint each day for 7 days significantly suppressed growth, lean and fat energy deposition, and total heat energy production compared with exposure to the acute stress for only 1 or 3 days (Laugero and Moberg, unpublished). As predicted by the model for chronic stress (Fig. 1.6), repeated exposure to the acute stressor resulted in an accumulated biological cost, similar to what we would predict if the animal was exposed to a prolonged, unrelenting stressor. Undoubtedly, it is such exposure to a series of acute stressors that accounts for most chronic stress in animals.

Using the concept of the total biological cost of stress, we can envisage how two acute stressors acting simultaneously on an animal could individually cause a shift in biological resources to result in an accumulative cost to

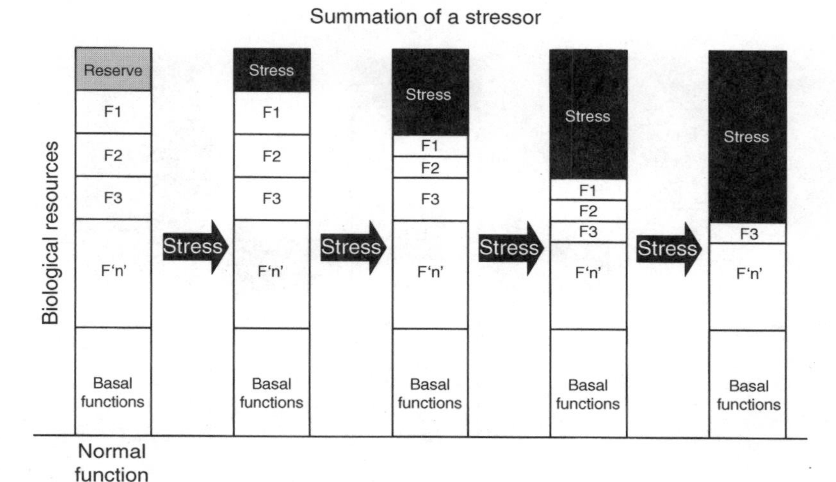

Fig. 1.6. A hypothetical scheme of the accumulative biological cost for an animal experiencing repeated exposure to the same stressor. While the first exposure may not result in the diversion of sufficient biological resources to affect other functions (F1–F′n′), with repeated exposure to the same stressor, the accumulative cost in biological resources may be sufficient to have a severe impact on or eliminate other biological functions (reprinted with permission from Moberg, 1999).

the animal that would have an impact on other functions, causing distress. For example, if an animal is already experiencing significant stress and then is exposed to the biological effects of a second stressor (Fig. 1.7), distress may result. To test this hypothesis, we restrained growing mice for 4 h per day for 7 days. On days 6 and 7 of the restraint treatment, the mice were injected with lipopolysaccharide (LPS) as a model for the stress of a mild infection. The combined cost of the two stressors (restraint plus LPS) suppressed growth to a greater extent and consumed more metabolic resources than either stressor acting alone (Laugero and Moberg, unpublished). In this case, the total biological cost of the two stressors was a summation of the biological resources needed to cope with each of the individual stressors. Since the period of distress would continue even after the stressors ceased and until the animals could restore biological functions to pre-stress levels, the magnitude of the stress response and its impact on function are sufficiently great to meet the criteria of a chronic stress.

Focusing on the biological cost of stress, we can argue that acute or chronic stress results in distress when the stress response shifts sufficient resources to impair other biological functions. When this occurs, the animal enters the prepathological state, is at risk of developing a pathological state and experiences distress. If the biological cost of the stress does not require the amount of resources that would have an impact on other functions, the

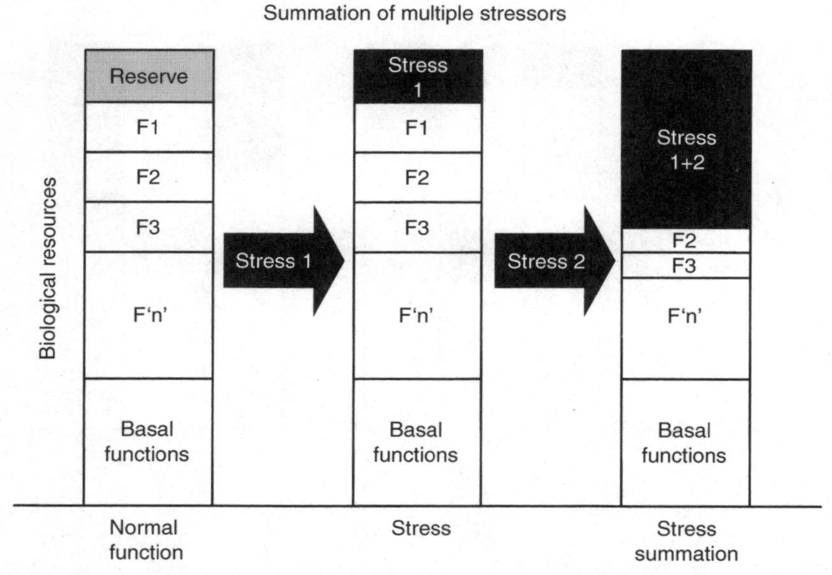

Fig. 1.7. A hypothetical scheme showing how the biological cost of two stressors can summate to impair normal function. As illustrated, exposure to only one stressor might not require the diversion of biological resources needed for other functions. If, however, two of these stressors occur simultaneously, the total cost may have a severe impact on other biological functions (reprinted with permission from Moberg, 1999).

animal is not at risk and does not experience distress. Using this definition of distress, we can see how the little hassles of life, good or bad, do not cause distress. This explains why the biological responses to various stimuli, good or bad stress, may be similar but the eventual effects on the animal differ. For example, the increase in cortisol observed during play and pain may be similar, but the increase in cortisol becomes critical to the animal's welfare only if the increase in the hormone is elevated for a sufficient time to cause a significant biological cost by shifting energy away from other biological functions. If this happens, it causes distress. If the response to the increase in cortisol does not affect other functions, then the stress response is just one of these little things in life.

Threat of subclinical stress

In the past, I have suggested that subclinical stress may increase the risk of distress (Moberg, 1985, 1999). A subclinical stress does not shift sufficient resources to affect normal biological functions and as a result does not cause distress. Because there is no change in function, there is no clinical indication of the stress; thus, the term 'subclinical stress'. However, such subclinical stress may consume sufficient resources to make the animal vulnerable to the effect of a second subclinical stress (Fig. 1.8). Either stressor by itself would have no effect on biological function, but their accumulated biological cost would result in distress.

This subclinical stress model might explain why natural animal populations suddenly decline when exposed to an environmental stressor that normally would be of no consequence to the animals. For example, a fish population exposed to low levels of an environmental toxin might not display any obvious clinical signs of this low-level stress. Yet, biological resources would need to be used for the fish to cope with this stressor. Upon being exposed to a second stressor, such as an infectious agent or reduction of oxygen levels in the water, the fish would not have sufficient reserves to cope with this second stressor and could survive only by shifting resources away from other functions. Since reproductive success in fish is closely related to energy reserves, we can envisage how meeting the total biological cost of these two subclinical stressors could deprive the fish of the energy needed for normal reproductive success. The end result would be an unexplained decline in the population. In this manner, subtle environmental stressors may well affect population dynamics and explain the unexpected decline of populations.

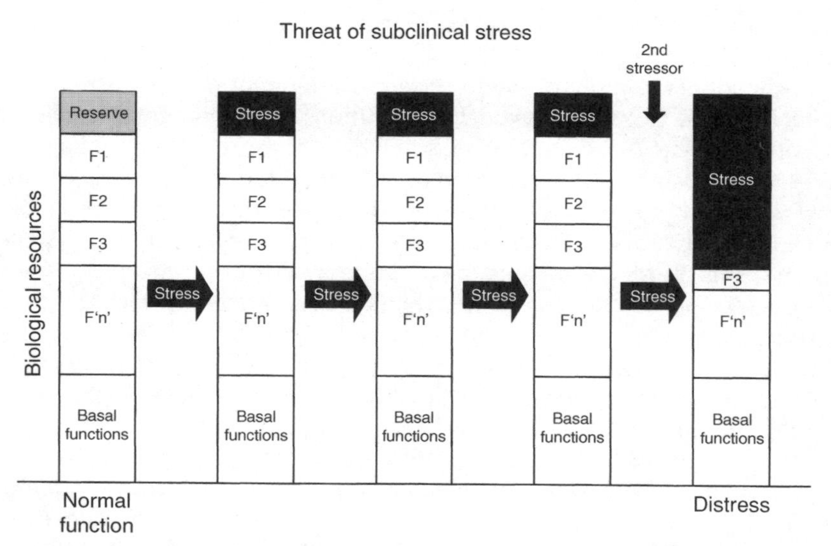

Fig. 1.8. A hypothetical scheme showing how subclinical stress may deplete biological resources without impairing other biological functions. This depletion of biological reserves makes the animal vulnerable to a second stressor whose additional biological cost can only be met by diverting resources from other biological functions.

There are limited data supporting the argument that subclinical stressors can make animals vulnerable to other stressors. A number of years ago, Holmes and co-workers (1979) observed that mallard ducks that consumed food contaminated with petroleum products showed no apparent symptoms of distress and no clear changes in body weight. However, when the ducks were exposed to a second mild, cold stressor, those consuming the petroleum-contaminated food had higher mortality. Likewise, Atlantic salmon previously exposed to predators had significantly higher mortality when exposed to an osmotic stressor than the fish not exposed to the predator (Järvi, 1989). In both of these examples the animals were able to cope with a subclinical stressor, but no longer had the biological reserves to meet the cost of the second stressor.

Animals confined in the laboratory or in intensive livestock operations are perhaps even more vulnerable to the subtle effects of subclinical stress (Moberg, 1996, 1999; Ladewig, Chapter 8, this volume). By its very nature, confinement restricts or eliminates behavioural options, results in greater human intrusion into the animals' lives and forces greater peer interactions or, possibly, even removes the animal from any social contact with its peers. Compounding these social aspects of confinement are the effects of the numerous husbandry practices such as vaccination, restraint, movement between cages and novel diets. Attempting to respond to these various minor stressors could result in a significant biological cost to the animal. If not carefully managed, these costs will result in a subclinical stress load that may place the animal at risk from what otherwise might be considered as rather minor inconsequential stressors.

In laboratory animals, one of the most important effects of subclinical stress might be to cause unexpected alterations in experimental results. For example, if an animal experiencing subclinical stress resulting from a recent series of husbandry activities (such as handling, removal to a clean cage, etc.) was then injected with an antigen intended for antibody production, the additional stress of the immune reaction might result in unanticipated distress. This distress might alter the immune response and cause unforeseen and even erroneous results. Previously, I discussed how a single 4-h period of restraint impaired a mouse's metabolic condition for 2 days. Yet, periods of relatively brief restraint are frequently used in research with no consideration of the lingering effects, which in this case could affect how the animal responds to other end-points being studied 1 or 2 days later.

If we have difficulty measuring stress and distress, detection of the subtle effects of subclinical stress presents an even greater problem. In addition, we are still struggling to understand fully the subtle effects of subclinical stress on the biology and welfare of the animal. In spite of these problems, it is essential for us to address this issue. As we begin to recognize the importance of stress in wild animals, we may find that subclinical stress contributes to unexplained declines in animal populations. With respect to confined animals, recognition of subclinical stress must become an important component of development of husbandry practices for the management of stress and distress in animals.

Management of stress and distress

Since stress in life is unavoidable, we can never expect to develop conditions that will always keep our animals stress free. Frequently in animal welfare discussions it is implied that any stress is unacceptable and that stress must be avoided at all costs. This is not possible, and we should understand that there is nothing inherently bad about stress, unless it results in distress. Since stress is a part of domestic animal production, maintaining animals in zoos, or using laboratory animals, we must develop husbandry practices to manage stress in these animals just as we manage their reproduction or nutrition (Moberg, 1992, 1993). The key to developing management strategies for stress is to focus on reducing the biological cost of stress. Our general strategy for managing stress should be to minimize the cost of any stress and certainly never to allow the cost of stress to rise beyond the level of a subclinical stress. Even then, we must strive to keep the period of subclinical stress as brief as possible. There are a number of strategies available to us to accomplish this goal.

Animal behaviour offers one attractive approach to managing stress. Any stress response begins with the perception of a stressor (Fig. 1.1). Convincing the animal that no threat exists or diverting its attention from the stimulus may eliminate the stress response. For example, placing visual barriers in cages housing groups of monkeys to reduce visual contact between dominant and subordinate monkeys can reduce stress. Comparable approaches may be applicable to other specific stresses. Another potential approach to reducing stress is to engage the animals in other behaviours during stress. Dantzer and Mormede (1981) found that the amount of cortisol secreted during the frustration of experiencing intermittent feeding schedules could be reduced by providing pigs with chains suspended into their pens from the ceilings. The pigs would chew and pull on the chains during the period of frustration, and the activity seemed to alleviate some of the stress. It would seem that electing similar adjunctive behaviours might be beneficial in reducing the biological costs of other stressors. Remembering that a behavioural response to a stressor probably comes at a lower cost to the animal than activation of the other stress defence systems, behavioural solutions may be a preferable approach for us to use.

Another approach to managing stress is to capitalize on the modifiers of the stress response (Fig. 1.3). I believe that manipulation of such modulating factors can be used as tools to ameliorate the stress response of animals. Both genetics and early experience serve as examples of this approach. In turkeys, selection for altered stress responses can improve the survival of animals exposed to cold stress (Brown and Nestor, 1973). Stress-susceptible strains of swine may show different physiological responses during environmental stress than non-susceptible breeds (Marple *et al.*, 1972). Genetic selection of domestic animals to make them more adaptable to confinement offers a potential tool for the management of stress (Pottinger, Chapter 14, this volume).

Likewise, since an animal's previous experience may influence its biological response to production conditions, manipulation of experience may serve as a management tool to ameliorate the stress response. In swine, Hemsworth *et al.* (1981) and Hemsworth and Barnett (Chapter 15, this volume) observed that the quality of human handling influences growth in young pigs, behaviour and the blood concentrations of the stress-sensitive corticosteroid hormones. Pigs exposed to unpleasant handling treatment grow more slowly and appear to exhibit indications of elevated stress responses. In domestic animal production, we have only begun to use such modifications of the stress response to alter how our animals cope with stress, but I believe that this area offers one of the most productive areas for future research on stress management.

Even if stress can be kept in the subclinical range, care must be taken not to expose the animals to additional minor stressors whose additional biological cost could induce a prepathological state, and thus distress. Simple husbandry precautions can prevent such a significant cost of stress from occurring. For example, pigs experience the greatest mortality if they are loaded for transportation just before feeding or if they are transported during high environmental temperatures (Honkavaara, 1989). Likewise, the longer the time of transportation, the greater the effect on the animals. This latter example further demonstrates the accumulative biological cost of multiple stressors over time. By simply separating these potentially stressful conditions and allowing the animals to recoup their biological reserve, the stress burden to the animal is reduced. Many of these additive conditions are obvious to the thoughtful animal technician or stockman, but more work is needed to understand fully the more subtle effects of stressors associated with animal production and their effects on the stress load of the animals.

Conclusion

The key to determining when stress affects an animal's welfare is the biological cost of the stress. When the biological cost of coping with the stressor diverts resources away from other biological functions, such as maintaining immune competence, reproduction or growth, the animal experiences distress. During distress, this impairment of function places the animal in a prepathological state that renders it vulnerable to a number of pathologies (Fig. 1.1).

Measurement of distress remains a problem. Measurements of the endocrine, physiological and behavioural responses to a stressor are valuable, but none of these end-points can serve as a litmus test for distress. Their use is compromised by interanimal differences in the nature of the stress response, by each type of stressor electing a unique biological response and by these systems' response to stimuli, whether or not the stressor actually induces distress.

Only a change in normal function during stress results in distress. This change in function may result from a disruption of a critical biological event, such as ovulation, when the biological cost of the stress diverts resources from other functions. Such a biological cost occurs when the animal experiences a severe stressor or when the accumulative cost of several minor stressors is sufficient to impair normal biological functions.

In the subsequent chapters, the biological responses to stress and the impact of these responses on biological function are discussed in more depth. Authors also address the critical issue of how to manage stress in animals to reduce the cost of stress in order to prevent distress and to protect the welfare of the animals under our care.

References

Blecha, F., Pollmann, D.S. and Nichols, D.A. (1983) Weaning pigs at an early age decreases cellular immunity. *Journal of Animal Science* 56, 396–400.

Bohus, B., Koolhaas, J.M., Nyakas, C., Steffens, A.B., Fokkema, D.S. and Scheurink, A.J.W. (1987) Physiology of stress: a behavioral view. In: Wiepkema, P.R. and van Adrichem, P.W.M. (eds) *Biology of Stress in Farm Animals: An Integrative Approach.* Martinus Nijhoff Publishers, Boston, pp. 57–70.

Brown, K.I. and Nestor, K.E. (1973) Some physiological responses of turkeys selected for high and low adrenal response to cold stress. *Poultry Science* 52, 1948–1954.

Cannon, W.B. (1929) *Bodily Changes in Pain, Hunger, Fear and Rage: An Account of Recent Researches into the Function of Emotional Excitement.* Appleton, New York.

Colborn, D.R., Thompson, D.L., Jr, Roth, T.L., Capehart, J.S. and White, K.L. (1991) Responses of cortisol and prolactin to sexual excitement and stress in stallions and geldings. *Journal of Animal Science* 69, 2556–2562.

Dantzer, R. and Mormede, P. (1981) Pituitary–adrenal consequences of adjunctive activities in pigs. *Hormones and Behavior* 15, 386–395.

Dubos, R. (1959) *The Mirage of Health.* Doubleday, New York.

Dunn, A. (1988) Nervous system–immune system interaction: an overview. *Journal of Receptor Research* 8, 589–607.

Engle, G.I. (1967) A psychological setting of somatic disease: the 'giving up–given up' complex. *Proceedings of the Royal Society of Medicine* 60, 553–555.

Hemsworth, P.H., Barnett, J.L. and Hansen, C. (1981) The influence of handling by humans on the behaviour, growth, and corticosteroids in the juvenile female pig. *Hormones and Behavior* 15, 396–403.

Henry, J.P. (1976) Mechanisms of psychosomatic disease in animals. *Advances in Veterinary Sciences and Comparative Medicine* 20, 115–145.

Henry, J.P. (1992) Biological basis of the stress response. *Integrative Physiological and Behavioral Science* 27, 66–83.

Holmes, W.N., Gorsline, J. and Cronshaw, J. (1979) Effects of mild cold stress on the survival of seawater-adapted mallard ducks (*Anas Platyrhynchos*) maintained on food contaminated with petroleum. *Environmental Research* 20, 425–444.

Honkavaara, M. (1989) Influence of selection phase, fasting and transport on porcine stress and on the development of PSE pork. *Journal of Agricultural Science* 61, 415–423.

Järvi, T. (1989) Synergistic effect on mortality in Atlantic salmon, *Salmo salar*, smolt caused by osmotic stress and presence of predators. *Environmental Biology of Fishes* 26, 149–152.

Marple, D.N., Aberle, E.D., Forrest, J.C., Blake, W.H. and Judge, M.D. (1972) Endocrine responses of stress susceptible and stress resistant swine to environmental stressors. *Journal of Animal Science* 35, 576–579.

Mason, J.W. (1968) 'Over-all' hormonal balance as a key to endocrine organization. *Psychosomatic Medicine* 30, 791–808.

Mason, J.W. (1975) Emotion as reflected in patterns of endocrine integrations. In: Levi, L. (ed.) *Emotions – Their Parameters and Measurements.* Raven, New York, pp. 143–181.

Mason, W.A., Mendoza, S.P. and Moberg, G.P. (1991) Persistent effects of early social experience on physiological responsiveness. In: Ehara, A., Kimura, T., Takenaka, D. and Iwamoto, M. (eds) *Primatology Today.* Elsevier Sciences Publishers, Amsterdam, pp. 469–471.

McEwen, B. and Stellar, E. (1993) Stress and the individual. *Archives of Internal Medicine* 153, 2093–2101.

Moberg, G.P. (1985) Biological response to stress: key to assessment of animal well-being? In: Moberg, G.P. (ed.) *Animal Stress.* American Physiological Society, Bethesda, Maryland, pp. 27–49.

Moberg, G.P. (1987a) Influence of the adrenal axis upon the gonads. In: Clarke, J. (ed.) *Oxford Reviews of Reproductive Biology.* Oxford University Press, New York, pp. 456–496.

Moberg, G.P. (1987b) Problems in defining stress and distress in animals. *Journal of the American Veterinary Medicine Association* 191, 1207–1211.

Moberg, G.P. (1992) Stress: diagnosis, cost and management. In: Mench, J.A., Phil, D., Mayer, S. and Krulisch, L. (eds) *The Well-being of Agricultural Animals in Biomedical and Agricultural Research.* Scientists Center for Animal Welfare, Bethesda, Maryland, pp. 58–61.

Moberg, G.P. (1993) Developing management strategies to reduce stress in swine: a new approach utilizing the biological cost of stress. In: Batterham, E.S. (ed.) *Proceedings of the Australasia Pig Science Association*, Vol. 4. Australasia Pig Science Association, Attwood, Australia, pp. 116–126.

Moberg, G.P. (1996) Suffering from stress: an approach for evaluating the welfare of an animal. In: Sandoe, P. and Hurnik, F. (eds) *Proceedings of Welfare of Domestic Animals Concepts, Theories and Methods of Measurement. Acta Agriculturae Scandinavica*, Sect. A, *Animal Science* (Suppl. 27), 46–49.

Moberg, G.P. (1999) When does stress become distress? *Laboratory Animals* 28, 22–26.

Nangalama, A.W. and Moberg, G.P. (1991) Interaction between cortisol and arachidonic acid on the secretion of LH from ovine pituitary tissue. *Journal of Endocrinology* 131, 87–94.

Naylor, A.M., Porter, D.W. and Lincoln, D.W. (1990) Central administration of corticotrophin-releasing factor in the sheep: effects on secretion of gonadotropins, prolactin and cortisol. *Journal of Endocrinology* 124, 117–125.

Rivier, C. (1995) Luteinizing-hormone-releasing hormone, gonadotropins, and gonadol steroids in stress. *Annals of the New York Academy of Sciences* 771, 187–191

Selye, H. (1950) *Stress.* Acta, Montreal.

Weiss, J.M. (1972) Influence of psychosocial variables on stress-induced pathology. In: Porter, R. and Knight, J. (eds) *Physiology, Emotion and Psychosomatic Illness.* Elsevier, Amsterdam, pp. 253–265.

Some Issues in the Interpretation of Behavioural Responses to Stress

2

J. Rushen

Dairy and Swine Research and Development Center, Agriculture and Agri-Food Canada, Lennoxville, Quebec, Canada

Introduction

Champions of behavioural indicators of stress argue that behavioural responses are often correlated with physiological or immune responses, and so can be used to predict the effect of stress on the biological functioning of the animal. Unfortunately, it is sometimes assumed that the interpretation of behavioural responses to stress is unproblematic. In this chapter, I try to show that interpretation of behavioural responses is fraught with difficulty, and that it is primarily problems of interpretation that limit our ability to use behavioural indicators of stress.

Anyone familiar with behavioural science will be aware of the distinction that is often drawn between causal explanations, i.e. the here-and-now things that make animals do what they do, and functional explanations, which try to explain why the behaviour has evolved by showing how the performance of the behaviour functions to increase the reproductive fitness of the animals that perform it. Although it is important to understand the causes of stress-related behaviours, I want to suggest that some recent studies of functional aspects of behaviour, especially predator avoidance behaviour, may also provide useful information for understanding how animals respond to stress.

There is now a sizeable literature on the behavioural responses of animals to stress, and I could not hope to provide a thorough review in the space available. Toates (1995), Møller *et al.* (1998) and Ramos and Mormède (1998) provide useful overviews. In this chapter, I focus on some issues in the interpretation of stress responses. The main messages of this chapter can be summarized as follows:

1. The control of the behaviour of animals in response to stress is complex and we cannot interpret behavioural responses, nor use them as indicators of stress, until we understand the underlying causes of the behaviour. This understanding can come from investigating both the motivation and the neurophysiological bases of the behaviour.
2. Behavioural responses to some types of stress, especially acute stressors that evoke fear in the animals, are derived at least in part from predator avoidance behaviours, and recent studies on the functional significance of these behaviours can help us to understand how animals respond to such acute stressors.
3. Because the behaviours that animals show during stress are performed to help the animal deal with the stress, the types of responses are often specific for a particular type of stressor. Thus, it is unlikely that there are 'general' behavioural stress responses that animals show regardless of the type of stressor. This makes it hard to use behavioural responses to judge the relative severity of quite different types of stressors. However, it does mean that behavioural responses can be very informative about the source of the stress.

Neuroendocrine integration of behaviour and physiology

Understanding the causes of behaviour will be advanced once we understand the neurophysiological circuits and the neurochemical systems that control the behaviour. The correlations that can sometimes be found between animals' behavioural and physiological responses to stress are not surprising in view of the recent findings that both behavioural and physiological responses to stress appear to be controlled, at least in part, by the same central neuroendocrine systems. Most of the research has focused on corticotropin-releasing hormone (CRH) because of the obviously central role it plays in regulating hypothalamic–pituitary–adrenal (HPA) activity, and because of the demonstrations of its involvement in human anxiety and depressive disorders (e.g. Arborelius *et al.*, 1999).

A number of studies have shown that, in rats, intracerebroventricular (i.c.v.) injections of CRH result in a number of behavioural changes that are similar to those seen under conditions of acute stress. For example, Morimoto *et al.* (1993) stressed rats by placing them in a novel cage and noted that, as well as exhibiting various signs of sympathetic nervous system activation such as increases in blood pressure and heart rate, the rats also showed a marked increase in locomotor behaviour. These same effects could be stimulated by i.c.v. injections of CRH, suggesting that it was the central secretion of CRH as a result of the stress that elicited both the peripheral physiological and behavioural responses. This interpretation was supported by the fact that both the physiological responses and the increase in locomotor activity when the rats were placed in a novel cage were substantially reduced by prior injections of a CRH antagonist. Interestingly, injections of a CRH antagonist into non-stressed rats caused no behavioural or physiological changes, suggesting that CRH has no 'basal' activity and is important only during periods of stress.

Research has clarified the neurocircuitry underlying the effects of CRH on the HPA axis (e.g. Herman and Cullinan, 1997). However, less is known of the motivational mechanisms by which CRH secretion triggers behavioural stress responses, or the relationship with the animal's emotional reactions. Deak *et al.* (1999) used a conditioned fear paradigm in an attempt to assess how CRH antagonists influence behavioural responses to stress. In this paradigm, rats receive a foot shock in a particular cage and then show conditioned fear responses (e.g. freezing) when they are returned to this cage. A CRH antagonist given when the rats are returned to the cage reduces the duration of this freezing. The researchers asked whether the CRH antagonist was directly suppressing the freezing behaviour itself or was instead reducing the (presumed) fear that motivated the freezing. They found that injection of a CRH antagonist given when the rats first received the shock also suppressed the freezing when the rats were returned to the cage, even though no CRH antagonist was administered at that time. This suggests that the CRH antagonist does not simply suppress the motor patterns of freezing behaviour, and the authors suggest that it affects the level of fear itself. However, before this interpretation can be accepted, we need to know that the antagonist given at the time of the conditioning did not interfere with the learning process or the animal's memory.

The role of CRH secretion in eliciting behavioural responses and the relationship between behavioural and physiological responses to acute stress can be complicated. For example, rats that receive an electric shock from a probe will often attempt to bury the electric probe in their bedding material. Korte *et al.* (1994) found that i.c.v. injections of a CRH antagonist reduced the amount of time that the rats spent burying the probe, suggesting a reduction in their level of fear. However, despite this effect on the behavioural responses, the CRH antagonist did not affect the shock-induced secretion of corticosterone or adrenocorticotropic hormone. Korte *et al.* (1994) speculate that although CRH may be involved in triggering both the behavioural and HPA responses, it does so by acting on different populations of receptors; the dose of CRH antagonist used in their experiment blocks only the receptors involved in the behavioural response. Korte *et al.* (1994) also noticed a number of other behavioural effects of the CRH antagonist including increased grooming and rearing behaviour, but it was not clear how these related to the stress response.

CRH also appears to be involved differently in the different behavioural responses to stress. For example, Buwalda *et al.* (1997) examined the effects of chronic infusions of CRH given over 10 days. The CRH injections initially resulted in increased locomotor activity, which is commonly found when rats are stressed (e.g. Morimoto *et al.*, 1993). However, this effect disappeared after 2 days, suggesting some adaptation to the CRH injections. This is not surprising since rats also adapt to chronic stress. However, the authors also examined the effect on a second behavioural response. In this case, the rats were placed in a maze in which some arms were closed and others were open. When the maze is lifted up, the rats tend to avoid the open arms, presumably because of their fear of falling. Buwalda *et al.* (1997) found that rats infused with CRH were even less likely to enter the open arms of the maze after 1 week, several days after the CRH effect on locomotor behaviour had disappeared. Thus CRH secretion may have different effects upon different fear-related behaviours.

Compared with the research on rats, comparatively little work has examined the neuroendocrine control of stress-related behaviours in farm animals. Both Johnson *et al.* (1994) and Salak-Johnson *et al.* (1997) report that i.c.v. injections of CRH in pigs elicit increased locomotor activity and stimulate vocalization. However, Salak-Johnson *et al.* (1997) reported that complex sequences of oral/nasal behaviours were reduced. Such behaviours are usually considered to be stereotypic behaviours and are thought to indicate that the animals are experiencing some stress. However, these behaviours, at least in pigs, are now thought to indicate high levels of feeding motivation (Rushen *et al.*, 1993), and CRH reduces feeding motivation (Parrott, 1990).

Although in this chapter I have dealt only with CRH, we must remember that the HPA axis is controlled, or affected, by a whole host of neuroendocrine systems (e.g. Jessop, 1999). We should similarly expect equally complex neuroendocrine control of behavioural responses to stress. In addition, behavioural responses to stress may also be influenced by many of the immune system changes that accompany exposure to stressors (Maier and Watkins, 1999).

Manipulation of neuroendocrine systems and the observation of the resulting behavioural effects does provide some useful information on the causal mechanisms underlying behavioural responses to stress. However, we should not assume that the relationship between behaviour and neuroendocrine systems is a simple one-way affair. The above discussion on CRH focused on unidirectional causation, i.e. where secretion of CRH resulted in changes in behaviour. However, we must avoid the simplistic interpretation that the direction of causation always goes one way, i.e. neuroendocrine changes produce behavioural changes. I will briefly consider another area where behaviour and physiology are linked, this time in a situation where the causal direction is two way, with behaviour apparently having a direct effect on neuroendocrine changes.

One instance of the interaction between behavioural and physiological responses to stress that has gained considerable attention in the research on human medicine but has been relatively overlooked in the area of applied animal ethology is the relationship between stress, sleep and growth hormone (GH) secretion. There is now evidence that both CRH and growth hormone-releasing hormone (GHRH), one of the main stimulants of GH secretion, have an influence on sleep. In fact there seems to be a reciprocal interaction between GHRH and CRH in controlling sleep. Numerous studies (reviewed in Friess *et al.*, 1995; Steiger *et al.*, 1998) show that CRH administration can reduce REM (rapid eye movement sleep), SWS (slow wave sleep) and GH secretion, while GHRH administration increases sleep (especially SWS) and GH secretion and reduces HPA activity.

The effect of CRH on sleep suggests that stress on an animal is likely to result in sleep disturbances, which in turn may be apparent in reduced GH secretion. Indeed, the relationship between stress, sleep and GH secretion has been demonstrated, at least in humans. Sakkas *et al.* (1998) report that SWS in depressed patients is reduced along with a reduction in the secretion of GH during the first period of sleep.

However, sleep also appears to influence the secretion of both CRH and GHRH. Studies on rats and humans have shown that GH tends to be released during sleep (e.g. Friess *et al.*, 1995). There is evidence of a link in some species between the various phases of sleep and GH secretion (Åström, 1995). The results are partly contradictory, but in humans GH secretion appears to occur during SWS. For example, Follenius *et al.* (1988) reported that people show falling GH levels during REM sleep. When people are deprived of sleep for one night, both SWS and plasma GH concentrations are increased during the following night (Beck *et al.*, 1975). The results from rats are, however, contrary to those obtained from humans. For example, rats selectively deprived of REM sleep rather than SWS show a decrease in serum GH (Toppila *et al.*, 1996).

Sleep is also related to endocrine systems involved in the responses to stress such as the sympathetic nervous system or the HPA axis. In humans, cortisol secretion appears to be suppressed by sleep onset and is high towards the end of sleep (Friess *et al.*, 1995). However, the different phases of sleep may also be related differently to HPA activity. Vgontzas *et al.* (1997) report that, in people, urinary free cortisol and adrenaline concentrations are positively correlated with the amount of REM sleep. Sleep is often increased during periods of infectious disease, and there is evidence that enhanced sleep may help some animals, e.g. rabbits, to recuperate from infection (Toth *et al.*, 1993). Intense stress can lead to increased REM sleep in rats when the stress is terminated (Rampin *et al.*, 1991), and the reciprocal link between sleep and HPA activity suggests that sleep may play a similar role in helping animals to adapt to or recuperate from stress. It is also clear that disturbances to sleep could have wide-ranging physiological consequences.

Despite the potential importance of this link, there has been relatively little research done on the relationship between stress and sleep in domestic animals. In fact, little is known of the sleep patterns of most domestic animals, except for the pioneering work of Ruckebusch (1974), which has not been adequately followed up. Ruckebusch (1974) described both SWS and REM sleep in cattle and showed that both tended to occur while the cattle were lying down. However, when cattle were prevented from lying down, SWS could still be detected when the cattle were standing, but REM sleep could not. Thus, preventing cattle from lying down appears selectively to reduce REM sleep. It is not yet known how stress, sleep and GH secretion interact in cattle. However, Munksgaard and Løvendahl (1993) found that restricting the times when cattle can lie down results in reduced concentrations of GH, suggesting a similar relationship between sleep and GH secretion in cattle to that found in other species.

Veissier *et al.* (1989) suggest that change in the circadian patterns of activity and rest could indicate how well animals are adapted to their environments. In view of the important link between sleep, stress and endocrine secretion, this is an important aspect of the relationship between behavioural and physiological responses to stress which has not been fully explored.

Complexity in the motivational control of stress-related behaviours: the open-field test

Understanding the neuroendocrine basis of behaviour is only one way of approaching the question of causation of behaviour. Another approach is to try to understand the motivational system underlying the behaviour. A major argument of this chapter is that interpretation of behavioural indicators of stress is and will remain difficult primarily because of the complexity of the motivational systems underlying behaviour. That is, unless we have some understanding of the motivational causes of the behavioural responses that animals make under stress, we will not be able to use these responses as a valid 'stress measure'. Considerable research by ethologists into the motivational bases of behaviour has shown that the behaviour of an animal at any one time is the result of an interaction between different motivational systems that often appear to 'compete' for control of the animal's behaviour (see McFarland, 1989). The important point for this discussion is that the behavioural responses that animals make to any given stressor will reflect a mix of different motivations.

The difficulty in interpreting behavioural responses to stress can be illustrated by the work on one very popular behavioural measure of an animal's responses to stress, the open-field test. Although the open-field test was originally used with rodents, I shall illustrate the problems with this test by referring mainly to work that has been done on farm animals, especially cattle.

The open-field test was originally developed to measure rather poorly defined characteristics of rats, such as 'fear' or 'emotionality' (Hall, 1936). A single rat was placed in a large, novel area and the rat's responses, which usually involved defaecation or increased activity, were interpreted as reflecting the animal's degree of fear, in response either to novelty or to the 'openness' of the enclosure. However, detailed analysis of the test showed the problems of this unidimensional interpretation (e.g. Archer, 1973; Ramos and Mormède, 1998). Careful experimental work suggested instead that the responses shown probably reflect a mix of motivations such as freezing in response to fear, exploration, escape attempts and specific responses to social isolation (Archer, 1973). More recently it has been shown that open-field activity of rats can be affected by factors apparently unrelated to the degree of fear, such as aerobic capacity (Friedman *et al.*, 1992). More recently, researchers have developed more sophisticated methods for analysing the multidimensional quality of stress responses in rodents (e.g. Ramos and Mormède, 1998).

Despite these problems, the open-field continues to be used as a measure of farm animals' responses to stress. For example, the open-field test has long been used to measure cattle's responses to fear-provoking situations (Kilgour, 1975). Often this involves placing a cow or calf in a novel area for a few minutes and then recording some aspect of its behaviour thought to best indicate the degree of fear the animal shows. In many cases, the open-field area is divided into a number of squares and the researchers count the number of squares that the animal enters. This research, however, has the same problems of interpretation as the rodent research. For example, Warnick *et al.* (1977) interpreted the degree of activity that calves show in an open-field as the degree of nervousness of the calves, while others (e.g. Dantzer *et al.*, 1983; Dellmeier *et al.*, 1985, 1990) interpreted the same degree of activity in terms of the level of 'locomotor' motivation. Which interpretation is correct? Probably both.

As is the case with rodents (Archer, 1973; Ramos and Mormède, 1998), it is most likely that the behaviour of calves in an open-field, or indeed in any 'stress test', reflects a changing mix of motivation rather than a single source of motivation such as fearfulness or locomotor motivation. Recently, de Passillé *et al.* (1995) used factor analysis as a means of analysing the mix of motivations that might underlie the behaviour of calves in an open-field. Even when only seven different behaviours were observed (sniffing/licking, walking, running, jumping, vocalization, defaecation and standing immobile), it was still necessary to use three factors to account for 60% of the variance in these behaviours. Based on the behaviours that had the highest correlations, these three factors were tentatively labelled as 'fear' (vocalization and defaecation), exploration (sniffing and licking) and locomotion (running and jumping). The correlations between these three factors were very small, suggesting that they are effectively independent sources of motivation. Furthermore, the amount of total activity that the calf showed correlated with all three factors. Thus, a calf could show a lot of activity because it had either a high level of fearfulness, or a high level of exploration, or a high level of locomotion, or a mixture of all three. Thus it would be very dangerous simply to measure total activity and then to interpret the calves' activity in this test as simply reflecting the level of fearfulness.

While factor analysis can be a useful way of examining the complex motivational structure underlying an animal's behaviour, the result can depend a great deal on the sort of assumptions made in carrying out the test. Hence the results are best thought of as hypotheses about the motivational structure of behaviour. These hypotheses, however, need to be tested before we can be certain as to their correct interpretation. De Passillé *et al.* (1995) attempted to test their interpretation of calves' open-field behaviour by experimentally varying the degree of motivation of each calf. For example, the factor analysis suggested that the amount of defaecation/vocalization might indicate the degree of fear the calves showed in response to the novelty of the enclosure. This was tested by examining how these aspects of the behaviour of the calf changed in response to three factors that were likely to alter the degree of novelty of the enclosure and hence the degree of fear of the calves: prior experience of the enclosure, the presence of a novel object and the age of the calves. As expected, the amount of defaecation/vocalization was higher for younger calves and was increased by adding a novel object, while allowing the calves to become familiar with the arena reduced the occurrence of these behaviours. Thus the amount of defaecation/vocalization could be a useful way of assessing the degree of fear that calves show in response to novelty.

Although older calves also showed more running and jumping, these behaviours were not affected by the degree of familiarity with the enclosure or by the presence of a novel object, suggesting that these type of activities cannot be used to measure calves' degree of fearfulness. In fact, there is evidence that the occurrence of these specific locomotor behaviours may well reflect primarily locomotor motivation (e.g. Dellmeier *et al.*, 1985; Jensen *et al.*, 1998).

It is likely that any behavioural response to stress will reflect a mix of motivations. However, this study shows that the use of multivariate statistical analysis, combined with experimental manipulations, can help us to untangle the causes of behaviour, interpret the behaviour that animals make in an open-field and choose the most appropriate type of measure to assess the degree of fear.

A functional approach: behavioural responses to acute stress that are derived from anti-predator behaviour

Many studies of the behavioural effects of stress involve exposing animals to acute stressors, such as forced swimming in cold water, electric shock, close restraint in a tube, or the open-field mentioned above. In the context of the normal life of the animal, these could be considered fairly 'artificial'. In such situations, the motivational basis of the behaviour is not evident and the function of the behavioural responses often difficult to understand. Blanchard and Blanchard (1988) criticized this use of artificial stressors.

They argue that stress responses evolved to help animals to deal adaptively with particular stressors that they are likely to encounter in their natural life. Hence, understanding the responses is best achieved if we use more 'natural' stressors, i.e. stressors that the animal is likely to have evolved to deal with. Arguing that many stress responses are in fact derived from anti-predator responses, Blanchard *et al.* (1998) developed a paradigm for studying stress responses of rodents that involved exposing rats to cats. They found that such chronic exposure resulted in marked increases in crouching by the rat and reductions in normal ongoing behaviours such as locomotion, grooming and rearing.

Vigilance

However, in this paradigm, the exposure to the cat, which was placed on top of the rat's cage, is fairly extreme. In reality, anti-predator behaviour is likely to involve a series of graded responses, and the first line of defence is to be able to detect the predator before it is too close. For most prey animals this is achieved by routine scanning of the environment, a phenomenon that has become known as 'vigilance'. To maximize the chance of avoiding predators, animals would need to maximize the time spent being vigilant. This scanning can interfere with other behaviours, most notably grazing, so most animals would be under some selection pressure to optimize the time spent scanning so as to reduce the risk of predation but still allow enough time to perform competing activities. Thus, the degree of vigilance would be affected not only by the animal's own assessment of the risk of predation but also by the strength of motivation to perform competing behaviours. Studies of vigilance may therefore provide a useful paradigm for studying the influence of different levels of perceived stress under different conditions of motivational conflict.

Considerable research has been carried out on vigilance behaviour in free-living animals, and the results show promise for elucidation of some of the causal mechanisms underlying behavioural responses to stress. One phenomenon that has attracted considerable attention is that the degree of vigilance shown by individual animals in a group is inversely related to the size of the group. This has now been reported in many species of birds and mammals (Elgar, 1989; Roberts, 1996), although the effect is sometimes small (e.g. Blumstein, 1996) or absent (e.g. Rose and Fedigan, 1995). A plausible functional explanation is that the additional group members increase the chance of predator detection so that, in a larger group, any one individual can reduce vigilance time but still have the same chance of detecting a predator. However, the increased group size may not only increase the chance of predator detection but also reduce the actual risk of predation, and there is not yet sufficient evidence to distinguish these possibilities (Lima, 1995; Roberts, 1996). Nevertheless, the effect of group size on vigilance does indicate that an animal's anti-predator behaviour can be quite sensitive to environmental factors that affect the risk of predation. This supports Blanchard and Blanchard's (1988) suggestion that animals make some assessment of risk when responding behaviourally to threats, and that this risk assessment is an essential part in explaining behavioural responses to some kinds of acute stressors.

Besides group size, other factors that influence the degree of detection of predators or the risk of predation have been shown to influence the degree of vigilance by animals. For example, vigilance has been shown to be higher at places where predators or other groups are more likely to be encountered, such as at waterholes or where group home ranges overlap (e.g. Burger and Gochfield, 1992; Rose and Fedigan, 1995), or when animals are relatively far from cover or a refuge (Pöysä, 1994; Frid, 1997). This suggests that animals may be quite sensitive to factors that affect the degree of risk. The risk of actual predation may not be the only factor that influences the degree of vigilance. For example, a loud noise increases the amount of visual scanning performed by rats while eating (Krebs *et al.*, 1997).

Studies of vigilance also provide a useful paradigm for obtaining information on how different sources of stress or risk interact. For example, Frid (1997) examined vigilance in Dall's sheep. The degree of vigilance was found to decrease as group size increased and as distance to cliffs (which the sheep use to escape from predators) decreased. Frid considered two models of how these two factors might together influence vigilance. According to the simplest model, the effect of the two factors might be additive, i.e. the magnitude of the effect of one factor on vigilance would be independent of the magnitude of the other factor. A more realistic model, however, is that the magnitude of the effect of one factor on vigilance would depend on the magnitude of other factors. Frid found evidence in support of this second, interactive model: the effect of changes in group size was much smaller when the animals were close to cliffs. The explanation proposed is that if animals have a low risk of predation because they are close to a refuge or to cover, then the reduced chance of detecting a predator when group sizes are small should be less of a problem than when animals face a high risk of predation by being far from cover or a refuge. Although this is a relatively simple example, it does illustrate how studies of vigilance can be used to examine how multiple stressors might interact in affecting the behaviour of the animals.

Communication

It is often suggested that communicative behaviours may provide some measure of the internal or subjective state of an animal, and so may be useful as measures of stress. For example, Griffin (1981) suggested that the study of communicative behaviour could provide a 'window on the minds of animals' (p. 149). Weary and Fraser (1995a) provide an excellent review of some of the issues involved in using signals as a means of assessing animals' needs, while Dawkins (1993) provides a more general review of signalling behaviour. The reader is referred to these articles.

There have been many attempts to use signalling or communicative behaviour to assess animals' responses to various stressors, and usually these involve examining vocalizations. For example, vocalization has been used to assess piglets' responses to castration (White *et al.*, 1995; Weary *et al.*, 1998), cattle's responses to branding (Schwartzkopf-Genswein *et al.*, 1998) or preslaughter handling (Grandin, 1998), and sheep's responses to castration and tail docking (Mellor and Murray, 1989). The assumption behind these studies is that there is variability in the vocalization given by the animals, in terms of either the likelihood of vocalization occurring, or the rate, amplitude or some aspect of the acoustic structure of the calls, and that some aspect of this variability provides reliable information about the inner state of the animal.

The view that communicative behaviour is essentially honest in the sense that it provides reliable information about an animal's internal state, especially its motivational state, is implicit in the classical ethological studies of animal displays (e.g. Tinbergen, 1969). Detailed analyses of the displays of animals were undertaken to dissect the particular mixture of motivations acting on an animal at any one time. Perhaps the most widely known example is Leyhausen's (1956) analysis of the facial expressions of cats as reflecting a changing balance of fear and aggression. The underlying assumption was that natural selection should favour accurate signals, i.e. signals that reliably communicate information about an animal's intentions or likely behaviour. However, more detailed analyses of the evolutionary pressures on animals, and particularly analyses focused on understanding the costs and benefits to individuals rather than species or groups of animals, show that, in some cases, natural selection can be expected to favour 'dishonest' communication, i.e. where animals convey inaccurate information about their needs or motivations in order to manipulate the behaviour of other animals to their advantage (e.g. Krebs and Dawkins, 1984). Indeed, there are a number of instances (reviewed in Weary and Fraser, 1995a) where animal communication appears to be quite 'deceptive' in that it contains inaccurate information about an animal's needs, intentions or internal state. More detailed analysis has shown that, in some cases, natural selection will favour 'honest' communication, but only under certain circumstances, e.g. where the sender and receiver are related genetically, or where there is some 'cost' to performing the signal (Zahavi, 1977; Maynard Smith, 1991; discussed in Weary and Fraser, 1995a).

I do not have the space here to deal with this central issue in animal communication in any detail. What is important for the present chapter is that we cannot simply assume that animal communication or signals will provide reliable information about the motivational or subjective state of the animal. Instead this must be demonstrated. Attempts to test the 'honesty' of animal signals involve either correlating the signal with some other measure of stress or experimentally manipulating the state of the animal and observing which aspects of the signals change.

How can we test the 'honesty' of animal communication, particularly when the animal is exposed to a stressor? In a number of cases, the extent of vocalizing by animals in conditions assumed to be stressful does correlate with physiological indices of stress. For example, as mentioned above, CRH injections can stimulate vocalization and HPA activity in pigs (Johnson *et al.*, 1994), while opioid antagonists can increase HPA activity and vocalization during restraint stress (Rushen and Ladewig, 1991). In open-field tests, individual differences in HPA activity and vocalization appear to be correlated (von Borell and Ladewig, 1992). There are fewer examples where an animal's state has been experimentally manipulated; the best example of these is the work done by Weary and Fraser (1995b) and Weary *et al.* (1996) on the vocalization of pigs. These authors examined the vocalizations shown by young pigs when they are separated from their mother. They tested the extent to which these vocalizations gave reliable information about need by increasing the level of hunger of some piglets by making them miss a milk ejection. They also compared piglets that were thriving, in the sense that they were above average weight for their litter and had good growth rates, with 'non-thriving' piglets, i.e. that were small and had poor growth rates. They found similar differences in both cases. Piglets that were not thriving or that had missed a milk ejection gave more vocalizations, more high frequency vocalizations, longer calls and calls with a greater increase in frequency. These acoustic properties were likely to increase the energetic cost of performing the signals, and since the calls were from offspring to their mothers (i.e. between genetic relatives) they argued that these calls filled the criteria necessary for honest signalling.

Finally, they examined what information was transmitted to the receiver of the calls by examining how sows responded to playback of the calls. Sows responded to the calls by orienting towards and approaching the loudspeaker, and they responded more strongly to calls that were recorded from needy piglets (i.e. smallest, slowest growing piglets that had missed a milk ejection and were in a cold environment). These experiments show that the vocalizations of piglets separated from their mothers do contain reliable information about the level of need of the piglets, and that the sows use this information in judging how to respond to the calls.

Thus, the experiments show that vocalization by pigs does contain some information about the level of risk to the piglets and the degree of stress they are under. However, research with a number of species shows that calls contain other types of information besides the level of risk.

Firstly, vocalizations can show a high degree of variability, and some of this variability is used to identify individuals rather than to communicate the emotional state of an individual. This individual recognition appears particularly important for the contact calls that many mammals give when separated from a group or from particular individuals within the group (Cheney *et al.*, 1996; Janik and Slater, 1998).

Secondly, signalling behaviour is affected by the presence of an audience. If communication is performed with the intent of influencing the behaviour of others, it would not be surprising if the absence or presence of an audience has an effect. Evans and Marler (1991, 1992) examined the alarm calls that cockerels gave in response to the sight of an aerial predator and found that the rate of calls given was significantly higher when the cockerel could see a hen, although the breed or degree of familiarity of the hen did not seem to make a difference.

Thirdly, research with primates suggests that alarm calls may not just provide information about the degree of risk, but also about the type of risk. This was first shown in intriguing experiments by Seyfarth *et al.* (1980), which are discussed and reviewed in Cheney and Seyfarth (1990). Vervet monkeys have a number of predators, most notably leopards, snakes and aerial predators such as eagles. The predator avoidance behaviour that vervets show is specific to the particular type of predator. For example, vervets react to the presence of eagles by looking up in the sky or running under bushes, to the presence of snakes by standing bipedally and looking at the ground, and to leopards by running up trees. Vervets also give acoustically different alarm calls when faced with different types of predators. Playback experiments demonstrated that these calls contain information about the type of predator rather than just the degree of risk: vervets show the appropriate predator avoidance behaviour when they hear a particular type of call. These findings indicate that vocal signals may not so much contain information about the internal state of the sender, but information about objects in the environment, i.e. semantic information. The importance of this finding for the present discussion is that the calls that animals give when faced with a stressful challenge may not so much indicate the degree of stress that an animal is under (or the degree of risk of predation) but rather the type of risk.

Similar findings have been reported for Barbary macaques (Fischer, 1998). These monkeys give an acoustically different alarm call when they are disturbed by a person from that when they are disturbed by a dog. The way that the monkeys perceive these calls was tested using the habituation–dishabituation paradigm. In this paradigm, a recording of one of the calls is repeatedly played back to the monkeys until they stop reacting (habituation). Once this has occurred, the other type of call is played. If the monkeys perceive the two types of calls as distinct, they respond to the second type of call (dishabituation). The receivers of the two types of alarm calls do classify these calls differently, suggesting that they contain information about the type of predator. Such semantic information appears not to be restricted to alarm calls. The scream vocalizations of rhesus monkeys given to elicit support in the face of an attack from another monkey appear to contain information about the rank difference between the monkey and its attacker (Gouzoules *et al.*, 1998).

It is not yet clear if such semantic calls occur in mammals other than primates. Marmots also show a variety of different alarm calls, but these do not appear to be systematically associated with a particular type of predator (Blumstein and Armitage, 1997). In fact, playback experiments suggest that variation in the acoustic structure of calls, particularly the rate of calling, conveys information about the degree of risk: marmots hearing playbacks were more likely to flee to their burrows if the alarm calls occurred at a higher rate. Blumstein and Armitage (1997) suggest that marmot calls may not indicate the type of predator because the predator avoidance behaviour of marmots (fleeing to the burrow) is the same regardless of the type of predator. In contrast, red squirrels do produce an acoustically different alarm call in response to large predatory birds from that in response to ground predators such as dogs or approaching people (Greene and Meagher, 1998). However, it is not yet known if this conveys specific semantic information to other squirrels or whether it reflects other factors. For example, the calls given in response to aerial predators are acoustically structured to make them difficult to localize, which may be more important for an aerial predator than for a ground predator. Nevertheless, the discovery of semantic alarm calls suggests that animals may, under some circumstances, vary their response depending on the type of danger they are facing rather than just the degree of danger.

The context specificity of vocalizations poses a problem to researchers wishing to use them as an indicator of the degree of stress an animal is experiencing. For example, Watts and Stookey (1999) examined vocalization by beef cattle and found increased vocalization (as well as a greater frequency range, higher maximum frequency and higher peak sound level) when cattle were branded. Restraint of the animals did not appear to increase vocalization, and the authors conclude that vocalization occurs in response to relatively severe stressors. However, cattle vocalize readily when isolated from other animals (Boissy and Le Neindre, 1997). Does this indicate that isolation is a very severe stressor (according to the logic of Watts and Stookey) or that some types of stressors are more potent at eliciting vocalization than are other types of stressor? If vocalizations are primarily contact calls, they may be given most frequently in contexts that involve social isolation.

The above discussion on alarm calls suggests that behavioural responses may be specific to the particular type of challenge. This could be seen as a disadvantage if we are interested in using behaviour to measure the relative severity of very different types of stressors. However, the specificity of behavioural responses to stress does have one big advantage: the responses can provide information about the particular cause or source of the stress.

Conclusions

The attractiveness of behavioural indices of stress lies in the fact that they are quicker and appear technically easier to obtain than physiological measures. In addition, they are considered to reflect more directly the animals' feelings or emotions. There are arguments in favour of these views. Unfortunately, much use of behavioural indices of stress has been based on an over-simplistic interpretation of the behaviours and a lack of evidence of the validity of the measures. This is best exemplified by the use of the open-field test. My belief is that interpretation of behavioural responses to stress is far from simple, and that until we understand more of the causal mechanisms underlying the behaviour we will not be able to use behavioural indices of stress with any confidence. Understanding the neurocircuitry underlying the behaviour is an essential step. The role played by CRH in controlling many behavioural responses to stress gives us some confidence in the use of such behaviours to assess stress. However, we should avoid making the assumption that the neuroendocrine control of stress-related behaviour is a simple matter. Furthermore, understanding just how neuroendocrine systems control the behaviour, and the implications this has for the emotional state of the animal during stress, will require that we understand how the neurocircuitry and the motivational controls of the behaviour interact. The two-way interaction between sleep and the secretion of GH and CRH has been underappreciated in studies of animal welfare, and it is clear that we need to broaden the range of behaviours and physiological systems that we consider when we examine animals under stress.

Finally, the great improvement in our capacity to dissect the functional role of behaviour that has accompanied the development of behavioural ecology and sociobiology provides some novel perspectives. In this chapter, I briefly discussed some of the functional approaches taken in investigations of both communication and anti-predator behaviour, which appear to have implications for our understanding of stress responses. The theoretical studies of the evolution of 'honest' communication make it clear that we cannot assume that communicative behaviours will provide a valid measure of the animals' emotional state; this must be directly tested. The notion that stress responses have evolved to deal with or adapt to particular stressors means that the type of behaviour shown will probably be context specific, making it unlikely that we will find general stress responses.

References

Arborelius, L., Owens, M.J., Plotsky, P.M. and Nemeroff, C.B. (1999) The role of corticotropin-releasing factor in depression and anxiety disorders. *Journal of Endocrinology* 160, 1–12.

Archer, J. (1973) Tests for emotionality in rats and mice: a review. *Animal Behaviour* 21, 205–235.

Åström, C. (1995) Interaction between sleep and growth hormone. *Acta Neurologica Scandanavica* 92, 281–296.

Beck, U., Brezinova, V., Hunter, W.M. and Oswald, I. (1975) Plasma growth hormone and slow wave sleep increase after interruption of sleep. *Journal of Clinical Endocrinology and Metabolism* 40, 812–815.

Blanchard, D.C. and Blanchard, R.J. (1988) Ethoexperimental approaches to the biology of emotion. *Annual Review of Psychology* 39, 43–68.

Blanchard, R.J., Nikulina, J.N., Sakai, R.R., McKittrick, C., McEwans, B. and Blanchard, D.C. (1998) Behavioral and endocrine change following chronic predatory stress. *Physiology and Behavior* 63, 561–569.

Blumstein, D.T. (1996) How much does social group size influence golden marmot vigilance? *Behaviour* 133, 1133–1151.

Blumstein, D.T. and Armitage, K.B. (1997) Alarm calling in yellow-bellied marmots: 1. The meaning of situational variable alarm calls. *Animal Behaviour* 53, 143–171.

Boissy, A. and Le Neindre, P. (1997) Behavioral, cardiac and cortisol responses to brief peer separation and reunion in cattle. *Physiology and Behavior* 61, 693–699.

Burger, J. and Gochfield, M. (1992) Effect of group size on vigilance while drinking in the coati, *Nasua narica* in Costa Rica. *Animal Behaviour* 44, 1053–1057.

Buwalda, B., de Boer, S.F., Van Kalkeren, A.A. and Koolhas, J.M. (1997) Physiological and behavioral effects of chronic intracerebroventricular infusion of corticotropin-releasing factor in the rat. *Psychoneuroendocrinology* 22, 297–309.

Cheney, D.L. and Seyfarth, R.M. (1990) *How Monkeys See the World.* University of Chicago Press, Chicago.

Cheney, D.L., Seyfarth, R.M. and Palombit, R. (1996) The function and mechanisms underlying baboon 'contact' barks. *Animal Behaviour* 52, 507–518.

Dantzer, R., Mormède, P., Bluthe, R.M. and Soissons, J. (1983) The effect of different housing conditions on behavioural and adrenocortical reactions in veal calves. *Reproduction, Nutrition, Développement* 23, 501–508.

Dawkins, M. (1993) Are there general principles of signal design? *Philosophical Transactions of the Royal Society. Series B: Biological Sciences* 340, 251–255.

de Passillé, A.M., Rushen, J. and Martin, F. (1995) Interpreting the behaviour of calves in an open-field test: a factor analysis. *Applied Animal Behaviour Science* 45, 201–213.

Deak, T., Kien, N.T., Ehrlich, A.L., Watkins, L.R., Spencer, R.L., Maier, S.F., Licino, J., Wong, M.-L., Chrousos, G.P., Webster, E. and Gold, P.W. (1999) The impact of the nonpeptide corticotropin-releasing hormone antagonist antalarmin on behavioral and endocrine responses to stress. *Endocrinology* 140, 79–86.

Dellmeier, G.R., Friend, T.H. and Gbur, E.E. (1985) Comparison of four methods of calf confinement. II, Behavior. *Applied Animal Behaviour Science* 60, 1102–1109.

Dellmeier, G., Friend, T. and Gbur, E.E. (1990) Effects of changing housing on open-field behavior of calves. *Applied Animal Behaviour Science* 26, 215–230.

Elgar, M.A. (1989) Predator vigilance and group size in mammals and birds: a critical review of the empirical evidence. *Biological Reviews* 64, 13–33.

Evans, C.S. and Marler, P. (1991) On the use of video images as social stimuli in birds: audience effects on alarm calling. *Animal Behaviour* 41, 17–26.

Evans, C.S. and Marler, P. (1992) Female appearance as a factor in the responsiveness of male chickens during antipredator behaviour and courtship. *Animal Behaviour* 43, 136–145.

Fischer, J. (1998) Barbary macaques categorize shrill barks into two types of calls. *Animal Behaviour* 55, 799–807.

Follenius, M., Brandenburger, G., Simon, C. and Schlienger, J.L. (1988) REM sleep in humans begins during decreased secretory activity of the anterior pituitary. *Sleep* 11, 546–555.

Frid, A. (1997) Vigilance by female Dall's sheep: interactions between predation risk factors. *Animal Behaviour* 53, 799–808.

Friedman, W.A., Garland, T. and Dohm, M. (1992) Individual variation in locomotor behavior and maximal oxygen consumption in mice. *Physiology and Behavior* 52, 97–104.

Friess, E., Wiedemann, K., Steiger, A. and Holsboer, F. (1995) The hypothalamic-pituitary-adrenocortical system and sleep in man. *Advances in Neuroimmunology* 5, 111–125.

Gouzoules, H., Gouzoules, S. and Tomaszycki, M. (1998) Agonistic screams and the classification of dominance relationships: are monkeys fuzzy logicians? *Animal Behaviour* 55, 51–60.

Grandin, T. (1998) The feasibility of using vocalization scoring as an indicator of poor welfare during cattle slaughter. *Applied Animal Behaviour Science* 56, 121–128.

Greene, E. and Meagher, T. (1998) Red squirrels, *Tamiasciurus hudsonicus*, produce predator-class specific alarm calls. *Animal Behaviour* 55, 511–518.

Griffin, D.R. (1981) *The Question of Animal Awareness*, 2nd edn. Rockefeller University Press, New York.

Hall, C.S. (1936) Emotional behavior in the rat. III. The relationship between emotionality and ambulatory activity. *Journal of Comparative Psychology* 22, 345–352.

Herman, J.P. and Cullinan, W.E. (1997) Neurocircuitry of stress: central control of the hypothalamo-pituitary–adrenocortical axis. *Trends in Neuroscience* 20, 78–84.

Janik, V.M. and Slater, P.J.B. (1998) Context-specific use suggests that bottlenose dolphin signature whistles are cohesion calls. *Animal Behaviour* 56, 829–838.

Jensen, M.B., Vestergaard, K.S. and Krohn, C.C. (1998) Play behaviour in dairy calves kept in pens: the effect of social contact and space allowance. *Applied Animal Behaviour Science* 56, 97–108.

Jessop, D.S. (1999) Central non-glucocorticoid inhibitors of the hypothalamo-pituitary–adrenal axis. *Journal of Endocrinology* 160, 169–180.

Johnson, R.W., von Borell, E., Anderson, L.L., Kojic, L.D. and Cunnick, J.E. (1994) Intracerebroventricular injection of corticotropin-releasing hormone in the pig: acute effects on behavior, adrenocorticotropin secretion, and immune suppression. *Endocrinology* 135, 642–648.

Kilgour, R. (1975) The open-field test as an assessment of the temperament of dairy cows. *Animal Behaviour* 23, 615–624.

Korte, S.M., Korte-Bouws, G.A.H., Bohus, B. and Koob, G.F. (1994) Effect of corticotropin-releasing factor antagonist on behavioral and neuroendocrine responses during exposure to defensive burying paradigm in rats. *Physiology and Behavior* 56, 115–120.

Krebs, H., Weyers, P., Macht, M., Weijers, H.-G. and Janke, W. (1997) Scanning behavior of rats during eating under stressful noise. *Physiology and Behavior* 62, 151–154.

Krebs, J.R. and Dawkins, R. (1984) Animal signals: mind reading and manipulation. In: Krebs, J.R. and Davis, N.R. (eds) *Behavioural Ecology: an Evolutionary Approach*, 2nd edn. Blackwell, Oxford, pp. 380–402.

Lima, S.L. (1995) Back to the basics of anti-predatory vigilance: the group size effect. *Animal Behaviour* 49, 11–20.

Leyhausen, P. (1956) *Verhaltensstudien an Katzen*. Paul Parey Verlag, Berlin.

Maier, S.F. and Watkins, L.R. (1999) Bidirectional communication between the brain and the immune system: implications for behaviour. *Animal Behaviour* 57, 741–751.

Maynard Smith, J. (1991) Honest signaling: the Philip Sidney game. *Animal Behaviour* 42, 1034–1035.

McFarland, D.J. (1989) *Problems of Animal Behaviour.* Longman Scientific and Technical, Harlow, UK.

Mellor, D.J. and Murray, L. (1989) Effects of tail docking and castration on behaviour and plasma cortisol concentrations in young lambs. *Research in Veterinary Science* 46, 387–391.

Møller, A.P., Milinski, M. and Slater, P.J.B.E. (1998) *Stress and Behavior (Advances in the Study of Behavior).* Vol. 27. Academic Press, London.

Morimoto, A., Nakamori, T., Morimoto, K., Tan, N. and Murakami, N. (1993) The central role of corticotropin-releasing factor (CRF-41) in psychological stress in rats. *Journal of Physiology* 460, 221–229.

Munksgaard, L. and Løvendahl, P. (1993) Effects of social and physical stressors on growth hormone levels in dairy cows. *Canadian Journal of Animal Science* 73, 847–853.

Parrott, R.F. (1990) Central administration of corticotropin releasing factor in the pig: effects on operant feeding, drinking and plasma cortisol. *Physiology and Behavior* 47, 519–524.

Pöysä, H. (1994) Group foraging, distance to cover and vigilance in the teal, *Ana crecca. Animal Behaviour* 48, 921–928.

Ramos, A. and Mormède, P. (1998) Stress and emotionality: a multidimensional and genetic approach. *Neuroscience and Biobehavioral Reviews* 22, 33–57.

Rampin, C., Cespuglio, R.C.N. and Jouvet, M. (1991) Immobilisation stress induces a paradoxical sleep rebound in rat. *Neuroscience Letters* 126, 113–118.

Roberts, G. (1996) Why individual vigilance declines as group size increases. *Animal Behaviour* 51, 1077–1086.

Rose, L.M. and Fedigan, L.M. (1995) Vigilance in white-faced capuchins, *Cebus capucinus*, in Costa Rica. *Animal Behaviour* 49, 63–70.

Ruckebusch, Y. (1974) Sleep deprivation in cattle. *Brain Research* 78, 495–499.

Rushen, J. and Ladewig, J. (1991) Stress-induced hypoalgesia and opioid inhibition of pigs' responses to restraint. *Physiology and Behavior* 50, 1093–1096.

Rushen, J., Lawrence, A.B. and Terlouw, E.M.C. (1993) The motivational basis of stereotypies. *Stereotypic Animal Behaviour: Fundamentals and Applications to Animal Welfare.* CAB International, Wallingford, UK, pp. 41–64.

Sakkas, P.N., Soldatos, C.R., Bergiannaki, J.D., Paparrigopoulus, T.J. and Stefanis, C.N. (1998) Growth hormone secretion during sleep in male depressed patients. *Progress in Neuro-Psychopharmacology and Biological Psychiatry* 22, 467–483.

Salak-Johnson, J.L., McGlone, J.J., Whisnant, C.S., Norman, R.L. and Kraeling, R.R. (1997) Intracerebroventricular porcine corticotropin-releasing hormone and cortisol effects on pig immune measures and behavior. *Physiology and Behavior* 61, 15–23.

Schwartzkopf-Genswein, K.S., Stookey, J.M., Crowe, T.G. and Genswein, B.M. (1998) Comparison of image analysis, exertion force, and behavior measurements for use in the assessment of beef cattle responses to hot-iron and freeze branding. *Journal of Animal Science* 76, 972–979.

Seyfarth, R.M., Cheney, D.L. and Marler, P. (1980) Vervet monkey alarm calls: semantic communication in a free-ranging primate. *Science* 210, 801–803.

Steiger, A., Antonijevic, I.A., Bohlhalter, S., Friebos, R.M., Friess, E. and Murck, H. (1998) Effects of hormones on sleep. *Hormone Research* 49, 125–130.

Tinbergen, N. (1969) *The Study of Instinct.* Clarendon Press, Oxford.

Toates, F. (1995) *Stress: Conceptual and Biological Aspects.* John Wiley and Sons, New York.

Toppila, J., Asikainen, M., Alanko, L., Turek, F.W., Stenburg, D. and Porkka-Heiskanen, T. (1996) The effect of REM sleep deprivation on somatostatin and growth hormone-releasing hormone gene expression in the rat hypothalamus. *Journal of Sleep Research* 5, 115–122.

Toth, L.A., Tolley, E.A. and Krueger, J.M. (1993) Sleep as a prognostic indicator during infectious disease in rabbits. *Proceedings of the Society for Experimental Biology and Medicine* 203, 179–192.

Veissier, I., Le Neindre, P. and Trillat, G. (1989) The use of circadian behaviour to measure adaptation of calves to changes in their environment. *Applied Animal Behaviour Science* 22, 1–12.

Vgontzas, A.N., Bixler, E.O., Papanicolaou, D.A., Kales, A., Stratakis, C.A., Vela-Bueno, A., Gold, P.W. and Chrousos, G.P. (1997) Rapid eye movement sleep correlates with the overall activities of the hypothalamic–pituitary–adrenal axis and sympathetic system in healthy humans. *Journal of Clinical Endocrinology and Metabolism* 82, 3278–3280.

von Borell, E. and Ladewig, J. (1992) Relationship between behaviour and adrenocortical response pattern in domestic pigs. *Applied Animal Behaviour Science* 34, 195–206.

Warnick, V.D., Arave, C.W. and Mickelsen, C.H. (1977) Effects of group, individual and isolated rearing of calves on weight gain and behavior. *Journal of Dairy Science* 60, 947–953.

Watts, J.M. and Stookey, J.M. (1999) Effects of restraint and branding on rates and acoustic parameters of vocalization in beef cattle. *Applied Animal Behaviour Science* 62, 125–135.

Weary, D.M. and Fraser, D. (1995a) Signaling need: costly signals and animal welfare assessment. *Applied Animal Behaviour Science* 44, 159–169.

Weary, D. and Fraser, D. (1995b) Calling by domestic piglets: reliable signals of need? *Animal Behaviour* 50, 1047–1055.

Weary, D.M., Lawson, G.L. and Thompson, B.K. (1996) Sows show stronger responses to isolation calls of piglets associated with greater levels of piglet need. *Animal Behaviour* 52, 1247–1253.

Weary, D.M., Braithwaite, L.A. and Fraser, D. (1998) Vocal response to pain in piglets. *Applied Animal Behaviour Science* 56, 161–172.

White, R.G., DeShazer, J.A., Tressler, C.J., Borcher, G.M., Davey, S., Waninge, A., Parkhurst, A.M., Milanuk, M.J. and Clemens, E.T. (1995) Vocalization and physiological response of pigs during castration with or without a local anesthetic. *Journal of Animal Science* 73, 381–386.

Zahavi, A. (1977) The cost of honesty (further remarks on the handicap principle). *Journal of Theoretical Biology* 67, 603–605.

Neuroendocrine Responses to Stress

3

R.L. Matteri, J.A. Carroll and C.J. Dyer

Animal Physiology Research Unit, Agricultural Research Service, United States Department of Agriculture, Animal Sciences Research Center, University of Missouri, Columbia, Missouri, USA

In his seminal work on the general adaptation syndrome, Hans Selye observed marked changes in the sizes of endocrine tissues subsequent to stressor exposure (Selye, 1939). Selye's observations have been confirmed many times over, and it is clear that endocrine responses constitute an integral component of the stress response (Van de Kar *et al.*, 1991; Stratakis and Chrousos, 1995). Stress can affect the hormonal control of metabolism (Weissman, 1990; Wenk, 1998), reproduction (Moberg, 1991; Rivier and Rivest, 1991; Rivest and Rivier, 1995), growth (Stratakis *et al.*, 1995) and immunity (Peterson *et al.*, 1991; Sheridan *et al.*, 1998). Since hormone signalling plays a vital role in the maintenance of homeostasis, virtually every endocrine system responds in some fashion to specific stressors. The overall effect on the animal's adaptive response to stress is an integration of multiple, and often interactive, hormone responses that directly affect physical health and well-being.

A complete discussion of the vast and complex literature on endocrine responses to stress is beyond the scope of the present chapter. Whenever possible, the readers are referred to more comprehensive reviews on selected topics. In the subsequent discussion, we will introduce and briefly review the general effects of stress on the major neuroendocrine systems. Additionally, we will present relatively new and important concepts in the neuroendocrinology of stress relating to appetite control.

Neuroendocrine systems

Background

Neuroendocrinology can be defined as the study of communication between the central nervous system and endocrine glands. The term 'neuro-endocrine' classically refers to hormone signalling involving the hypothalamus, pituitary gland and peripheral body systems. The hypothalamus is a bilaterally symmetric region on the inferior aspect of the brain above the anterior pituitary gland (Fig. 3.1; Braunstein, 1995). While not precisely defined by anatomical borders, the anterior to posterior range of the hypothalamus is considered to extend approximately from the optic chiasm to the mammillary bodies. The lateral boundaries extend roughly to the optic tracts. From superior to inferior, the hypothalamus extends from the floor of the third ventricle to the median eminence (lower-most region), where the pituitary stalk is formed. Axon terminals of parvicellular hypothalamic neurons release hormones into a capillary network (hypophysial portal vasculature) that traverses the pituitary stalk from the median eminence to the anterior pituitary gland. The parvicellular neurons produce hormones that regulate the function of the anterior pituitary gland. Magnocellular neurons extend directly through the pituitary stalk into the posterior pituitary gland. The magnocellular neurons produce vasopressin and oxytocin, the hormones released from the posterior pituitary gland.

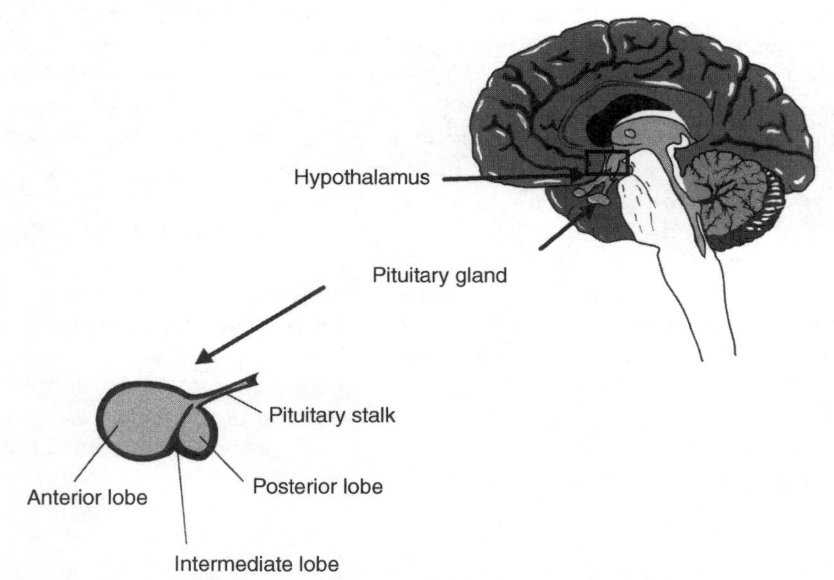

Fig. 3.1. Location of the hypothalamus and pituitary gland (with anterior, intermediate and posterior pituitary lobes indicated).

The pituitary gland comprises three components: (i) the anterior lobe; (ii) the posterior lobe; and (iii) the intermediate lobe (Fig. 3.1; Asa *et al.*, 1995). The anterior pituitary gland contains specialized cells that produce and secrete growth hormone, adrenocorticotropic hormone (ACTH), thyroid-stimulating hormone (TSH), luteinizing hormone (LH), follicle-stimulating hormone (FSH), and prolactin (Fig. 3.2). The chemical structures and major functions of the anterior pituitary hormones have been described elsewhere (Matteri, 1994). The role of the intermediate lobe is species specific, and in the human as well as a small number of other mammals the intermediate lobe is considered rudimentary. The posterior lobe of the pituitary is an extension of the floor of the third ventricle and maintains physical contact with the hypothalamus via the pituitary stalk. The posterior lobe acts as a storage site for vasopressin and oxytocin. Vasopressin, also known as antidiuretic hormone (ADH), enhances water reabsorption by the kidney, constricts smooth muscles surrounding blood vessels and participates in stress-induced ACTH secretion (see below). Oxytocin stimulates uterine contractions during labour and milk ejection from mammary tissue.

Hypothalamic–pituitary–adrenal (HPA) axis

One of the best known and consistent neuroendocrine responses to stress is activation of the HPA axis, resulting in the secretion of steroid hormones from the adrenal gland (Fig. 3.3). This relationship between stress and

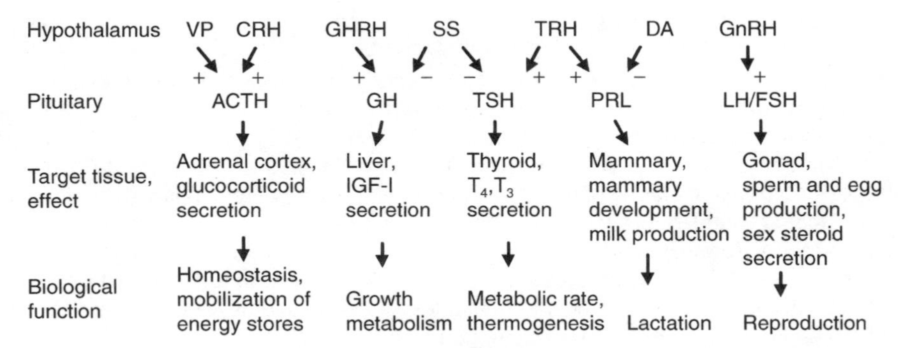

Fig. 3.2. Schematic representation of hypothalamic–pituitary neuroendocrine axes, and major biological actions. Hypothalamic factors: CRH, corticotropin-releasing hormone; DA, dopamine; GHRH, growth hormone-releasing hormone; GnRH, gonadotropin-releasing hormone; SS, somatostatin; TRH, thyrotropin-releasing hormone; VP, vasopressin. +, stimulatory hypothalamic factor; −, inhibitory hypothalamic factor. Pituitary hormones: ACTH, adrenocorticotropic hormone; FSH, follicle-stimulating hormone; GH, growth hormone; LH, luteinizing hormone; PRL, prolactin; TSH, thyroid-stimulating hormone; IGF, insulin-like growth factor.

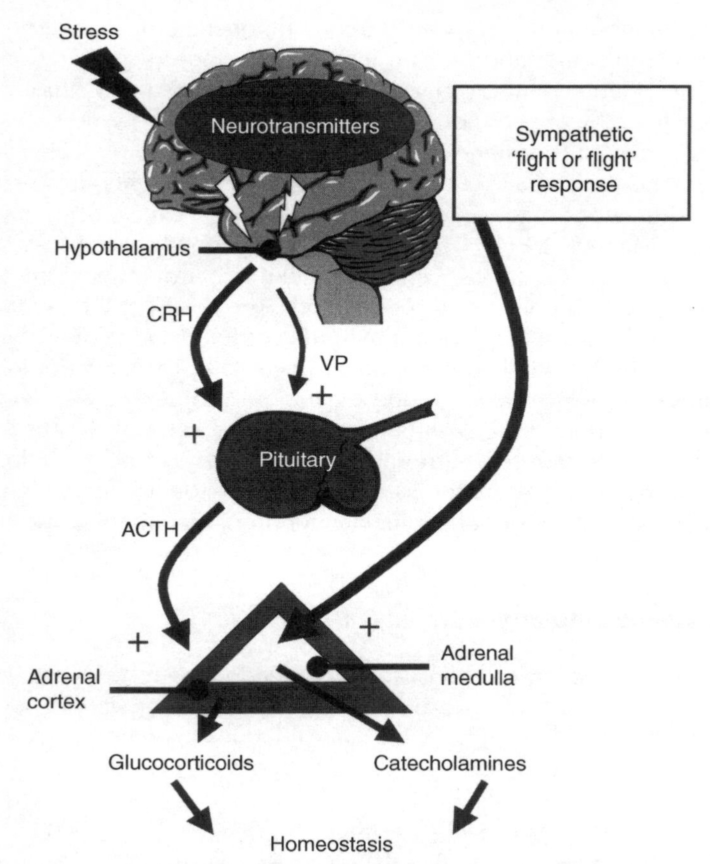

Fig. 3.3. Schematic diagram of the major components of the hypothalamic–pituitary–adrenal axis. External stimuli perceived as a stressor initiate a cascade of events which leads to activation of the sympathetic division of the nervous system and stimulation of corticotropin-releasing hormone (CRH) and vasopressin (VP) release from hypothalamic neurons. Activation of the sympathetic pathway initiates release of catecholamines from the adrenal medulla which act on various target organs and tissues. CRH and VP stimulate the release of adrenocorticotropic hormone (ACTH) from anterior pituitary corticotrophs which in turn stimulates glucocorticoid release from the adrenal cortex. Glucocorticoids act on a variety of target tissues and organs to maintain homeostasis.

adrenocortical activation was one of the first recognized in the study of the endocrinology of stress (Selye, 1939). Early investigators concluded that the regulation of glucocorticoid secretion from the adrenal gland depended on a linkage between the hypothalamus and the pituitary gland. Harris (1948) suggested that neurons of the hypothalamus regulate the secretion of hormones from the anterior pituitary. Harris's work led to further investigation into the hypothalamic–pituitary linkage, which demonstrated that factors produced in hypothalamic neurons did indeed regulate the secretion of ACTH from the anterior pituitary (Guillemin and Rosenberg, 1955; Saffran *et al.*, 1955; Porter and Jones, 1956). ACTH is produced from a larger molecule known as pro-opiomelanocortin (POMC), which is also a precursor of β-endorphin, lipotropin and melanotropin (Matteri, 1994). POMC-related peptides are synthesized in specialized anterior pituitary cells known as corticotrophs. ACTH stimulates the synthesis and release of steroids from the adrenal cortex by promoting the uptake of cholesterol and its enzymatic conversion to cortisol and corticosterone, the glucocorticoid hormones. Cortisol is the primary glucocorticoid in humans and most mammals, whereas in the rodent corticosterone is the primary glucocorticoid.

As their name implies, glucocorticoids play an important role in gluconeogenesis by stimulating the liver to convert fat and protein to intermediate metabolites that are ultimately converted to glucose for energy. Glucocorticoids also support this response by potentiation of the synthesis and action of epinephrine (adrenaline), a catecholamine released by the adrenal medulla during the stress response. Adrenaline stimulates gluconeogenesis and lipolysis, which mobilize energy stores for vigorous 'fight or flight' activity. Maintaining a sufficient, yet not excessive, concentration of glucocorticoids is necessary in order to maintain homeostasis. Chronic elevation of glucocorticoids results in protein catabolism, hyperglycaemia, immune suppression, susceptibility to infection and depression. McEwen and Sapolsky (1995) also noted that mental performance and hippocampal volume were reduced following chronic elevation of corticosteroids. Considering the potent deleterious actions of chronically elevated glucocorticoids, an important function of these steroids is to curtail the HPA response to stress. This occurs through negative feedback inhibition, where glucocorticoids inhibit further HPA response at the levels of the brain and pituitary (McEwen, 1979; Fink *et al.*, 1991).

Activation of the pituitary component of the HPA axis can be mediated by several neuroendocrine hormones. In accordance with conventional nomenclature, Saffran and colleagues (1955) named the ACTH regulator corticotropin-releasing factor (CRF). Today, this neurohormone is commonly referred to as corticotropin-releasing hormone (CRH) by many scientists, though the use of CRF remains in the scientific vocabulary. During this same time period, others were conducting *in vivo* stress studies with vasopressin (VP) in the rat (Martini and Morpurgo, 1955; McCann, 1957). Initially VP was thought to be the putative CRH. This led to a controversy in the scientific community as to which substance, CRH or VP, was the primary factor responsible for the regulation of ACTH secretion. The controversy continued throughout the 1960s and 1970s, although there was increasing evidence that VP was indeed not the primary stimulator of ACTH secretion (Arimura *et al.*, 1967; Portanova and Sayers, 1973). Finally, in 1981, Vale and colleagues characterized the chemical structure and sequenced ovine CRH as a 41-amino acid hypothalamic peptide with intrinsic ACTH-releasing activity. Following the characterization of CRH were several studies demonstrating the physiological role of CRH as a primary regulator of ACTH secretion (Rivier and Plotsky, 1986).

It is now apparent that CRH and VP work independently, as well as in concert, to mediate glucocorticoid secretion at the level of the pituitary and adrenal glands. The magnitude and duration of the glucocorticoid response are secretogogue-dependent. An intriguing aspect of ACTH regulation is the ability of VP to increase the potency of CRH. In several species (rat, human, porcine, bovine), it has been demonstrated that VP possesses the ability to potentiate CRH-induced ACTH secretion (Liu *et al.*, 1983; Rivier and Vale, 1983; Watabe *et al.*, 1988; Carroll *et al.*, 1993; Minton and Parsons, 1993). However, what has not been fully elucidated is the need for multiple ACTH secretogogues.

While Selye's work sparked an interest in stress regulation throughout the scientific community, his hypothesis (that regardless of the stressor the body would respond in the same physiological manner) was an oversimplification of what we now know to be a very complex and integrated system that regulates glucocorticoid secretion. There are at least two other regulatory factors in addition to CRH and VP that have been reported to induce ACTH secretion from the anterior pituitary: adrenaline (Giguere and Labrie, 1983) and oxytocin (Link *et al.*, 1993). High affinity receptors for both adrenaline (Petrovic *et al.*, 1983) and oxytocin (Antoni, 1986) have been identified in the rat pituitary gland. The biological need for multiple stimulators of ACTH secretion has not been fully elucidated; however, the existence of multiple stimulators does highlight the relative importance of this system in the maintenance of homeostasis. Ultimately, an increase in plasma concentration of ACTH stimulates release of glucocorticoids from the adrenal cortex.

In mammals, the adrenal gland, like the pituitary gland, comprises multiple endocrine organs: the adrenal cortex and the adrenal medulla (Fig. 3.3). The adrenal medulla is encapsulated by the adrenal cortex; its primary secretions are the catecholamines epinephrine (adrenaline), norepinephrine (noradrenaline) and dopamine. The adrenal cortex comprises three layers: (i) the zona glomerulosa; (ii) the zona fasciculata; and (iii) the zona reticularis. In general, the adrenal cortex is responsible for the synthesis and release of three classes of adrenocortical steroid hormones (i.e. mineralocorticoids, glucocorticoids and androgens). Mineralocorticoids, predominately aldosterone, are synthesized in the zona glomerulosa, whereas the major products of the zona fasciculata and zona reticularis are the glucocorticoids and the androgens. While the role of adrenal androgens is limited to effects on reproductive performance, glucocorticoids and mineralocorticoids are essential for survival. Mineralocorticoids are essential for the maintenance of sodium balance and extracellular fluid volume, and glucocorticoids elicit a variety of effects on the metabolism of carbohydrates and protein.

Regulation of the stress response at the level of the adrenal gland is no less complex than regulation at the level of the hypothalamus or anterior pituitary gland. While glucocorticoid secretion from the adrenal cortex is mediated primarily by the endocrine action of ACTH from the anterior pituitary, CRH and VP may also regulate glucocorticoid production and secretion via paracrine (local) actions within the adrenal gland. The ability of CRH to increase adrenal blood flow and the localization of CRH and CRH mRNA in the adrenal gland suggest an intra-adrenal role for this neurohormone (Minamino *et al.*, 1988; Usui *et al.*, 1988; Muglia *et al.*, 1994). Indeed, CRH has been reported to directly stimulate glucocorticoid secretion from the adrenal gland in humans (Fehm *et al.*, 1988; Parker *et al.*, 1995), rats (Mazzocchi *et al.*, 1989) and cattle (Jones and Edwards, 1990; Carroll *et al.*, 1996). Additionally, CRH has been identified in the adrenal medulla of humans and several other mammalian species (Suda *et al.*, 1984), and has been reported to be capable of stimulating local ACTH production from chromaffin cells in the presence or absence of VP (Markowska *et al.*, 1993; Mazzocchi *et al.*, 1994, 1997). It has been suggested that VP may be capable of sustaining catecholamine and steroid secretion by acting directly through VP receptors in the adrenal cortex, or indirectly by acting on medullary VP receptors that stimulate local ACTH secretion (Markowska *et al.*, 1993; Mazzocchi *et al.*, 1994, 1997). In cultured bovine adrenocortical cells, VP stimulates cortisol in a time-dependent manner (Carroll *et al.*, 1996). In terms of potency, ACTH remains recognized as the predominant physiological regulator of cortisol production, although many investigations continue to assess the relative roles of angiotensin II, cytokines, growth factors, VIP, VP and CRH. Thus, the traditional view of HPA activation is continually undergoing modification as novel approaches are employed to elucidate the complex and integrative aspects of glucocorticoid synthesis and secretion.

The adrenal medulla also plays a role in the overall regulation of the HPA axis at several levels, including the brain and the pituitary gland. During embryonic development, the adrenal medulla originates from the neural crest and becomes a rather specialized sympathetic ganglion that secretes its products directly into the circulatory system rather than into a synaptic cleft. Thus, while the adrenal medulla is part of the sympathetic division of the autonomic nervous system, it functions as an endocrine and paracrine organ. While both adrenaline and noradrenaline increase alertness and awareness, adrenaline seems to be associated more with anxiety and fear (Fell *et al.*, 1985; Abelson *et al.*, 1996). In humans, adrenaline accounts for the majority of catecholamine output from the adrenal, while in the cat and other species noradrenaline is the primary catecholamine secreted. Catecholamines influence the HPA axis at many levels, including stimulating neurohormone release from the hypothalamus (Plotsky *et al.*, 1989), ACTH release from the pituitary gland (Axelrod and Reisine, 1984; Dinan, 1996) and cortisol release from the adrenal cortex (Dinan, 1996).

Thus, the increase in catecholamine synthesis and secretion during acute stress (Kvetnansky *et al.*, 1971; Axelrod, 1972; Kvetnansky, 1973) probably facilitates activation of the HPA axis.

HPA axis activation is therefore an important and complex adaptive response to stress. At this time, it is apparent that the original concept of an 'all or none' stress response is an over-simplification with regard to the HPA axis. In the mid-1980s Plotsky and co-workers demonstrated that specific stressors would elicit specific ACTH secretogogues. During haemorrhage, CRH, VP, oxytocin and catecholamines (all known stimulators of ACTH secretion) are released; however, during hypotension, CRH is the only secretogogue released (Plotsky *et al.*, 1985a, b). Others have also provided evidence that specific stressors elicit specific patterns of neurohormonal activation (Mason, 1974; Seggie and Brown, 1982), which also supports the concept that activation of the HPA axis is stressor specific. The effectiveness of glucocorticoid-mediated negative feedback on the HPA axis also varies among specific stressors (Plotsky *et al.*, 1993). It would appear that the brain has the ability to distinguish between stressors and to release ACTH secretogogues depending on the physiological response needed to cope with the current stressor. Similarly, plasma concentrations of catecholamines, which regulate the HPA response, are elevated by acute stress; but the magnitude of the increase is dependent on the intensity of the stressors (Natelson *et al.*, 1981; Goldstein *et al.*, 1983). While acute immobilization stress in the rat causes a subsequent reduction in adrenal medullary adrenaline, noradrenaline levels are unaffected. Thus, the body mounts a specific HPA response of a magnitude necessary to maintain or return to homeostasis. Central interpretative regulation in the brain, coupled with the intercommunication that exists between the sympathetic nervous system and the HPA, works in concert to maintain homeostasis.

Somatotrophic axis

The term 'somatotrophic axis' is generally used to refer to the integrated neural and endocrine mechanisms that control growth hormone (GH) production/secretion and the subsequent physiological responses to the secreted GH. Specialized cells of the anterior pituitary gland called somatotrophs produce and release GH. A variety of hormonal inputs can affect the somatotroph (Bertherat *et al.*, 1995). Growth hormone-releasing hormone (GHRH) and somatostatin are important hypothalamic factors that exert stimulatory and inhibitory effects on GH secretion, respectively (Fig. 3.2). One of the effects of GH is the stimulation of the liver to produce and release insulin-like growth factor-I (IGF-I). The growth and development of a variety of peripheral tissues are dependent on IGF-I. GH also exerts direct effects on numerous peripheral tissues (Holly and Wass, 1989).

Stress-induced reductions of GH and IGF-I secretion have been reported in rats (Armario *et al.*, 1987; Straus, 1994; Peisen *et al.*, 1995); however, data from other vertebrate species indicate that the somatotrophic axis responds to stress by concurrently increasing GH and decreasing IGF-I secretion (Vance *et al.*, 1992; Kakizawa *et al.*, 1995; Bruggeman *et al.*, 1997; Carroll *et al.*, 1998; McCusker, 1998). These endocrine responses act to divert energy from growth to survival. The increase in circulating GH antagonizes the effects of insulin by direct GH receptor-mediated actions on peripheral target tissues, thus reserving blood glucose. A reduction in IGF-I is thought to minimize growth during times of distress, further preserving energy for purposes of survival. While evidence is limited, it is interesting to note that positive emotional experiences may be associated with decreased GH secretion (Berk *et al.*, 1989).

In rats, crowding reduces GH secretion by altering regulatory signals from the hypothalamus (Armario *et al.*, 1987). Acute psychological stress transiently increases GH secretion in non-rodent species (Rushen *et al.*, 1993; Cataldi *et al.*, 1994; Gerra *et al.*, 1998). In humans, a variety of psychological stress tests elevate GH secretion within 30 min, with a significantly enhanced response in individuals with higher aggressive tendencies (Gerra *et al.*, 1998). Fifteen minutes of restraint by nose snare increases GH secretion in prepubertal gilts, with no attenuation when repeated daily for 9 days (Rushen *et al.*, 1993). The acute increase in GH secretion following restraint stress appears to be mediated by an increase in GHRH release (Cataldi *et al.*, 1994). Concurrent activation of the HPA axis during stress could enhance the GH response, as acute glucocorticoid treatment increases GH secretion (Casanueva *et al.*, 1990). While acute restraint elevates GH secretion in pigs within 0.5 h (Farmer *et al.*, 1991; Rushen *et al.*, 1993), IGF-I concentrations are rapidly reduced following application of the stressor (Farmer *et al.*, 1991). The physiological mechanism for this acute reduction of circulating IGF-I concentrations in the face of elevated GH secretion remains to be determined; however, a rapid reduction in IGF-I production can be induced by cortisol in cultured target cells (McCarthy *et al.*, 1990).

The effects of nutritional stress (undernutrition, fasting) on the somatotrophic axis are well documented (Vance *et al.*, 1992; Straus, 1994). Inadequate nutrition reduces both GH and IGF-I secretion in rodents (Straus, 1994). In other species undernutrition induces a concurrent elevation in GH secretion and suppression of circulating IGF-I (Ketelslegers *et al.*, 1995). Undernutrition reduces liver GH receptors and/or signal transduction at the GH receptor (Straus, 1994; Villares *et al.*, 1994; Ketelslegers *et al.*, 1995), thus contributing to suppressed IGF-I secretion in concert with high levels of GH. In piglets, weaning results in elevated GH secretion and reduced IGF-I and IGF-II secretion (Carroll *et al.*, 1998), the expected effects of undernutrition. As noted above, the concurrent elevation in GH and suppression of IGF secretion is an important adaptive response that diverts energy substrate from growth to survival.

Acute exposure to high temperature significantly elevates, while chilling reduces, GH secretion (Weeke and Gunderson, 1983). Chronic heat stress increases both basal and secretagogue-stimulated GH secretion in lactating sows (Barb *et al.*, 1991). Exposure to lower temperatures reduces IGF-I secretion (Ozawa *et al.*, 1994). The reduction in IGF secretion may be an adaptive response to an environment that requires increased energy expenditure to maintain body temperature rather than growth. Consistent with this concept, chronic exposure to temperatures at the upper end of the thermoneutral zone favours IGF secretion (Ma *et al.*, 1992). The effects of the thermal environment on the somatotrophic axis may be dependent on the developmental status of the animal. In very young pigs, housing at temperatures somewhat below the thermoneutral zone does not appear to affect GH secretion or somatotroph function (Matteri and Becker, 1994, 1996a). Interestingly, while housing at low temperature does not affect circulating levels of IGF-I or IGF-II in the young pig, lower levels of hepatic IGF-I, IGF-II and GH receptor mRNA are observed (Carroll *et al.*, 1999).

Lactotrophic axis

Most pituitary hormones respond to stimulatory input from releasing factors; however, the primary regulation of pituitary prolactin (PRL) secretion is thought to be mediated by the suppressive effects of hypothalamic dopamine, which reaches the pituitary gland through the hypophysial portal system (Fig. 3.2). It should be noted that the well-documented negative regulation of PRL does not preclude stimulatory control by thyrotropin-releasing hormone (TRH), neurophysin, substance P and other factors (Kuan *et al.*, 1990; Henriksen *et al.*, 1995; Shin *et al.*, 1995; Watanobe and Sasaki, 1995). The best-documented function of PRL is to stimulate milk synthesis and secretion; however, a variety of other functions may exist (Bole-Feysot *et al.*, 1998). In rodents, but not other species, PRL has a trophic effect directly on the corpus luteum (CL). PRL also appears to play a role in CL maintenance in the pig by preventing uterine luteolytic substances from reaching the ovaries (Bazer *et al.*, 1991). PRL also has been implicated in the control of salt–water balance, immunity, growth, development and metabolism (Nicoll, 1980; Weigent, 1996). Stress activation of the lactotrophic axis is a consistent observation.

Acute psychological stress elevates PRL secretion in a variety of species (Klemcke *et al.*, 1987; Jurcovicova *et al.*, 1990; Kirschbaum *et al.*, 1993; Matthews and Parrott, 1994; Juszczak, 1998). The effect of stress on PRL secretion is recognized as a factor that must be taken into consideration for the accurate clinical diagnosis of hyperprolactinaemia (Muneyyirci-Delale *et al.*, 1989). Behavioural factors may influence PRL secretion in response to stress. Passive, rather than active, coping responses to stress are associated with increased PRL secretion (Theorell, 1998). Aggressive tendencies such as fighting with peers, oppositive behaviour and rule violation in adolescent boys appear to be related to a higher level of basal PRL secretion (Gerra *et al.*, 1998). Since there are several hundred potential functions of PRL in vertebrates (Bole-Feysot *et al.*, 1998), assigning a specific adaptive purpose to stress-induced PRL secretion is difficult. There is speculation, however, that PRL may enhance the acquisition of active avoidance behaviour, induce analgesia and offer protection against chronic stress effects (Drago *et al.*, 1989).

Prolactin secretion is positively related to environmental temperature (Wettemann and Tucker, 1974). Although the inhibitory effect of dopamine on PRL secretion at the pituitary gland is well recognized, recent evidence suggests that stimulatory dopamine signalling within the ventromedial hypothalamus is involved with elevated PRL release due to heat exposure (Colthorpe *et al.*, 1998). Pretreatment of sheep with an antagonist to a specific dopamine receptor (D1 subtype) inhibits the PRL response to high ambient temperature, yet does not affect acute PRL secretion following psychological stress (Colthorpe *et al.*, 1998). Circulating levels of PRL are reduced within 2 h of cold exposure in humans (Leppaluoto *et al.*, 1988). Acute and chronic elevations in ambient temperature increase PRL secretion. This response is consistent and conserved across a variety of species. The increase in PRL secretion subsequent to long-term heat exposure is accompanied by a marked increase in pituitary PRL available for release (Matteri and Becker, 1994, 1996b). The underlying mechanism is unknown, but may involve an expansion of the lactotroph population, as can occur during oestrogen treatment (Pasolli *et al.*, 1992).

While most of the PRL responses to stress are characterized by increases in hormone secretion, reduced PRL levels may occur as a consequence of chronic stress, such as that encountered during prolonged illness (Van den Berghe and de Zegher, 1996). Whether this decline in PRL is attributable to the disease *per se* or to the associated decrease in food intake is not known. Fasting reduces PRL secretion in pigs, rats and humans (Tegelman *et al.*, 1986; Bergendahl *et al.*, 1989; Rojkittikhun *et al.*, 1993).

Gonadotrophic axis

Luteinizing hormone (LH) and follicle-stimulating hormone (FSH) are collectively termed gonadotropins due to their positive effects on gonadal structure and function. These hormones are produced in specialized cells of the anterior pituitary gland called gonadotrophs. The secretion and synthesis of LH and FSH are positively regulated by hypothalamic gonadotropin-releasing hormone (GnRH; Fig. 3.2). While gonadotrophs synthesize both

LH and FSH, a variety of mechanisms involving hormones such as activin, inhibin, follistatin and sex steroids exist that allow separate control of the secretion of both gonadotropins (Farnworth, 1995; Schwartz, 1995). LH induces ovulation and exerts trophic effects on the corpus luteum in the female. FSH supports maturation of ovarian follicles, maintains the size of the ovary and stimulates oestrogen production. In the male, LH stimulates the production of androgens from testicular Leydig cells, and FSH stimulates Sertoli cell function and is needed for sperm production.

Acute psychological and thermal stress can elicit a short-lived increase in LH, but not FSH, secretion (Siegel *et al.*, 1981; Sakamoto *et al.*, 1991). Chronic stress, however, is recognized as a cause of decreased gonadotropin secretion and reproductive failure (Moberg, 1991; Rivier and Rivest, 1991). As Hans Selye observed in 1939, 'It appears that in cases of emergency, the pituitary tends to produce more adrenocorticotropic and less gonadotropic hormone than under normal conditions. The reason for this is probably that, under certain conditions, an abundant supply of the life-maintaining principle of the adrenal cortex is a more imminent necessity than the preservation of normal sex function' (Selye, 1939). The neural and endocrine mechanisms underlying stress-induced reproductive failure are complex and have yet to be fully elucidated. Evidence exists for suppressive effects of glucocorticoids, vasopressin, ACTH, opioids and CRH on gonadotropin secretion (Matteri *et al.*, 1984; Moberg, 1991; Ferin, 1993; Rivest and Rivier, 1995; Dobson and Smith, 1995; Xiao *et al.*, 1996). The role of physiological levels of glucocorticoids in suppressing gonadotropin secretion is somewhat controversial, but is supported by recent evidence in sheep (Adams *et al.*, 1999; Daley *et al.*, 1999). Glucocorticoids may also exert direct inhibitory effects on gonadal steroid secretion and sensitivity of target tissues to sex steroids (Magiakou *et al.*, 1997).

The administration of CRH has been shown to inhibit GnRH release in rats, monkeys and women (Petraglia *et al.*, 1987; Gindoff and Ferin, 1987; Barbarino *et al.*, 1989; Chrousos *et al.*, 1998). In rats, neutralization of CRH or blockade of CRH action prevents the inhibition of LH secretion during stress (Rivier *et al.*, 1986). The suppressive effect on GnRH neurons may be mediated directly by CRH or by CRH induction of POMC-related peptides (Rivier and Rivest, 1991; Chrousos *et al.*, 1998; Tellam *et al.*, 1998). The evidence for a role for CRH in the inhibition of reproduction during stress, however, is not completely consistent among studies. The CRH-induced reduction in LH secretion observed in ovariectomized rhesus monkeys (Gindoff and Ferin, 1987) is not apparent in intact animals (Norman, 1994). The suppression of LH secretion by restraint stress in CRH-deficient mice demonstrates that other stress-related factors certainly are involved in the inhibition of reproduction (Jeong *et al.*, 1999). In sheep, central CRH administration appears to increase LH secretion, but the effect is dependent on the presence of sex steroids (Caraty *et al.*, 1997). Although evidence exists to support a role for CRH in the inhibition of reproduction during stress under specific experimental conditions, the underlying mechanisms are complex and may be dependent on animal species and a variety of physiological factors.

Nutritional stress poses a serious challenge to homeostasis. The importance of nutritional intake in maintaining reproductive function is well established (Schillo, 1992; Brown, 1994; Chilliard *et al.*, 1998). Inadequate nutrition delays or prevents the onset of puberty, interferes with normal cyclicity in the female and results in hypogonadism and infertility in males. A consistent observation across species is that undernutrition results in decreased gonadotropin secretion, an event mediated by reduced hypothalamic GnRH release (Schillo, 1992; Brown, 1994). A complex array of neural and neuroendocrine signals convey information of nutritional status to the reproductive system (Pettigrew and Tokach, 1993; Maeda *et al.*, 1996; Polkowska, 1996). Metabolites can also exert potent effects on endocrine systems (Widmaier, 1992), and have been implicated in the control of gonadotropin secretion (Pettigrew and Tokach, 1993). More recent evidence suggests that leptin, a hormone produced in fat tissue, is required for reproduction (Houseknecht *et al.*, 1998). Reduced leptin secretion due to inadequate nutrition may play an important role in mediating the associated disruption of reproductive neuroendocrine function (Houseknecht *et al.*, 1998; Nagatani *et al.*, 1998).

The sensitivity of the gonadotrophic axis to environmental temperature is well recognized. Heat stress has repeatedly been shown to exert inhibitory effects on gonadotropin secretion. High ambient temperature attenuates the post-castration rise in gonadotropins (Flowers and Day, 1990). Gonadotropin secretion is reduced by heat exposure, which reflects an inhibition of hypothalamic GnRH release (Donoghue *et al.*, 1989; Barb *et al.*, 1991; Gilad *et al.*, 1993). Responsiveness of the pituitary gland to GnRH can also be reduced by heat stress (Gilad *et al.*, 1993). Relative to the effects of heat, little information exists on the effects of cool temperatures on gonadotropin secretion. While acute exposure to cold water results in a brief elevation in LH secretion, this effect may reflect a general acute stress response (Sakamoto *et al.*, 1991). Long-term exposure to temperatures below the thermoneutral zone reduces LH and FSH secretion and the amount of releasable hormone in the pituitary gland (Wada, 1993; Matteri and Becker, 1996b). The underlying mechanism of reduced gonadotropin secretion may reflect inadequate nutrition relative to increased metabolic demand at colder temperatures. As leptin appears to be required for GnRH and gonadotropin secretion (Houseknecht *et al.*, 1998), a reduction in leptin production during cold stress (see section below on appetite control) would be expected to suppress the reproductive neuroendocrine axis.

Thyrotrophic axis

The principal components of the endocrine thyroid (thyrotrophic) axis, summarized in Fig. 3.2, are hypothalamic thyrotropin-releasing hormone (TRH), pituitary thyroid-stimulating hormone (TSH) and the thyroid hormones (tri- and tetra-iodothyronine, T_3 and T_4, respectively). As with other hypothalamic hormones, TRH is released into the hypophysial portal vessels, where it is carried to the anterior pituitary. The pituitary cell type that produces and secretes TSH in response to TRH is called the thyrotroph. As potent metabolic regulators, thyroid hormones play a major role in controlling body temperature and metabolism (Danforth and Burger, 1984).

Relative to other anterior pituitary hormones, less is known with regard to the effects of psychological stress on TSH secretion. The effect of acute psychological stress on TSH secretion may vary between species. While short-term stress produces a transient increase in TSH secretion in humans (Richter *et al.*, 1996), a decrease in secretion is observed in rodents (Goya *et al.*, 1995; Marti *et al.*, 1996). In rats, crowding reduces TSH secretion without altering pituitary sensitivity to secretory stimulation (Armario *et al.*, 1987). Acute psychological stress induces an increase in TSH secretion in humans that occurs within 20 min (Richter *et al.*, 1996). Maximal increases in T_3 and T_4 secretion occur within 10 min of acute stress in pigs (Farmer *et al.*, 1991). The physiological benefit of stress-altered TSH secretion has not been elucidated.

Several hours of cold stress increase TSH and thyroid hormone levels in rodents (Goya *et al.*, 1995) but not humans (Leppaluoto *et al.*, 1988). Perhaps the smaller size and associated thermal capacity of small animals result in more rapid changes in core temperature during cold stress which then trigger activation of the thyroid axis. In general, thyroid hormone production responds to changes in thermogenic demand imposed by chronic changes in the thermal environment. Cool temperatures increase the activity of the thyroid axis (Arancibia *et al.*, 1996). Housing in a cool thermoneutral environment greatly increases pituitary thyrotroph responsiveness to secretory stimulation as compared with a warm thermoneutral environment (Matteri and Becker, 1994). Exposure to cold temperatures increases TRH gene expression and secretion (Joseph-Bravo *et al.*, 1998).

Nutritional stress decreases overall activity of the thyroid axis. The elevation in HPA activity during chronic undernutrition may contribute to suppressed activity of the thyroid axis. Infusion of cortisol at levels similar to those observed during fasting suppresses TSH secretion (Samuels and McDaniel, 1997). In addition to reducing thyrotroph function, undernutrition diminishes hypothalamic TRH release, thyroid hormone production and levels of peripheral thyroid hormone receptors (Tegelman *et al.*, 1986; Nordio *et al.*, 1989; Everts *et al.*, 1996; Tagami *et al.*, 1996; Diano *et al.*, 1998). The decrease in thyroid axis function at so many levels reflects an important adaptive response to undernutrition (Sul and Wang,

1998). A reduction in metabolic rate and associated energy use has a positive survival value at times when the food supply is limited (Chilliard *et al.*, 1998).

Neuroendocrine control of appetite during the stress response

In the sections above, we have briefly touched upon the major impacts of stress on classical neuroendocrine systems. Various stressors also influence neural and neuroendocrine mechanisms involved with appetite control. The impact of stress on appetite has long been recognized (Forbes, 1995); however, recent advances in our understanding of appetite control have revealed neural and endocrine linkages that offer novel explanations for the suppression of appetite. As feed intake is necessary for the growth and survival of all animals, it is important for us to understand how common stressors reduce feed intake at the biochemical level, with the hope of someday being able to prevent or diminish appetite loss and subsequent reduction in the growth, health and well-being of animals. While the primary emphasis of this section relates to interactions between stress and appetite, the reader should also be aware that regulation of food intake is currently a topic of great interest within the human clinical fields associated with diseases such as diabetes, obesity, cancer and anorexia nervosa.

Regions of the hypothalamus have long been associated with the control of essential homeostatic functions including water balance, thermoregulation and appetite. Great strides have been taken towards elucidating the complex interplay of neuroendocrine signals involved in appetite regulation, particularly within the leptin–neuropeptide Y (NPY) axis. Although many different stressors influence feed intake, we are only just beginning to investigate the endocrine changes occurring due to a particular stress. Within the area of endocrine control of appetite, changes occurring due to hypothermic activation of the metabolic axis, stress activation of the HPA axis and disease activation of the immune axis have been best characterized. These are outlined below, following a brief review of the leptin–NPY axis.

Review of appetite regulation

One of the most potent stimulators of feed intake in animals is neuropeptide Y, a 36 amino acid peptide found throughout the peripheral nervous system but also produced in the brain (Tatemoto *et al.*, 1982). In the hypothalamus, NPY mRNA is produced in cell bodies within the arcuate nucleus, whereas the mature NPY peptide is released at terminals mainly within the paraventricular (upper lateral) nucleus (Pelletier, 1990). Feed restriction increases both transcription and peptide levels of NPY in the hypothalamus of rodents (Sahu *et al.*, 1988) and sheep (McShane *et al.*, 1993). Central (intracerebroventricular) administration of NPY induces an immediate increase in appetite in sheep (Miner *et al.*, 1989) and pigs (Parrott *et al.*, 1986), as well as in many other animals.

Leptin (Zhang *et al.*, 1994) is a 16 kDa protein produced in several tissues, including adipose. Adipocytes produce and secrete leptin in quantities directly and positively correlated with the adiposity of the animal and thus indirectly correlated with body weight (for a comprehensive review of leptin, see Houseknecht and Portocarrero, 1998). While roles for leptin have recently been implicated in the areas of placental nutrient transfer and fetal growth (Hassink *et al.*, 1997), as well as in angiogenesis (Rocio Sierra-Honigmann *et al.*, 1998), a primary role for leptin is the endocrine regulation of NPY production and release in the brain. Leptin administered either centrally or peripherally decreases feed intake (and subsequently decreases body weight), presumably at least partly through its actions on NPY release. There are leptin receptors within the arcuate (ARC, lower medial) area of the hypothalamus of sheep (Dyer *et al.*, 1997a), and they are co-localized with NPY neurons in murine ARC (Mercer *et al.*, 1996). Animals that lack functional leptin (such as the *ob/ob* 'obese' mouse) or do not produce functional leptin receptors (such as *db/db* 'diabetic' mice and *fa/fa* Zucker 'fatty' rats) over-express NPY, are hyperphagic and obese.

In normal animals, leptin and NPY work in concert to maintain energy balance. Feedback mechanisms exist which would further ensure homeostasis. Leptin receptors appear to be down-regulated by leptin and highly expressed when leptin levels are low (Dyer *et al.*, 1997a; Mercer *et al.*, 1997; Baskin *et al.*, 1998). Increases in mRNA expression of both leptin and NPY receptors in sheep adipose tissue occur in response to a peripheral injection of NPY (Dyer *et al.*, 1997b).

A variety of other peptides, such as galanin and cholecystokinin, influence appetite (Lee *et al.*, 1994; Hirschberg, 1998). The recently discovered neuropeptide orexins (ORX) also increase feed intake in adult rats (Sakurai *et al.*, 1998) and in weanling pigs (Dyer *et al.*, 1999). Like NPY, orexin mRNA expression increases in rats which have been feed-restricted (Sakurai *et al.*, 1998). Very little is known about the orexins or their receptors, and there is no evidence yet to indicate any role for leptin in regulating ORX function. Our laboratory has found positive relationships between body weight and orexin mRNA expression, and between orexin mRNA and type 2 orexin receptor mRNA expression in weanling pigs (unpublished). Although we know too little at this point to speculate further on the possible role of orexins in appetite control, work is continuing in this direction and hopefully we will soon know more about this new family of neuropeptides.

Effects of thermal environment on appetite

It makes sense that an individual adapts to changes in thermal environment by adjusting energy expenditure to maintain normal body temperature (see section on thyrotrophic axis above). At the molecular level, however, we are far from understanding the complex neuroendocrine mechanisms that occur during episodes of climatic change. The suppressive effect of heat stress on appetite is well recognized (Forbes, 1995), but virtually nothing is known about the underlying endocrine mechanisms. The limited information that exists about neuroendocrine–appetite interactions during thermal stress relates to cold exposure. Since the appetite regulators leptin and NPY also have metabolic effects, responses of the leptin–NPY axis to ambient temperature are of particular interest. In addition to stimulating feed intake, central administration of NPY suppresses sympathetic outflow to brown adipose tissue (BAT) and reduces uncoupling protein-1 (UCP-1) gene expression (Egawa *et al.*, 1991; Billington *et al.*, 1994). UCP-1 uncouples mitochondrial respiration, causing an increase in energy utilization for non-shivering thermogenesis. While a decrease in UCP-1 expression would minimize energy loss and therefore be beneficial during feed restriction, it would not be beneficial and could potentially be fatal during a period of hypothermia.

Recent work suggests that a response in NPY expression to hypothermia depends on the duration of exposure, but hypothermia consistently decreases leptin expression. Trayhurn *et al.* (1995) reported a complete disappearance of leptin mRNA within hours in adipose tissue of cold-exposed mice, and provided evidence that the cold-induced suppression of leptin mRNA expression is mediated by catecholamine activation of the fat cell β-adrenoceptor. In a similar study, plasma leptin concentrations declined in normal rats exposed to 4°C for 24 h, but not in cold-exposed *fa/fa* rats which lack the fully functional form of the leptin receptor (Hardie *et al.*, 1996). The authors of this study proposed that the lack of leptin modulation in the *fa/fa* rats in response to cold may reflect reduced β-adrenoceptor numbers in the adipose tissue of these animals (Muzzin *et al.*, 1991).

McCarthy *et al.* (1993) exposed rats to 4°C for 2.5 or 18 h, and observed that NPY protein levels were significantly elevated (80–170%) in various hypothalamic areas, but were *reduced* by 21% in the ARC after 2.5 h and comparable with controls after 18 h. While changes in NPY mRNA levels were not investigated, it was proposed that the increase in NPY protein content was not due to increased synthesis, as much as to inhibited release. In a later experiment, Mercer and colleagues (1997) found that exposure of mice to 4°C for 24 h caused an induction of both leptin receptor and NPY mRNA expression, but actual NPY release was not measured.

In contrast to the results of acute-term exposure experiments such as those described above, chronic cold exposure in rats (4°C for 21 days) increased feed intake and decreased plasma leptin concentrations but did not change hypothalamic NPY protein levels or mRNA expression (Bing *et al.*, 1998). In this experiment, cold-exposed rats weighed 14% less than controls after 21 days, despite a 10% increase in daily food intake which continued from the third day of exposure. Leptin levels continued to correlate to animal body weight throughout the experiment. BAT activity (as measured by UCP-1 mRNA expression) was increased by day 2 and was elevated to 335% of control values by day 21. The authors noted that, although the profile associated with cold-induced hyperphagia (decreased leptin but no increase in NPY) differs from starvation-induced hyperphagia (where NPY is elevated), the difference may in fact be physiologically appropriate. The inhibition of thermogenesis caused by an induction of NPY activity would be extremely detrimental to an animal facing hypothermia.

Although the mechanisms are still largely unknown, it appears that leptin acts centrally to inhibit feeding *in a manner independent of NPY modulation*, and stimulates sympathetically driven BAT activity and thermogenesis. The hyperphagia seen in cold-exposed rats is less than the marked increases in feed intake seen when comparable body weight is lost through feed restriction, diabetes or lactation. In those situations, NPY release is elevated and BAT activity is suppressed. Expression of other orexogenic (appetite-stimulating) hormones has not been measured in hypothermic conditions; possible contributors to hyperphagia without suppressing BAT activity may include the orexins, or as yet undiscovered factors.

The hypothermia scenario described above may apply to animals with BAT, but it is unknown how these mechanisms may be altered in animals that lack BAT (e.g. swine). Once pigs reach 1 week of age, they respond to cooler temperatures by ingesting more milk (Forbes, 1995). Sometimes, the increase in voluntary consumption is greater than necessary to maintain weight. Reduced nocturnal temperature increases intake and growth by about 10% in piglets. In older growing pigs, Holme and Coey (1967) found that cold-stressed animals not only ate more than controls but also gained more weight. This phenomenon has been observed in both younger (Suguhara *et al.*, 1970) and older (Jensen *et al.*, 1969) growing pigs. Comparisons across studies must be made with great care, as species differences, degree of cold stress and duration of cold stress must all be considered.

Effects of generalized stress (induction of HPA axis) on appetite

Only recently has enough information been gathered on how the factors of the HPA axis (in particular CRH and glucocorticoids) affect feed intake regulators to form a theory as to how feed intake is decreased during times of general or perceived stress. Once again, complicated interplays between multiple hormones make elucidation of the pathway(s) involved extremely difficult.

CRH, produced in the hypothalamus, has a direct effect of decreasing feed intake in both normal and NPY-deficient mice (NPY –/– knockout), suggesting that this effect is not mediated through NPY (Hollopeter *et al.*, 1998). Leptin treatment increases hypothalamic CRH content and decreases appetite in rats (Uehara *et al.*, 1998). The suppression of appetite produced by leptin treatment is attenuated by co-administration of a CRH antagonist (Uehara *et al.*, 1998). Glucocorticoids increase plasma leptin levels in rats (Newcomer *et al.*, 1998). Makino *et al.* (1998) postulated that the increased leptin secretion does not decrease feed intake exclusively through decreasing NPY levels, but by increasing sensitivity to CRH. There are receptors for CRH within the ventromedial hypothalamus (VMH), an area which is rich in leptin receptors. In a series of experiments using adrenalectomized, feed-restricted and corticosterone-treated rats, Makino and colleagues (1998) rendered convincing evidence that the decreased feed intake associated with high glucocorticoid levels is at least partially induced by increasing leptin secretion, thereby increasing the sensitivity of the VMH to CRH by inducing CRH receptor mRNA expression. The exact mechanism by which CRH acts remains unknown.

Effects of immunological stress on appetite

Disease challenges elicit potent and well-defined stress responses involving a variety of immune system hormones referred to as cytokines (Curfs *et al.*, 1997). The most studied of these hormones with regard to stress are the pro-inflammatory cytokines interleukin-1β (IL-1β), interleukin-6 (IL-6) and tumour necrosis factor-α (TNF-α), which are involved in the acute-phase response to inflammatory stressors (Koj, 1996; Maule and VanderKooi, 1999). Other prominent pro-inflammatory cytokines include IL-8 and interferon-γ (Koj, 1998). Cytokines can affect many components of neuroendocrine function (Mandrup-Poulsen *et al.*, 1995; Maule and VanderKooi, 1999).

One of the first symptoms of any disease is decreased appetite, often coinciding with fever. Evidence has begun to accumulate that pro-inflammatory cytokines play a role in disease-induced anorexia, as reviewed by Johnson (1997, 1998). IL-6, IL-1β and TNF-α have been implicated in disease-related anorexia. Both peripheral and central injections of recombinant IL-1β induce anorexia (Dantzer and Kelley, 1989), whereas only central administration of IL-6 will induce anorexia (Schobitz *et al.*, 1995; Weingarten, 1996). TNF-α, like IL-1β, suppresses feed intake when administered either centrally (Warren *et al.*, 1997; Plata-Salaman *et al.*, 1998) or peripherally (Mahoney *et al.*, 1988). IL-1β is much more effective than TNF-α when given centrally (Johnson *et al.*, 1997; Segreti *et al.*, 1997); however, TNF-α may have an important peripheral role in directly stimulating leptin secretion from adipocytes (Finck *et al.*, 1998). All of these cytokines are produced within the central nervous system (Johnson *et al.*, 1997) and are inducible by central (Ilyin *et al.*, 1998) or peripheral (Laye *et al.*, 1994) lipopolysaccharide (LPS) administration. Unfortunately, very little is known about the mechanism(s) by which these cytokines act within the brain to inhibit feeding. When LPS was administered intracerebroventricularly in rats, IL-1β, IL-1 receptor type 1, and TNF-α mRNA expression in the hypothalamus, as well as anorexia, was induced; but no change in NPY mRNA expression was observed (Ilyin *et al.*, 1998). Potent activation of the HPA axis by cytokines (Mandrup-Poulsen *et al.*, 1995; Rivier, 1995) may also contribute to appetite suppression (see preceding section on the effects of generalized stress).

Summary

Neuroendocrine and endocrine responses to stress play an integral role in the maintenance of homeostasis. In general, endocrine responses to stress work towards inhibiting non-essential functions such as growth and reproduction, in favour of maintenance and survival. Substantial evidence suggests that neuroendocrine responses to stress can be specific and graded, rather than 'all or none'. Acute responses have important adaptive functions and are vital to coping and survival. Long-term chronic stressors elicit endocrine responses that may actually contribute to morbidity and mortality. The effects of stress on appetite are well recognized, but we are just beginning to understand the complex hormonal interplay mediating these effects. Integrated, multidisciplinary approaches that fully utilize emerging technologies and methods of molecular biology, neurology and endocrinology will lead future advances in our understanding of the biology of stress and related animal well-being.

References

Abelson, J.L., Weg, J.G., Nesse, R.M. and Curtis, G.C. (1996) Neuroendocrine responses to laboratory panic: cognitive intervention in the doxapram model. *Psychoneuroendocrinology* 21, 375–390.

Adams, T.E., Sakurai, H. and Adams, B.M. (1999) Effect of stress-like concentrations of cortisol on estradiol-dependent expression of gonadotropin-releasing hormone receptor in orchidectomized sheep. *Biology of Reproduction* 60, 164–168.

Antoni, F.A. (1986) Oxytocin receptors in rat adenohypophysis: evidence from radioligand binding studies. *Endocrinology* 119, 2393–2395.

Arancibia, S., Rage, F., Astier, H. and Tapia-Arancibia, L. (1996) Neuroendocrine and autonomous mechanisms underlying thermoregulation in cold environment. *Neuroendocrinology* 64, 257–267.

Arimura, A., Saito, T., Bowers, C. and Schally, A. (1967) Pituitary–adrenal activation in rats with hereditary hypothalamic diabetes insipidus. *Acta Endocrinologica* 54, 155–165.

Armario, A., Garcia-Marquez, C. and Jolin, T. (1987) Crowding-induced changes in basal and stress levels of thyrotropin and somatotropin in male rats. *Behavioral and Neural Biology* 48, 334–343.

Asa, S.L., Kovacs, K. and Melmed, S. (1995) The hypothalamic–pituitary axis. In: Melmed, S. (ed.) *The Pituitary*. Blackwell Science, Cambridge, Massachusetts, pp. 3–44.

Axelrod, J. (1972) Dopamine-β-hydroxylase: regulation of its synthesis and release from nerve terminals. *Pharmacological Reviews* 24, 233–243.

Axelrod, J. and Reisine, T.D. (1984) Stress hormones: their interaction and regulation. *Science* 224, 452–459.

Barb, C.R., Estienne, M.J., Kraeling, R.R., Marple, D.N., Rampacek, G.B., Rahe, C.H. and Sartin, J.L. (1991) Endocrine changes in sows exposed to elevated ambient temperature during lactation. *Domestic Animal Endocrinology* 8, 117–127.

Barbarino, A., De Marinis, L., Tofani, A., Della Casa, S., D'Amico, C., Mancini, A., Corsello, S.A., Sciuto, R. and Barini, A. (1989) Corticotropin-releasing hormone inhibition of gonadotropin release and the effects of opioid blockade. *Journal of Clinical Endocrinology and Metabolism* 68, 523–528.

Baskin, D.G., Seeley, R.J., Kuijper, J.L., Lok, S., Weigle, D.S., Erickson, J.C., Palmiter, R.D. and Schwartz, M.W. (1998) Increased expression of mRNA for the long form of the leptin receptor in the hypothalamus is associated with leptin hypersensitivity and fasting. *Diabetes* 47, 538–543.

Bazer, F.W., Simmen, R.C. and Simmen, F.A. (1991) Comparative aspects of conceptus signals for maternal recognition of pregnancy. *Annals of the New York Academy of Sciences* 622, 202–211.

Bergendahl, M., Perheentupa, A. and Huhtaniemi, I. (1989) Effect of short-term starvation on reproductive hormone gene expression, secretion and receptor levels in male rats. *Journal of Endocrinology* 121, 409–417.

Berk, L.S., Tan, S.A., Fry, W.F., Napier, B.J., Lee, J.W., Hubbard, R.W., Lewis, J.E. and Eby, W.C. (1989) Neuroendocrine and stress hormone changes during mirthful laughter. *American Journal of Medical Science* 298, 390–396.

Bertherat, J., Bluet-Pajot, M.T. and Epelbaum, J. (1995) Neuroendocrine regulation of growth hormone. *European Journal of Endocrinology* 132, 12–24.

Billington, C.J., Briggs, J.E., Grace, H.M. and Levine, A.S. (1994) Neuropeptide Y in hypothalamic paraventricular nucleus: a center coordinating energy metabolism. *American Journal of Physiology (Regulatory, Integrative, and Comparative Physiology)* 35, R1765–R1770.

Bing, C., Frankish, H.M., Pickavance, L., Wang, Q., Hopkins, D.F., Stock, M.J. and Williams, G. (1998) Hyperphagia in cold-exposed rats is accompanied by decreased plasma leptin but unchanged hypothalamic NPY. *American Journal of Physiology (Regulatory, Integrative, and Comparative Physiology)* 43, R62–R68.

Bole-Feysot, C., Goffin, V., Edery, M., Binart, N. and Kelly, P.A. (1998) Prolactin (PRL) and its receptor: actions, signal transduction pathways and phenotypes observed in PRL receptor knockout mice. *Endocrine Reviews* 19, 225–268.

Braunstein, G.D. (1995) The hypothalamus. In: Melmed, S. (ed.) *The Pituitary*. Blackwell Science, Cambridge, Massachusetts, pp. 309–340.

Brown, B.W. (1994) A review of nutritional influences on reproduction in boars, bulls and rams. *Reproduction, Nutrition, Développement* 34, 89–114.

Bruggeman, V., Vanmontfort, D., Renaville, R., Portetelle, D. and Decuypere, E. (1997) The effect of food intake from two weeks of age to sexual maturity on plasma growth hormone, insulin-like growth factor-I, insulin-like growth factor-binding proteins, and thyroid hormones in female broiler breeder chickens. *General and Comparative Endocrinology* 107, 212–220.

Caraty, A., Miller, D.W., Delaleu, B. and Martin, G.B. (1997) Stimulation of LH secretion in sheep by central administration of corticotrophin-releasing hormone. *Journal of Reproduction and Fertility* 111, 249–257.

Carroll, J.A., Gillespie, J.C., Willard, S.T., Kemper, C.N. and Welsh, T.H., Jr (1993) Utilization of corticotropin-releasing factor and vasopressin to assess the pituitary–adrenocortical system in cattle. *Journal of Animal Science* 71 (Suppl. 1), 207.

Carroll, J.A., Willard, S.T., Bruner, B.L., McArthur, N.H. and Welsh, T.H., Jr (1996) Mifepristone modulation of ACTH and CRH regulation of bovine adrenocorticosteroidogenesis *in vitro*. *Domestic Animal Endocrinology* 13, 339–349.

Carroll, J.A., Veum, T.L. and Matteri, R.L. (1998) Endocrine responses to weaning and changes in post-weaning diet in the young pig. *Domestic Animal Endocrinology* 15, 183–194.

Carroll, J.A., Buonomo, F.C., Becker, B.A. and Matteri, R.L. (1999) Interactions between environmental temperature and porcine growth hormone (pGH) treatment in neonatal pigs. *Domestic Animal Endocrinology* 16, 103–113.

Casanueva, F.F., Burguera, B., Muruais, C. and Dieguez, C. (1990) Acute administration of corticoids: a new and peculiar stimulus of growth hormone secretion in man. *Journal of Clinical Endocrinology and Metabolism* 70, 234–237.

Cataldi, M., Magnan, E., Guillaume, V., Dutour, A., Sauze, N., Mazzocchi, L., Conte-Devolx, B. and Oliver, C. (1994) Acute stress stimulates secretion of GHRH and somatostatin into hypophysial portal blood of conscious sheep. *Neuroscience Letters* 178, 103–106.

Chilliard, Y., Bocquier, F. and Doreau, M. (1998) Digestive and metabolic adaptations of ruminants to undernutrition, and consequences on reproduction. *Reproduction, Nutrition, Développement* 38, 131–152.

Chrousos, G.P., Torpy, D.J. and Gold, P.W. (1998) Interactions between the hypothalamic-pituitary-adrenal axis and the female reproductive system: clinical implications. *Annals of Internal Medicine* 129, 229–240.

Colthorpe, K.L., Anderson, S.T., Martin, G.B. and Curlewis, J.D. (1998) Hypothalamic dopamine D1 receptors are involved in the stimulation of prolactin secretion by high environmental temperature in the female sheep. *Journal of Neuroendocrinology* 10, 503–509.

Curfs, J.H., Meis, J.F. and Hoogkamp-Korstanje, J.A. (1997) A primer on cytokines: sources, receptors, effects, and inducers. *Clinical Microbiology Reviews* 10, 742–780.

Daley, C.D., Sakurai, H., Adams, B.M. and Adams, T.E. (1999) Effect of stress-like concentrations of cortisol on gonadotroph function in orchidectomized sheep. *Biology of Reproduction* 60, 158–163.

Danforth, E., Jr and Burger, A. (1984) The role of thyroid hormones in the control of energy expenditure. *Clinics in Endocrinology and Metabolism* 13, 581–595.

Dantzer, R. and Kelley, K.W. (1989) Stress and immunity: an integrated view of relationships between the brain and the immune system. *Life Sciences* 44, 1995–2008.

Diano, S., Naftolin, F., Goglia, F. and Horvath, T.L. (1998) Fasting-induced increase in type II iodothyronine deiodinase activity and messenger ribonucleic acid levels is not reversed by thyroxine in the rat hypothalamus. *Endocrinology* 139, 2879–2884.

Dinan, T.G. (1996) Serotonin and the regulation of the hypothalamic-pituitary-adrenal axis function. *Life Sciences* 58, 1683–1694.

Dobson, H. and Smith, R.F. (1995) Stress and reproduction in farm animals. *Journal of Reproduction and Fertility* 49 (Suppl.), 451–461.

Donoghue, D.J., Krueger, B.F., Hargis, B.M., Miller, A.M. and el Halwani, M. (1989) Thermal stress reduces serum luteinizing hormone and bioassayable hypothalamic content of luteinizing hormone-releasing hormone in hens. *Biology of Reproduction* 41, 419–424.

Drago, F., D'Agata, V., Iacona, T., Spadaro, F., Grassi, M., Valerio, C., Raffaele, R., Astuto, C., Lauria, N. and Vitetta, M. (1989) Prolactin as a protective factor in stress-induced biological changes. *Journal of Clinical Laboratory Analysis* 3, 340–344.

Dyer, C.J., Simmons, J.M., Matteri, R.L. and Keisler, D.H. (1997a) Leptin receptor mRNA is expressed in ewe anterior pituitary and adipose tissue and is differentially expressed in hypothalamic regions of well-fed and feed-restricted ewes. *Domestic Animal Endocrinology* 14(2), 119–128.

Dyer, C.J., Simmons, J.M., Matteri, R.L. and Keisler, D.H. (1997b) Effects of an intravenous injection of NPY on leptin and NPY-Y1 receptor mRNA expression in ovine adipose tissue. *Domestic Animal Endocrinology* 14(5), 325–333.

Dyer, C.J., Touchette, K.J., Carroll, J.A., Allee, G.L. and Matteri, R.L. (1999) Cloning of porcine prepro-orexin cDNA and effects of synthetic porcine orexin-B on feed intake in young pigs. *Domestic Animal Endocrinology* 16, 145–148.

Egawa, M., Yoshimatsu, H. and Bray, G.A. (1991) Neuropeptide Y suppresses sympathetic activity to intrascapular brown adipose tissue in rats. *American Journal of Physiology (Regulatory, Integrative, and Comparative Physiology)* 29, R328–R334.

Everts, M.E., de Jong, M., Lim, C.F., Docter, R., Krenning, E.P., Visser, T.J. and Hennemann, G. (1996) Different regulation of thyroid hormone transport in liver and pituitary: its possible role in the maintenance of low T3 production during nonthyroidal illness and fasting in man. *Thyroid* 6, 359–368.

Farmer, C., Dubreuil, P., Couture, Y., Brazeau, P. and Petitclerc, D. (1991) Hormonal changes following an acute stress in control and somatostatin-immunized pigs. *Domestic Animal Endocrinology* 8, 527–536.

Farnworth, P.G. (1995) Gonadotrophin secretion revisited. How many ways can FSH leave a gonadotroph? *Journal of Endocrinology* 145, 387–395.

Fehm, H.L., Holl, R., Spath-Schwalbe, E., Born, J. and Voigt, K.H. (1988) Ability of corticotropin releasing hormone to stimulate cortisol secretion independent from pituitary ACTH. *Life Sciences* 42, 679–686.

Fell, D., Derbyshire, D.R., Maile, C.J., Larsson, I.M., Ellis, R., Achola, K.J. and Smith, G. (1985) Measurement of plasma catecholamine concentrations: an assessment of anxiety. *British Journal of Anaesthesiology* 57, 770–774.

Ferin, M. (1993) Neuropeptides, the stress response, and the hypothalamo-pituitary-gonadal axis in the female rhesus monkey. *Annals of the New York Academy of Sciences* 697, 106–116.

Finck, B.N., Kelley, K.W., Dantzer, R. and Johnson, R.W. (1998) *In vivo* and *in vitro* evidence for the involvement of TNF-α in the induction of leptin by lipopolysaccharide. *Endocrinology* 139, 2278–2283.

Fink, G., Rosie, R., Sheward, W.J., Thomson, E. and Wilson, H. (1991) Steroid control of central neuronal interactions and function. *Journal of Steroid Biochemistry and Molecular Biology* 40, 123–132.

Flowers, B. and Day, B.N. (1990) Alterations in gonadotropin secretion and ovarian function in prepubertal gilts by elevated environmental temperature. *Biology of Reproduction* 42, 465–471.

Forbes, J.M. (1995) Environmental factors affecting intake. In: Forbes, J.M. (ed.) *Voluntary Food Intake and Diet Selection in Farm Animals.* CAB International, Wallingford, UK, pp. 332–353.

Gerra, G., Zaimovic, A., Giucastro, G., Folli, F., Maestri, D., Tessoni, A., Avanzini, P., Caccavari, R., Bernasconi, S. and Brambilla, F. (1998) Neurotransmitter-hormonal responses to psychological stress in peripubertal subjects: relationship to aggressive behavior. *Life Sciences* 62, 617–625.

Giguere, V. and Labrie, F. (1983) Additive effects of epinephrine and corticotropin-releasing factor (CRF) on adrenocorticotropin release in rat anterior pituitary cells. *Biochemical and Biophysical Research Communications* 110, 456–462.

Gilad, E., Meidan, R., Berman, A., Graber, Y. and Wolfenson, D. (1993) Effect of heat stress on tonic and GnRH-induced gonadotrophin secretion in relation to concentration of oestradiol in plasma of cyclic cows. *Journal of Reproduction and Fertility* 99, 315–321.

Gindoff, P.R. and Ferin, M. (1987) Endogenous opioid peptides modulate the effect of corticotropin-releasing factor on gonadotrophin release in the primate. *Endocrinology* 121, 837–842.

Goldstein, D.S., McCarty, R., Polinsky, R.J. and Kopin, I.J. (1983) Relationship between plasma norepinephrine and sympathetic neural activity. *Hypertension* 5, 552–559.

Goya, R.G., Sosa, Y.E., Console, G.M. and Dardenne, M. (1995) Altered thyrotropic and somatotropic responses to environmental challenges in congenitally athymic mice. *Brain, Behavior, and Immunity* 9, 79–86.

Guillemin, R. and Rosenberg, B. (1955) Humoral hypothalamic control of anterior pituitary: a study with combined tissue cultures. *Endocrinology* 57, 599–607.

Hardie, L.J., Raynor, D.V., Holmes, S. and Trayhurn, P. (1996) Circulating leptin levels are modulated by fasting, cold exposure and insulin administration in lean but not Zucker (fa/fa) rats as measured by ELISA. *Biochemical and Biophysical Research Communications* 223(3), 660–665.

Harris, G. (1948) Neural control of the pituitary gland. *Physiological Reviews* 28, 139–179.

Hassink, S.G., Delancy, E., Sheslow, D.V., Smithkirwin, S.M., O'Connor, D.M., Considine, R.V., Opentanova, I., Dostal, K., Spear, M.L., Leef, K., Ash, M., Spitzer, A.R. and Funanage, V.L. (1997) Placental leptin: an important new growth factor in intra-uterine and neonatal development. *Pediatrics* 100, E11–E16.

Henriksen, J.S., Saermark, T., Vilhardt, H. and Mau, S.E. (1995) Tachykinins induce secretion of prolactin from perifused rat anterior pituitary cells by interactions with two different binding sites. *Journal of Receptor and Signal Transduction Research* 15, 529–541.

Hirschberg, A.L. (1998) Hormonal regulation of appetite and food intake. *Annals of Medicine* 30, 7–20.

Hollopeter, G., Erickson, J.C., Seeley, R.J., Marsh, D.J. and Palmiter, R.D. (1998) Response of neuropeptide Y-deficient mice to feeding effectors. *Regulatory Peptides* 75-6, 383–389.

Holly, J.M. and Wass, J.A. (1989) Insulin-like growth factors; autocrine, paracrine or endocrine? New perspectives of the somatomedin hypothesis in light of recent developments. *Journal of Endocrinology* 122, 611–618.

Holme, D.W. and Coey, W.E. (1967) The effect of environmental temperature and method of feeding on the performance and carcass composition of bacon pigs. *Animal Production* 9, 209–218.

Houseknecht, K.L. and Portocarrero, C.P. (1998) Leptin and its receptors: regulators of whole-body energy homeostasis. *Domestic Animal Endocrinology* 15(6), 457–475.

Houseknecht, K.L., Baile, C.A., Matteri, R.L. and Spurlock, M.E. (1998) The biology of leptin, a review. *Journal of Animal Science* 76, 1405–1420.

Ilyin, S.E., Gayle, D., Flynn, M.C. and Plata-Salaman, C.R. (1998) Interleukin-1-beta system (ligand, receptor type I, receptor accessory protein and receptor antagonist), TNF-alpha, TGF-beta-1 and neuropeptide Y mRNAs in specific brain regions during bacterial LPS-induced anorexia. *Brain Research Bulletin* 45, 507–515.

Jensen, A.H., Becker, D.E. and Harmon, B.G. (1969) Response of growing/fattening swine to different housing environments during winter seasons. *Journal of Animal Science* 63, 1737–1758.

Jeong, K.H., Jacobson, L., Widmaier, E.P. and Majzoub, J.A. (1999) Normal suppression of the reproductive axis following stress in corticotropin-releasing hormone-deficient mice. *Endocrinology* 140, 1702–1708.

Johnson, R.W. (1997) Inhibition of growth by pro-inflammatory cytokines: an integrated view. *Journal of Animal Science* 75, 1244–1255.

Johnson, R.W. (1998) Immune and endocrine regulation of food intake in sick animals. *Domestic Animal Endocrinology* 15, 309–319.

Johnson, S.R., Gheusi, G., Segreti, S., Dantzer, R. and Kelley K.W. (1997) C3H/HeJ mice are refractory to lipopolysaccharide in the brain. *Brain Research* 752, 219–226.

Jones, C.T. and Edwards, A.V. (1990) Adrenal responses to corticotropin-releasing factor in conscious hypophysectomized calves. *Journal of Physiology* 430, 25–36.

Joseph-Bravo, P., Uribe, R.M., Vargas, M.A., Perez-Martinez, L., Zoeller, T. and Charli, J.L. (1998) Multifactorial modulation of TRH metabolism. *Cellular and Molecular Neurobiology* 18, 231–247.

Jurcovicova, J., Kvetnansky, R., Dobrakovova, M., Jezova, D., Kiss, A. and Makara, G.B. (1990) Prolactin response to immobilization stress and hemorrhage: the effect of hypothalamic deafferentations and posterior pituitary denervation. *Endocrinology* 126, 2527–2533.

Juszczak, M. (1998) Melatonin affects the oxytocin and prolactin responses to stress in male rats. *Journal of Physiology and Pharmacology* 49, 151–163.

Kakizawa, S., Kaneko, T., Hasegawa, S. and Hirano, T. (1995) Effects of feeding, fasting, background adaptation, acute stress, and exhaustive exercise on the plasma somatolactin concentrations in rainbow trout. *General and Comparative Endocrinology* 98, 137–146.

Ketelslegers, J.M., Maiter, D., Maes, M., Underwood, L.E. and Thissen, J.P. (1995) Nutritional regulation of insulin-like growth factors. *Metabolism: Clinical and Experimental* 44, 50–57.

Kirschbaum, C., Pirke, K.M. and Hellhammer, D.H. (1993) The 'Trier Social Stress Test' – a tool for investigating psychobiological stress responses in a laboratory setting. *Neuropsychobiology* 28, 76–81.

Klemcke, H.G., Neinaber, J.A. and Hahn, G.L. (1987) Stressor-associated alterations in porcine plasma prolactin. *Proceedings of the Society for Experimental Biology and Medicine* 186, 333–343.

Koj, A. (1996) Initiation of acute phase response and synthesis of cytokines. *Biochimica et Biophysica Acta* 1317, 84–94.

Koj, A. (1998) Termination of acute-phase response: role of some cytokines and anti-inflammatory drugs. *General Pharmacology* 31, 9–18.

Kuan, S.I., Login, I.S., Judd, A.M. and MacLeod, R.M. (1990) A comparison of the concentration-dependent actions of thyrotropin-releasing hormone, angiotensin II, bradykinin, and Lys-bradykinin on cytosolic free calcium dynamics in rat anterior pituitary cells: selective effects of dopamine. *Endocrinology* 127, 1841–1848.

Kvetnansky, R. (1973) Transsynaptic and humoral regulation of adrenal catecholamine biosynthesis in stress. In: Usdin, E. and Snyder, S.H. (eds) *Frontiers in Catecholamine Research*. Pergamon Press, New York, p. 223.

Kvetnansky, R., Weise, V.K., Gerwitz, G.P. and Kopin, I.J. (1971) Synthesis of adrenal catecholamines in rats during and after immobilization stress. *Endocrinology* 89, 46–49.

Laye, S., Parnet, P., Goujon, E. and Dantzer, R. (1994) Peripheral administration of lipopolysaccharide induces the expression of cytokine transcripts in the brain and pituitary of mice. *Molecular Brain Research* 27, 157–162.

Lee, M.C., Schiffman, S.S. and Pappas, T.N. (1994) Role of neuropeptides in the regulation of feeding behavior: a review of cholecystokinin, bombesin, neuropeptide Y, and galanin. *Neuroscience and Biobehavioral Reviews* 18, 313–323.

Leppaluoto, J., Korhonen, I., Huttenen, P. and Hassi, J. (1988) Serum levels of thyroid and adrenal hormones, testosterone, TSH, LH, GH and prolactin in men after a 2-h stay in a cold room. *Acta Physiologica Scandinavica* 132, 543–548.

Link, H., Dayanithi, G. and Gratzl, M. (1993) Glucocorticoids rapidly inhibit oxytocin-stimulated adrenocorticotropin release from rat anterior pituitary cells, without modifying intracellular calcium transients. *Endocrinology* 132, 873–878.

Liu, J.H., Muse, K., Contreras, P., Gibbs, D. and Vale, W. (1983) Augmentation of ACTH-releasing activity of synthetic corticotropin-releasing factor (CRF) by vasopressin in women. *Journal of Clinical Endocrinology and Metabolism* 57, 1087–1089.

Ma, L., Burton, K.A., Saunders, J.C. and Dauncey, M.J. (1992) Thermal and nutritional influences on tissue levels of insulin-like growth factor-I mRNA and peptide. *Journal of Thermal Biology* 17, 89–95.

Maeda, K., Nagatani, S., Estacio, M.A. and Tsukamura, H. (1996) Novel estrogen feedback sites associated with stress-induced suppression of luteinizing hormone secretion in female rats. *Cellular and Molecular Neurobiology* 16, 311–324.

Magiakou, M.A., Mastorakos, G., Webster, E. and Chrousos, G.P. (1997) The hypothalamic-pituitary-adrenal axis and the female reproductive system. *Annals of the New York Academy of Sciences* 816, 42–56.

Mahoney, S.M., Beck, S.A. and Tisdale, M.J. (1988) Comparison of weight loss induced by recombinant TNF with that produced by a cachexia-inducing tumor. *British Journal of Cancer* 57, 385–389.

Makino, S., Nishiyama, M., Asaba, K., Gold, P.W. and Hashimoto, K. (1998) Altered expression of type 2 CRH receptor mRNA in the VMH by glucocorticoids and starvation. *American Journal of Physiology (Regulatory, Integrative, and Comparative Physiology)* 44(4), R1138–R1145.

Mandrup-Poulsen, T., Nerup, J., Reimers, J.I., Pociot, F., Andersen, H.U., Karlsen, A., Bjerre, U. and Bergholdt, R. (1995) Cytokines and the endocrine system. I. The immunoendocrine network. *European Journal of Endocrinology* 133, 660–671.

Markowska, A., Rebuffat, P., Rocco, S., Gottardo, G., Mazzocchi, G. and Nussdorfer, G.G. (1993) Evidence that an extra hypothalamic pituitary corticotropin-releasing hormone (CRH/adrenocorticotropin (ACTH)) system controls adrenal growth and secretion in rats. *Cell and Tissue Research* 272, 439–445.

Marti, O., Gavalda, A., Jolin, T. and Armario, A. (1996) Acute stress attenuates but does not abolish circadian rhythmicity of serum thyrotropin and growth hormone in the rat. *European Journal of Endocrinology* 135, 703–708.

Martini, L. and Morpurgo, C. (1955) Neurohumoral control of the release of adrenocorticotrophic hormone. *Nature* 175, 1127–1128.

Mason, J.W. (1974) Specificity in the organization of neuroendocrine response profiles. In: Seeman, P. and Brown, G. (eds) *Frontiers in Neurology and Neuroscience Research*. University of Toronto Press, p. 68.

Matteri, R.L. (1994) Anterior pituitary hormones. In: *Encyclopedia of Chemical Technology*, vol. 13. John Wiley & Sons, New York, pp. 370–380.

Matteri, R.L. and Becker, B.A. (1994) Somatotroph, lactotroph and thyrotroph function in three-week-old gilts reared in a hot or cool thermal environment. *Domestic Animal Endocrinology* 11, 217–226.

Matteri, R.L. and Becker, B.A. (1996a) Lactotroph and somatotroph function in piglets reared in a constant hot environment. *Life Sciences* 58, 711–717.

Matteri, R.L. and Becker, B.A. (1996b) Gonadotroph function in 3-week-old gilts reared in a warm or cool thermal environment. *Life Sciences* 59, 27–32.

Matteri, R.L., Watson, J.G. and Moberg, G.P. (1984) Stress or acute adrenocorticotrophin treatment suppresses LHRH-induced LH release in the ram. *Journal of Reproduction and Fertility* 72, 385–393.

Matthews, S.G. and Parrott, R.F. (1994) Centrally administered vasopressin modifies stress hormone (cortisol, prolactin) secretion in sheep under basal conditions, during restraint and following intravenous corticotrophin-releasing hormone. *European Journal of Endocrinology* 130, 297–301.

Maule, A.G. and VanderKooi, S.P. (1999) Stress-induced immune-endocrine interaction. In: Balm, P.H.M. (ed.) *Stress Physiology in Animals*. Sheffield Academic Press Ltd, Sheffield, UK, pp. 205–245.

Mazzocchi, G., Rebuffat, P., Meneghelli, V. and Nussdorfer, G.G. (1989) Effects of the infusion with ACTH or CRH on the secretory activity of rat adrenal cortex. *Journal of Steroid Biochemistry* 32, 841–843.

Mazzocchi, G., Malendowicz, L., Markowaska, A. and Nussdorfer, G. (1994) Effect of hypophysectomy on corticotropin-releasing hormone and adrenocorticotropin immunoreactivities in the rat adrenal gland. *Molecular and Cellular Neurosciences* 5, 345–349.

Mazzocchi, G., Malendowicz, L., Rebuffat, P., Tortorella, C. and Nussdorfer, G. (1997) Arginine-vasopressin stimulates CRH and ACTH release by rat adrenal medulla, acting via the V1 receptor subtype and a protein kinase C-dependent pathway. *Peptides* 18, 191–195.

McCann, S. (1957) The ACTH-releasing activity of extracts of the posterior lobe of the pituitary *in vivo*. *Endocrinology* 60, 664–676.

McCarthy, H.D., Kilpatrick, A.P., Trayhurn, P. and Williams, G. (1993) Widespread increases in regional hypothalamic neuropeptide Y levels in acute cold-exposed rats. *Neuroscience* 54(1), 127–132.

McCarthy, T.L., Centrella, M. and Canalis, E. (1990) Cortisol inhibits the synthesis of insulin-like growth factor-I in skeletal cells. *Endocrinology* 126, 1569–1575.

McCusker, R.H. (1998) Controlling insulin-like growth factor activity and the modulation of insulin-like growth factor binding protein and receptor binding. *Journal of Dairy Science* 81, 1790–1800.

McEwen, B.S. (1979) Influences of adrenocortical hormones on pituitary and brain functions. *Monographs in Endocrinology* 12, 467–492.

McEwen, B.S. and Sapolsky, R.M. (1995) Stress and cognitive function. *Current Opinions in Neurobiology* 5, 205–216.

McShane, T.M., Petersen, S.L., McCrone, S. and Keisler, D.H. (1993) Influence of food restriction on neuropeptide-Y, proopiomelanicortin, and luteinizing hormone-releasing hormone gene expression in sheep hypothalami. *Biology of Reproduction* 49, 831–839.

Mercer, J.G., Hoggard, N., Williams, L.M., Lawrence, C.B., Hannah, L.T., Morgan, P.J. and Trayhurn, P. (1996) Coexpression of leptin expression and prepro-neuropeptide Y mRNA in arcuate nucleus of mouse hypothalamus. *Journal of Neuroendocrinology* 8, 733–735.

Mercer, J.G., Moar, K.M., Raynor, D.V., Trayhurn, P. and Hoggard, N. (1997) Regulation of leptin receptor and NPY gene expression in hypothalamus of leptin-treated obese (*ob/ob*) and cold-exposed lean mice. *FEBS Letters* 402, 185–188.

Minamino, N., Uehara, A. and Arimura, A. (1988) Biological and immunological characterization of corticotropin-releasing activity in the bovine adrenal medulla. *Peptides* 9, 37–45.

Miner, J.L., Della-Ferra, M.A., Paterson, J.A. and Baile, C.A. (1989) Lateral cerebroventricular injection of neuropeptide Y stimulates feeding in sheep. *American Journal of Physiology* 257, R383.

Minton, J.E. and Parsons, K.M. (1993) Adrenocorticotropic hormone and cortisol response to corticotropin-releasing factor and lysine vasopressin in pigs. *Journal of Animal Science* 71, 724–729.

Moberg, G.P. (1991) How behavioral stress disrupts the endocrine control of reproduction in domestic animals. *Journal of Dairy Science* 74, 304–311.

Muglia, L.J., Jenkins, N.A., Hilbert, D.J., Copeland, N.G. and Majzoub, J.A. (1994) Expression of the mouse corticotropin-releasing hormone gene *in vivo* and targeted inactivation in embryonic stem cells. *Journal of Clinical Investigation* 93, 2066–2072.

Muneyyirci-Delale, O., Goldstein, D. and Reyes, F.I. (1989) Diagnosis of stress-related hyperprolactinemia. Evaluation of the hyperprolactinemia rest test. *New York State Journal of Medicine* 89, 205–208.

Muzzin, P., Revelli, J.P., Kuhne, F., Gocayne, J.D., McCombie, W.R., Venter, J.C., Giacobino, J.P. and Fraser, C.M. (1991) An adipose tissue-specific beta-adrenergic receptor. Molecular cloning and down-regulation in obesity. *Journal of Biological Chemistry* 266(35), 24,053–24,058.

Nagatani, S., Guthikonda, P., Thompson, R.C., Tsukamura, H., Maeda, K.I. and Foster, D.L. (1998) Evidence for GnRH regulation by leptin: leptin administration prevents reduced pulsatile LH secretion during fasting. *Neuroendocrinology* 67, 370–376.

Natelson, B.H., Trapp, W.N., Adamus, J.E., Mittler, J.C. and Levine, B.E. (1981) Humoral indices of stress in rats. *Physiology and Behavior* 26, 1049–1054.

Newcomer, J.W., Selke, G., Melson, A.K., Gross, J., Vogler, G.P. and Dagogojack, S. (1998) Dose-dependent cortisol-induced increases in plasma leptin concentrations in healthy humans. *Archives of General Psychiatry* 55(11), 995–1000.

Nicoll, C.S. (1980) Ontogeny and evolution of prolactin's functions. *Federation Proceedings* 39, 2563–2566.

Nordio, M., Vaughan, M.K., Sabry, I. and Reiter, R.J. (1989) Undernutrition potentiates melatonin effects in maturing female rats. *Journal of Endocrinological Investigation* 12, 103–110.

Norman, R.L. (1994) Corticotropin-releasing hormone effects on luteinizing hormone and cortisol secretion in intact female rhesus macaques. *Biology of Reproduction* 50, 949–955.

Ozawa, A., Johke, T. and Hodate, K. (1994) Plasma insulin-like growth factor-1 response to cold exposure in barrows. *Endocrine Journal* 41, 725–730.

Parker, C.R., Jr, Stankovic, A.K. and Goland, R.S. (1995) Corticotropin-releasing hormone enhances steroidogenesis by cultured human adrenal cells. *77th Annual Meeting of the Endocrine Society*, Abstract no. P3-574.

Parrott, R.F., Heavens, R.P. and Baldwin, B.A. (1986) Stimulation of feeding in the satiated pig by introcerebroventricular injection of neuropeptide Y. *Physiology and Behavior* 36, 523–525.

Pasolli, H.A., Torres, A.I. and Aoki, A. (1992) Influence of lactotroph cell density on prolactin secretion. *Journal of Endocrinology* 134, 241–246.

Peisen, J.N., McDonnell, K.J., Musroney, S.E. and Lumpkin, M.D. (1995) Endotoxin-induced suppression of the somatotropic axis is mediated by interleukin-1 beta and corticotropin-releasing factor in the juvenile rat. *Endocrinology* 136, 3378–3390.

Pelletier, G. (1990) Ultrastructural localization of neuropeptide Y in the hypothalamus. *Annals of the New York Academy of Sciences* 611, 232–246.

Peterson, P.K., Chao, C.C., Molitor, T., Murtaugh, M., Strgar, F. and Sharp, B.M. (1991) Stress and the pathogenesis of infectious disease. *Reviews in Infectious Diseases* 13, 710–720.

Petraglia, F., Sutton, S., Vale, W. and Plotsky, P. (1987) Corticotropin-releasing factor decreases plasma luteinizing hormone levels in female rats by inhibiting gonadotropin-releasing hormone release into hypophysial-portal circulation. *Endocrinology* 120, 1083–1088.

Petrovic, S.L., McDonald, J.K., Snyder, G.D. and McCann, S.M. (1983) Characterization of β-adrenergic receptors in the rat brain and pituitary using a new high affinity ligand (^{125}I) iodocyanopindolol. *Brain Research* 261, 249–259.

Pettigrew, J.E. and Tokach, M.D. (1993) Metabolic influences on sow reproduction. *Pig News and Information* 14, 69N–72N.

Plata-Salaman, C.R., Ilyin, S.E. and Gayle, D. (1998) Brain cytokine mRNAs in anorectic rats bearing prostate adenocarcinoma tumor cells. *American Journal of Physiology (Regulatory, Integrative, and Comparative Physiology)* 44, R566–R573.

Plotsky, P., Bruhn, T. and Vale, W. (1985a) Evidence for multifactor regulation of the adrenocorticotropin secretory response to hemodynamic stimuli. *Endocrinology* 116, 633–639.

Plotsky, P., Bruhn, T. and Vale, W. (1985b) Hypophysiotropic regulation of ACTH secretion in response to insulin-induced hypoglycemia. *Endocrinology* 117, 323–329.

Plotsky, P.M., Cunningham, E.T. and Widmair, E.P. (1989) Catecholamine modulation of corticotrophin-releasing factor and adrenocorticotrophin secretion. *Endocrine Reviews* 10, 437–458.

Plotsky, P.M., Thrivikraman, K.V. and Meaney, M.J. (1993) Central and feedback regulation of hypothalamic corticotropin-releasing factor secretion. *Ciba Foundation Symposium* 172, 59–75.

Polkowska, J. (1996) Stress and nutritional influences on GnRH and somatostatin neuronal systems in the ewe. *Acta Neurobiologiae Experimentalis* 56, 797–806.

Portanova, R. and Sayers, G. (1973) Isolated pituitary cells: CRF-like activity of neurohypophyseal and related polypeptides. *Proceedings of the Society for Experimental Biology and Medicine* 143, 661–666.

Porter, J. and Jones, J. (1956) Effect of plasma from hypophyseal-portal vessel blood on adrenal ascorbic acid. *Endocrinology* 58, 62–67.

Richter, S.D., Schurmeyer, T.H., Schedlowski, M., Hadicke, A., Tewes, U., Schmidt, R.E. and Wagner, T.O. (1996) Time kinetics of the endocrine response to acute psychological stress. *Journal of Clinical Endocrinology and Metabolism* 81, 1956–1960.

Rivest, S. and Rivier, C. (1995) The role of corticotropin-releasing factor and interleukin-1 in the regulation of neurons controlling reproductive functions. *Endocrine Reviews* 16, 177–199.

Rivier, C. (1995) Influence of immune signals on the hypothalamic–pituitary axis of the rodent. *Frontiers in Neuroendocrinology* 16, 151–182.

Rivier, C.L. and Plotsky, P.M. (1986) Mediation by corticotropin-releasing factor (CRF) of adenohypophysial hormone secretion. *Annual Review of Physiology* 48, 475–494.

Rivier, C. and Rivest, S. (1991) Effect of stress on the activity of the hypothalamic-pituitary-gonadal axis: peripheral and central mechanisms. *Biology of Reproduction* 45, 523–532.

Rivier, C. and Vale, W. (1983) Interaction of corticotropin-releasing factor and arginine vasopressin on adrenocorticotropin secretion *in vivo*. *Endocrinology* 113, 939–942.

Rivier, C., Rivier, J. and Vale, W. (1986) Stress-induced inhibition of reproductive function: role of endogenous corticotropin-releasing factor. *Science* 231, 607–609.

Rocio Sierra-Honigmann, M., Nath, A.K., Murakami, C., Garcia-cardena, G., Papapetropoulos, A., Sessa, W.C., Madge, L.A., Schechner, J.S., Schwabb, M.B., Polverini, P.J. and Flores-Riveros, J.R. (1998) Biological action of leptin as an angiogenic factor. *Science* 281, 1683–1686.

Rojkittikhun, T., Uvnas-Moberg, K. and Einarsson, S. (1993) Plasma oxytocin, prolactin, insulin and LH after 24 h of fasting and refeeding in lactating sows. *Acta Physiologica Scandinavica* 148, 413–419.

Rushen, J., Schwarze, N., Ladewig, J. and Foxcroft, G. (1993) Opioid modulation of the effects of repeated stress on ACTH, cortisol, prolactin, and growth hormone in pigs. *Physiology and Behavior* 53, 923–928.

Saffran, M., Schally, A. and Benfey, B. (1955) Stimulation of the release of corticotropin from the adrenohypophysis by a neurohypophyseal factor. *Endocrinology* 57, 439–444.

Sahu, A., Kalra, P.S. and Kalra, S.P. (1988) Food deprivation and ingestion induced reciprocal changes in neuropeptide Y concentrations in the paraventricular nucleus. *Peptides* 9, 83–86.

Sakamoto, K., Wakabayashi, I., Yoshimoto, S., Masui, H. and Katsuno, S. (1991) Effects of physical exercise and cold stimulation on serum testosterone level in man. *Japanese Journal of Hygiene* 46, 635–638.

Sakurai, T., Amemiya, A., Ishii, M., Matsuzaki, I., Chemilli, R.M., Tanaka, H., Williams, S.C., Richardson, J.A., Kozlowski, G.P., Wilson, S., Arch, J.R.S., Buckingham, R.E., Haynes, A.C., Carr, S.A., Annan, R.S., McNulty, D.E., Liu, W.S., Terrett, J.A., Elshourbagy, N.A., Bergsma, D.J. and Yanagisawa, M. (1998) Orexins and orexin receptors: a family of hypothalamic neuropeptides and G protein-coupled receptors that regulate feeding behavior. *Cell* 92, 573–585.

Samuels, M.H. and McDaniel, P.A. (1997) Thyrotropin levels during hydrocortisone infusions that mimic fasting-induced cortisol elevations: a clinical research center study. *Journal of Clinical Endocrinology and Metabolism* 82, 3700–3704.

Schillo, K.K. (1992) Effects of dietary energy on control of luteinizing hormone secretion in cattle and sheep. *Journal of Animal Science* 70, 1271–1282.

Schobitz, B., Pezeshki, G., Pohl, T., Hemmann, U., Heinrich, P.C., Holsboer, F. and Reul, J.M. (1995) Soluble interleukin-6 (IL-6) receptor augments central effects of IL-6 *in vivo*. *Federation of American Societies for Experimental Biology Journal* 9, 659–664.

Schwartz, N.B. (1995) The 1994 Stevenson award lecture. Follicle-stimulating hormone and luteinizing hormone: a tale of two gonadotropins. *Canadian Journal of Physiology and Pharmacology* 73, 675–684.

Seggie, J. and Brown, G.M. (1982) Profiles of hormone stress response: recruitment or pathway specificity. In: Collu, R., Ducharme, J.R., Barbeau, A. and Tolis, G. (eds) *Brain Neurotransmitters and Hormones*. Raven Press, New York, p. 277.

Segreti, J., Gheusi, G., Dantzer, R., Kelley, K.W. and Johnson, R.W. (1997) Defect in interleukin-1 beta secretion prevents sickness behavior in C3H/HeJ mice. *Physiology and Behavior* 61, 873–878.

Selye, H. (1939) The effect of adaptation to various damaging agents in the female sex organs in the rat. *Endocrinology* 25, 615–624.

Sheridan, J.F., Dobbs, C., Jung, J., Chu, X., Konstantinos, A., Padgett, D. and Glaser, R. (1998) Stress-induced neuroendocrine modulation of viral pathogenesis and immunity. *Annals of the New York Academy of Sciences* 840, 803–808.

Shin, S.H., Heisler, R.L. and Lee, C.S. (1995) Neurophysin stimulates prolactin release from primary cultured rat pituitary cells. *Journal of Endocrinology* 144, 225–231.

Siegel, R.A., Weidenfeld, J., Feldman, S., Conforti, N. and Chowers, I. (1981) Neural pathways mediating basal and stress-induced secretion of luteinizing hormone, follicle-stimulating hormone, and testosterone in the rat. *Endocrinology* 108, 2302–2307.

Stratakis, C.A. and Chrousos, G.P. (1995) Neuroendocrinology and pathophysiology of the stress system. *Annals of the New York Academy of Sciences* 771, 1–18.

Stratakis, C.A., Gold, P.W. and Chrousos, G.P. (1995) Neuroendocrinology of stress: implications for growth and development. *Hormone Research* 43, 162–167.

Straus, D.S. (1994) Nutritional regulation of hormones and growth factors that control mammalian growth. *Federation of American Societies for Experimental Biology Journal* 8, 6–12.

Suda, T., Tomori, N., Tozana, F., Mouri, T., Demupa, H. and Shizume, K. (1984) Distribution and characterization of ir-CRF in human tissues. *Journal of Clinical Endocrinology and Metabolism* 59, 861–867.

Suguhara, M., Baker, D.H., Harman, B.G. and Jensen, A.H. (1970) Effect of ambient temperature on performance and carcass development in young swine. *Journal of Animal Science* 31, 59–62.

Sul, H.S. and Wang, D. (1998) Nutritional and hormonal regulation of enzymes in fat synthesis: studies of fatty acid synthase and mitochondrial glycerol-3-phosphate acyltransferase gene transcription. *Annual Reviews in Nutrition* 18, 331–351.

Tagami, T., Nakamura, H., Sasaki, S., Miyoshi, Y. and Nakao, K. (1996) Starvation-induced decrease in the maximal binding capacity for triiodothyronine of the thyroid hormone receptor is due to a decrease in the receptor protein. *Metabolism: Clinical and Experimental* 45, 970–973.

Tatemoto, K., Carlquist, M. and Mutt, V. (1982) Neuropeptide Y– a novel brain peptide with structural similarities to peptide YY and pancreatic polypeptide. *Nature* 296, 659–660.

Tegelman, R., Lindeskog, P., Carlstrom, K., Pousette, A. and Blomstrand, R. (1986) Peripheral hormone levels in healthy subjects during controlled fasting. *Acta Endocrinologica* 113, 457–462.

Tellam, D.J., Perone, M.J., Dunn, I.C., Radovick, S., Brennand, J., Rivier, J.E., Castro, M.G. and Lovejoy, D.A. (1998) Direct regulation of GnRH transcription by CRF-like peptides in an immortalized neuronal cell line. *Neuroreport* 9, 3135–3140.

Theorell, T. (1998) Prolactin – a hormone that mirrors passiveness in crisis situations. *Integrative Physiological and Behavioral Science* 27, 32–38.

Trayhurn, P., Duncan, J.S. and Rayner, D.V. (1995) Acute cold-induced suppression of ob (obese) gene expression in white adipose tissue of mice: mediation by the sympathetic system. *Biochemical Journal* 311(3), 729–733.

Uehara, Y., Shimizu, H., Ohtani, K., Sato, N. and Mori, M. (1998) Hypothalamic corticotropin-releasing hormone is a mediator of the anorexigenic effect of leptin. *Diabetes* 47, 890–893.

Usui, T., Nakai, Y., Tsukada, T., Jingami, H., Takahashi, H., Fukata, J. and Imura, H. (1988) Expression of adrenocorticotropin-releasing hormone precursor gene in placenta and other nonhypothalamic tissues in man. *Molecular Endocrinology* 2, 871–875.

Vale, W., Spiess, J., Rivier, C. and Rivier, J. (1981) Characterization of a 41-residue ovine hypothalamic peptide that stimulates secretion of corticotropin and beta-endorphin. *Science* 213, 1394–1397.

Vance, M.L., Hartman, M.L. and Thorner, M.O. (1992) Growth hormone and nutrition. *Hormone Research* 38, 85–88.

Van de Kar, L.D., Richardson-Morton, K.D. and Rittenhouse, P.A. (1991) Stress: neuroendocrine and pharmacological mechanisms. In: Jasmin, G. and Cantin, M. (eds) *Methods and Achievements in Experimental Pathology*, vol. 14. Karger, Basel, pp. 133–173.

Van den Berghe, G. and de Zegher, F. (1996) Anterior pituitary function during critical illness and dopamine treatment. *Critical Care Medicine* 24, 1580–1590.

Villares, S.M.F., Goujon, L., Maniar, S., Delehaye-Zervas, M.-C., Martini, J.-F., Kleinknecht, C. and Postel-Vinay, M.-C. (1994) Reduced food intake is the main cause of low growth hormone receptor expression in uremic rats. *Molecular and Cellular Endocrinology* 106, 51–56.

Wada, M. (1993) Low temperature and short days together induce thyroid activation and suppression of LH release in Japanese quail. *General and Comparative Endocrinology* 90, 355–363.

Warren, E.J., Finck, B.N., Arkins, S., Kelley, K.W., Scamurra, R.W., Murtaugh, M.P. and Johnson, R.W. (1997) Coincidental changes in behavior and plasma cortisol in unrestrained pigs after intra-cerebroventricular injection of tumour necrosis factor-alpha. *Endocrinology* 138, 2365–2371.

Watabe, T., Tanaka, K., Kumagae, M. and Itoh, S. (1988) Role of endogenous AVP in potentiating CRH-stimulated corticotropin secretion in man. *Journal of Clinical Endocrinology and Metabolism* 66, 1132–1137.

Watanobe, H. and Sasaki, S. (1995) Effect of thyroid status on the prolactin-releasing action of vasoactive intestinal peptide in humans: comparison with the action of thyrotropin-releasing hormone. *Neuroendocrinology* 61, 207–212.

Weeke, J. and Gunderson, H.J. (1983) The effect of heating and central cooling on serum TSH, GH, and norepinephrine in resting normal men. *Acta Physiologica Scandinavica* 117, 33–39.

Weigent, D.A. (1996) Immunoregulatory properties of growth hormone and prolactin. *Pharmacological Therapeutics* 69, 237–257.

Weingarten, H.P. (1996) Cytokines and food intake: the relevance of the immune system to the student of ingestive behavior. *Neuroscience and Behavior Reviews* 20, 163–170.

Weissman, C. (1990) The metabolic response to stress: an overview and update. *Anesthesiology* 73, 308–327.

Wenk, C. (1998) Environmental effects on nutrient and energy metabolism in pigs. *Archiv für Tierernährung* 51, 211–224.

Wettemann, R.P. and Tucker, H.A. (1974) Relationship of ambient temperature to serum prolactin in heifers. *Proceedings of the Society for Experimental Biology and Medicine* 146, 908–911.

Widmaier, E.P. (1992) Metabolic feedback in mammalian endocrine systems. *Hormone and Metabolic Research* 24, 147–153.

Xiao, E., Xia-Zhang, L., Thornell, D., Shalts, E. and Ferin, M. (1996) The luteinizing hormone but not the cortisol response to arginine vasopressin is prevented by naloxone and a corticotropin-releasing hormone antagonist in the ovariectomized rhesus monkey. *Neuroendocrinology* 64, 225–232.

Zhang, Y., Proenca, R., Maffei, M., Barone, M., Leopold, L. and Friedman, J.M. (1994) Positional cloning of the mouse obese gene and its human homologue. *Nature* 372, 425–432.

The Metabolic Consequences of Stress: Targets for Stress and Priorities of Nutrient Use

4

T.H. Elsasser,[1] K.C. Klasing,[2] N. Filipov[3] and F. Thompson[3]

[1]USDA, Agricultural Research Service, Growth Biology Lab, Beltsville, Maryland, USA; [2]Department of Animal Science, University of California, Davis, California, USA; [3]Department of Physiology, School of Veterinary Medicine, University of Georgia, Athens, Georgia, USA

Introduction

Inherent in any discussion on the effects of 'stress' on metabolism is the problem of defining how animals cope with stress. To lessen the stress, animals remove themselves from the 'discomfort', confront the stress, or adapt to it (Lefcourt, 1986). Behavioural, social, environmental, injury and disease stresses have some commonality in their capacity to alter an animal's metabolism. A common denominator of the responses to these stresses is the endocrine system (Davis, 1998). While most of the concepts developed in this chapter relate to the impact of stresses from infection and tissue trauma on metabolism, the effector and regulatory mechanisms affected and activated transcend several stressors.

The impact of stress on metabolism can be characterized as a gradient response with some positive correlation between the magnitude of the stress challenge(s) and the change in metabolism (Beisel, 1988). Variables associated with this impact can be quantified for comparative purposes through measurements of the efficiency of nutrient use for a specified purpose or assessment of a change in chemical composition or biochemical activity of various tissue beds in the body. The interpretation of metabolic efficiency in various stress paradigms requires caution and one should consider that the loss of growth efficiency in a young animal experiencing a disease stress is often offset by an increased efficiency of nutrient use for thermogenesis. The cost to growth is offset by enhanced immunosurveillance and higher biological priority for the animal to fight off the infection in order to survive (Powanda, 1980).

The host's own cognitive and non-cognitive responses to a stress dictate the patterns of chemical messengers that redirect the use of nutrients by various tissues. Animals seldom experience a single stressor that alone underwrites the overall impact of stress on metabolism. Rather, there is a constantly changing milieu of biochemical signals (neurally excitant and depressant amino acids, prostaglandins, neuropeptides, hormones and cytokines) that are needed to: (i) initiate responses; (ii) rebalance and stabilize the internal environment; and (iii) facilitate recovery of physiological processes. Some of the greatest challenges to an animal's metabolism can be generated during the initial stages of sudden-onset stress, particularly disease stress. Excessive tissue production and release of particular cytokines, which are normally tightly controlled and in extreme concentrations (tumour necrosis factor-α, TNF-α, in particular), often start a cascade of responses potentially deleterious to the animal's health. The intense reactions typical of acute bacterial infection or endotoxaemia can culminate in severe acute deficiencies in cardiopulmonary function and metabolic derangements largely associated with hypoglycaemia, acidosis and hypocalcaemia. Chronic effects of cytotoxic reactions in tissues can result in overproduction of oxygen and nitrogen free radicals, hypoxia, ischaemia, and losses of metabolic regulatory controls which, for example, impair insulin secretion from the pancreas or hormone secretion from the pituitary.

The release of stress-responsive hormones and cytokines is not the only factor that reshapes metabolism in the face of a perceived challenge. Contributing to the complexity of the stress response is the modulation of the numbers of receptors for each of these signals, receptor signal transduction factors, circulating hormone transport-binding proteins, organ perfusion/blood shunting responses, and cell-to-cell communication in organs like the liver (Kupffer cell and hepatocyte), pancreas (differential islet cells) and the pituitary (infiltrating macrophages and intercommunicating hormone-secreting pituitary cells). We have come to recognize that factors like food intake and nutritional status, photoperiod, age and sex can moderate the metabolic response to stress by modifying these regulatory processes.

Inappetence and lower feed intake, lethargy, reduced activity and fever are key indicators of many types of stresses, particularly those associated with infection or trauma. In some states of mild stress resulting in increased basal metabolic rates or maintenance energy requirements, the animal can obtain additional calories by increasing its intake. More often, however, especially in cases of clinically evident infections, feed intake declines and some calories have to be made up either by redirecting nutrients from some 'less important' tissues or by actively breaking down tissue stores to supply energy substrates and amino acids.

The gradient of metabolic response of tissues to stress need not be uniform throughout the body or across all tissue beds. Different tissue depots can be selectively curbed in their use of nutrients for growth or other productive functions. Conversely, tissue beds can be selectively recruited to make up for the nutrient and calorie deficits inherent in the cost of febrile response incurred in the face of reduced voluntary intake. This becomes particularly complex when the thermogenic requirements (of fever in addition to maintenance of core temperature) are balanced against calories available through the reduced voluntary feed intake often associated with stress responses. This chapter will outline the interrelationships between nutrition, the immune system and the endocrine system, which serve as targets for stress-related metabolic perturbations in animals. It becomes apparent that strategies to limit the extent of metabolic perturbations stemming from stress must embrace this cycle of interactions. Recognizing how defined biochemical reactions are altered in stress is a key to developing nutritional and pharmacological measures to limit metabolic illness in animals.

Ascribing the priority by which tissues receive and utilize nutrients

In this chapter, shifts in metabolism will be viewed in terms of how tissues use or provide nutrients during stress and how the priority with which nutrients are available to the different tissues of the body is determined. Metabolic shifts away from physiological processes resulting in a net increase in synthesized product (growth, lactation, etc.) occur during stress, particularly disease stress. A working model of the hierarchy of nutrient processing by tissues is depicted in Fig. 4.1 (Hammond, 1944). Hammond assessed metabolic priorities in terms of the impact of variations in plane of nutrition and nutrient availability on the development and growth of different tissues. Applying this to observations in domestic animals, Hammond ranked priority of nutrient use from highest to lowest, where tissues with the highest metabolic rate receive first or higher priority of nutrient use as compared with less metabolically active tissues. When there is a nutrient deficit, tissues like adipose are the first to lose priority. Hammond (1944) also demonstrated that earlier developing tissues *in utero* continued to be metabolically more active after birth than later developing tissues and continued to receive a greater opportunity to assimilate and process nutrients as animals matured. This model layer evolved to incorporate environmental and genetic factors (Hammond, 1952). In this latter model, energy and nutrition are accounted for and partitioned to different tissues through a prioritization scheme of use (Hammond, 1952; Touchberry, 1984) where neural utilization > visceral > bone > muscle > adipose. As stated by Touchberry (1984), '. . . it is quite evident that priorities are set for the utilization of the nutrients in the circulating blood according to the importance and activities of various body tissues'. This original Hammond thesis implied that the tissues with the

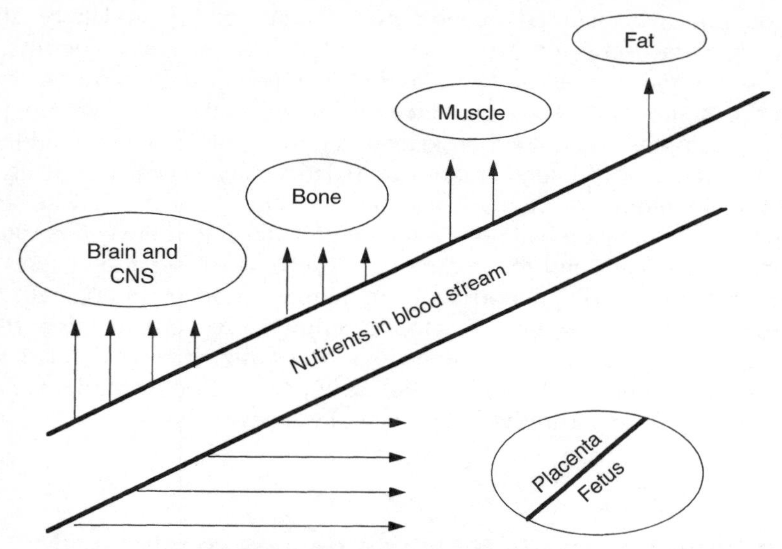

Fig. 4.1. Hammond's original scheme for the prioritization of nutrient use by different tissues in the animal body as ranked by metabolic rate (adapted from Hammond, 1944). In terms of metabolic priority, tissues with the greater number of arrows have the higher priority of use.

least effect on coordinating body functions and survival were the tissues permitted to receive fewer and fewer nutrients as the availability (plane of nutrition) decreased. Thus, nutrient use by adipose tissue is of low priority both because it is a late maturing tissue and because its relative accretion is highly dependent on nutrient 'excess'.

However, flexibility needs to be imparted to the ranking of adipose tissue accretion. We now know that fat accretion may be preferentially increased in some animals (hibernators and photoseasonal reproducers) in anticipation of future physiological needs such as hibernation or lactation, or in anticipation of harsh cold conditions (Hammond *et al.*, 1983). In addition, the role of adipose tissue to coordinate body functions may be far from passive. Recent information (see the review by Heiman *et al.*, 1998) suggests that adipose tissue may 'flip-flop' in roles as a pseudo-endocrine and pseudoimmune organ as indicated by its production of insulin-like growth factor-I (IGF-I) (Ramsay *et al.*, 1995; Kim *et al.*, 1998), leptin (Heiman *et al.*, 1998) and TNF-α (Morin *et al.*, 1998). Adipose tissue is implicated as a direct endocrine link regulating appetite and feed intake through the elaboration of adipocyte factors like leptin, which can interact with specific receptors for leptin in the hypothalamus in nuclei traditionally identified as feeding and satiety centres (Heiman *et al.*, 1998).

Endocrine–immune signal integration: dissecting systemic responses to reveal the localized tissue response

Most papers dealing with metabolism during stress characterize the changes in concentrations of numerous hormones, cytokines and metabolites circulating in plasma with respect to a particular stress paradigm. Correlations between an effector and a metabolite of cellular function are usually interpreted as cause and effect relationships, wherein the changes in plasma concentrations of the effector drive the host metabolic response. This view is simplistic and fails to account for the integrated effector responses that permit discrete cell responses, even those localized in specific regions within organs.

Regulation of cell function, and therefore metabolism, occurs not only through endocrine-type responses (the elaboration of a hormone or cytokine that is transported in the blood to change the functioning of a more distal organ or cell), but also more locally through paracrine (localized adjacent cell-to-cell chemical signals) and autocrine (where a cell regulates a part of its own response similar to ultrashort-loop feedback) regulation. The term 'metabolic cooperation' has been applied to the processes by which juxtaposed cells of organs exchange nutrients and biochemical signals to modulate cell metabolism in the localized area (Bettger and McKeehan, 1986). We proposed a set of interactions between the endocrine and immune systems, which are modulated by the nutritional status of the animal (Figs 4.2 and 4.3), that illustrates these multiple levels of control. In this model, cells integrate impinging signals to compartmentalize the metabolic responses to stress. In essence, the output of the endocrine system is moderated by the prevailing immune and nutritional status. Similarly, the immune system is enhanced and repressed according to nutrition and endocrine status. Both the endocrine and immune systems' responses affect and are affected by nutrition. Each of these is shaped by the stress inputs to cognitive and non-cognitive (immune) centres.

Shifts in metabolism during stress can be mediated both by different concentrations of effector molecules (hormones and cytokines) and by the temporal character and patterns of secretion and metabolic clearance of each hormone. While many hormones (such as insulin) have secretion patterns that, for the most part, occur in response to a stimulus (such as feeding or specific nutrient infusion), other hormones have secretion patterns with diurnal and ultradian rhythms (glucocorticoids, for example). In addition, other hormones, such as growth hormone (GH), show bursts of secretion (episodic secretion) that are characterized in terms of the concentration, pulse height, and duration of a secretory burst (as assessed by changes in measured plasma concentrations of a hormone), the frequency of the secretory bursts, and the kinetics associated with removal and inactivation of the hormone from intracellular fluids (Tannenbaum, 1991).

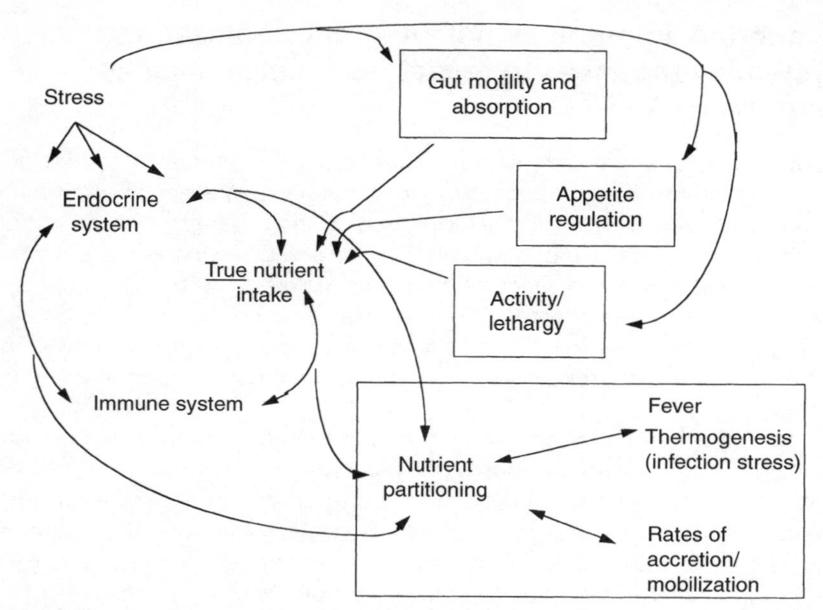

Fig. 4.2. Stress affects endocrine and immune system targets as well as nutrient intake and supply. The interactions between the endocrine and immune systems are further refined according to the availability of nutrients and the need or use for those nutrients. Ultimately, the total body response to stress can be measured as suboptimal rates of growth, losses in the efficiency of nutrient use for growth and increased cost of maintenance metabolizable energy.

A variety of factors regulate hormone and cytokine secretion by specific glands and cells (e.g. pituitary, pancreas, adrenals, lymphocytes and Kupffer cells). For the most part, these regulatory peptides and factors are synthesized and released in structures in the brain (the hypothalamus, in particular). The interaction of these neurotransmitters and regulatory peptides alters the sensitivity and responsiveness of various tissues to primary regulatory hormones and thus changes the output of a hormone or cytokine by a tissue or cell (Elsasser, 1979). 'Sensitivity' at the cell level is the lowest level of regulation that can initiate a cell response or the amount of change in level of regulatory effector that is needed to increase or decrease a cell's function past the point of initiation. The term 'responsiveness' refers to the magnitude of the output for a given level of stimulatory or inhibitory input; responsiveness necessarily embraces sensitivity in that it is the measurable increase in response through which sensitivity can be defined.

The hypothalamo-pituitary unit is a target for disease stress in infected animals (Elsasser *et al.*, 1991; Abebe *et al.*, 1993) which significantly alters metabolism. Our research focuses on the role of the pituitary in stress states, and has demonstrated that pituitary as well hypothalamic regulatory

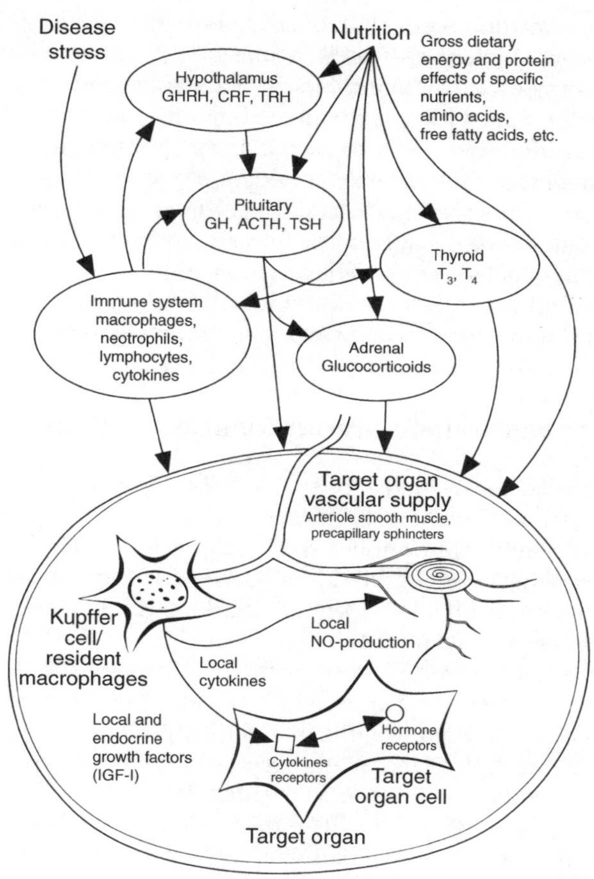

Fig. 4.3. The ultimate capability for integration of all metabolic regulatory signals resides in the target tissues. The cellular response is shaped not only by the concentrations of hormones, cytokines and nutrients reaching a given cell, but also through the timing and patterns of these factors, blood flow to the tissue, interactions with neighbouring cells, and changes in enzyme activity, receptors and transport mechanisms by which regulatory factors enter or modify the metabolic capabilities of cells.

mechanisms are impaired during stress due to parasitism. Parasitized calves have decreased circulating plasma concentrations of GH associated with fewer secretion pulses of lower magnitude and shorter duration (Elsasser *et al.*, 1986). They also have a reduced sensitivity and responsiveness to the growth hormone-releasing factor. Reduced GH output was partly due to significant increases in plasma and tissue (pancreatic and gut) concentrations of somatostatin (SS) (Elsasser *et al.*, 1990). In addition, these calves have higher basal plasma TNF-α concentrations. Additional support for our hypothesis was related to experimental results that demonstrated: (i) the ability of TNF-α to blunt increases in plasma GH subsequent to *in vivo* challenge with growth hormone-releasing peptides; (ii) the presence of specific receptors for TNF-α in pituitary cell homogenates; and (iii) the ability of TNF-α to abolish directly the stimulatory effects of growth hormone-releasing peptides on GH release into media *in vitro* (Elsasser *et al.*, 1991). An interesting observation by Abebe *et al.* (1993) is that under some disease stress situations, macrophages can infiltrate from the circulation to the anterior pituitary and selectively change the ability of GH-secreting cells to release hormone. This effect is apparently due to the localized release of immune cytokines from the macrophages in proximity to the somatotrophs.

Fine-tuning responses among tissues and cells

Metabolically active cells integrate numerous endocrine, immune and nutritional signals. The task of interpreting factors such as hormone or cytokine concentration, temporal patterns of effector signals and receptor activity (which determines sensitivity and responsiveness) resides within and between cell types that make up a target organ.

The interaction of peripheral, blood-borne and local factors that shape cell response and metabolism is presented in Fig. 4.3. The major effector hormones forming the checks and balances on metabolism are associated with: (i) a dominant anabolic action of the somatotropic axis (Salomon *et al.*, 1991; Beermann and Devol, 1992) which comprises GH, the IGF-I complexes (Etherton and Bauman, 1998) and peptides that regulate GH secretion such as SS and GHRH; (ii) the adrenal axis (Dayton and Hathaway, 1992) with the catabolic actions of adrenocorticotropic hormone (ACTH) and glucocorticoids; and (iii) the thyroid axis as associated with regulation of basal metabolism, transmembrane nutrient uptake by cells and regulatory input to the somatotropic axis (Rodriquez-Arnao *et al.*, 1993).

There may be additional modulating conditions that affect how hormonal, cytokine and nutritional information reaches cells. Modulation of hormone and cytokine action can occur through changes in regional blood flow patterns to organs and through intracellular states of oxidation and reduction (Jaeschke, 1995). Blood flow to specific organs can be altered dramatically in the acute phase response stage of infection or tissue trauma by changes in vasoconstriction and vasodilation as affected by the induced release of arachidonic acid metabolites (prostaglandins, prostacyclins and thromboxanes) and nitric oxide (NO), especially within arteriole smooth muscle and precapillary sphincters (Lancaster, 1992; Griffith and Stuehr, 1995). For example, the imbalance in the relative elaboration of arachidonic acid-derived thromboxane and prostacyclin is part of the reason that signs of pulmonary hypertension and peripheral hypotension are apparent in septic shock (Demling *et al.*, 1986).

That NO has both toxic and beneficial effects may be due to the character of the redox-activated forms of the molecule, since free radical nitric oxide is further converted to nitrosonium (NO^+) or nitroxyl anion (NO^-) altering its potential to complex with other molecules in the cell (Beckman and Koppenol, 1996; Stamler and Feelisch, 1996). It is interesting that IGF-I has been found to modulate NO-synthase activity in vascular tissue locations (Tsukahara *et al.*, 1994) and may account for some aspects of GH- and IGF-mediated effects on blood flow as well as nutritional regulation of regional blood flow. The relationship demonstrates the integrated links between nutrition, the endocrine system and the immune system and underscores the value of recognizing where and how stress can alter the balance in this relationship and affect metabolism. Contributors to this adjustment in metabolism are reflected in the fact that: (i) diet is the most prominent regulator of IGF-I production (Clemmons *et al.*, 1990); (ii) GH is the principal hormonal regulator of IGF-I production; and (iii) the protein and energy content of diets affect not only IGF-I production by promoting high affinity GH receptor binding (Breier *et al.*, 1988), but also cytokine and NO responses to stress (Kahl *et al.*, 1997).

Many hormones circulate bound to plasma binding proteins, fatty acids or circulating receptor fragments (Daughaday and Trivedi, 1991; Cohick and Clemmons, 1993). In addition to increasing plasma half-life of lower molecular weight peptides, transport proteins regulate the passage of hormones across endothelial barriers between blood and target tissues. Thus, the biological activity of hormones and cytokines can be further modulated by their partitioning between bound and free states. During different physiological and pathophysiological situations, specific plasma proteases can degrade hormone-binding proteins and modify the binding capacity of specific hormones. This is particularly evident for the IGF-I-binding proteins whose characteristics, patterns and concentrations are regulated by GH, nutrition, insulin and cytokines (Clemmons *et al.*, 1990; Clemmons and Underwood, 1991; Cohick and Clemmons, 1993). Patterns of these binding proteins and therefore the availability of the transported hormones to tissues are markedly altered during the stress due to infection (Elsasser *et al.*, 1995; Fan *et al.*, 1995).

Further integration of the stress response occurs in the cascade of intracellular messengers that transmit information from hormone and cytokine receptors to responding genes and epigenetic elements. At the receptor level, for example, modulation of sensitivity and responsiveness can occur through changes in: (i) receptor numbers; (ii) receptor–ligand binding affinity; (iii) post-receptor signal transduction messages via alteration of intracellular second and tertiary messenger production (such as cyclic AMP or calcium flux); (iv) phosphorylation activation of specific protein kinases; and (v) a host of additional factors and processes that ultimately turn on or off the transcription and translation of a specific series of genes to alter cell function in the further elaboration of protein products such as enzymes and energy-transforming reactants in mitochondria, etc. It is the disturbance in these biochemical reactions that triggers much of the pathophysiological response to stress that appears as aberrant metabolism.

Concentration or flux?

In terms of the impact of disease on metabolism, a concept often ignored is how the host response to the disease process can alter the timing and presentation of hormones and cytokines to tissue receptors, as well as the delivery of nutrients and oxygen and removal of metabolic wastes. Alterations in blood flow mediated by various effectors that regulate vascular constriction and dilation significantly impact tissue function by allowing more or less oxygen, nutrients and regulatory factors to reach cells in a given time (Huntington and Reynolds, 1987). Tissue extraction and uptake is influenced by blood flow, tissue transit time, and capillary and endothelial transport. These last two are in turn affected by the transport state of a hormone or nutrient (e.g. binding proteins), local pH, etc.

The challenge to the immune system is to clear a threat from the internal environment and, in doing so, signal a reduction in metabolism to assist survival. Such signals are produced by macrophages, neutrophils and lymphocytes, and include the systemically active pro-inflammatory cytokines IL-1, IL-6, TNF-α and interferon-γ. Sometimes the acute cytokine response is overpowering as in the case of acute bacteraemia, phasic parasitic eruptions, endotoxaemia and major tissue trauma. Overreaction by the immune system becomes a source of metabolic pathogenesis in its own right as cascading waves of effectors including cytokines, prostaglandins, prostacyclins and thromboxanes are released in disproportionate amounts and culminate in cardiovascular shock and acute organ failure, hypoglycaemia and hypocalcaemia (Sherry and Cerami, 1988; Beutler and Cerami, 1989; Fong and Lowry, 1990). The actions of cytokines that contribute to a redirection of metabolism towards tissue mobilization and catabolism are summarized in Table 4.1. In essence, the actions of pro-inflammatory cytokines perturb metabolism through direct effects on target tissue like muscle and adipose tissue, and through associated actions that increase the systemic and local production of catabolic endocrine hormones (e.g. glucocorticoids) while decreasing the action of anabolic hormones (e.g. insulin).

In stress responses, the priority of nutrient use by tissues is altered

It is simplistic to assume that the reverse of Hammond's nutrient partitioning scheme is true when tissues are affected by stress. Responses to stress range from mild decreases in growth rate in younger animals to the cachectic

Table 4.1. Actions of cytokine response to infection and endotoxin to modify and alter metabolic processes.

Promotes	Depresses or disrupts
Cartilage and bone remodelling	Cartilage production and long bone elongation
Osteoclast activity	Osteoblast activity
Hypocalcaemia	
Inflammatory response via:	Haematopoiesis
Macrophage, Kupffer cell and local tissue responses	Progenitor/stromal cells
Priming and initiation of cytokine–arachidonic acid cascade	T-, B-cell numbers
Tissue remodelling and replacement of senescent tissue	Myogenesis
Apoptosis	Pituitary GH release (species specific)
Hepatic acute phase response protein synthesis	Voluntary feed intake
Nitric oxide, superoxide and peroxide generation	Energy metabolism for growth and lactation
Fever/endogenous pyrogen release	Skeletal muscle protein synthesis
Antiviral/antiparasitic protection	Iron and zinc in plasma
ACTH/glucocorticoid secretion	Anabolism
Redistribution of organ perfusion shunting	Tissue-specific secondary and tertiary messenger and signal transduction mechanisms
Glycogenolysis	Fat accumulation
Energy metabolism for febrile response	
Amino acid transmembrane fluxes into specific tissues	

catabolism that results in muscle degradation and fat mobilization when stress is severe. Adipose tissue is often thought of as the first source of energy substrates to provide metabolic fuel; skeletal muscle provides amino acids, glutamine, etc. Actually, the liver and skeletal muscle may be called upon first during a stress response to provide glucose substrates through hormone (glucagon and catecholamine)-mediated breakdown of glycogen. Secondarily, other tissue sources such as adipose may be mobilized to provide energy substrates in the form of fatty acids.

Muscle and liver metabolism contribute important substrates that serve as alternative fuel sources in times of stress. For example, cardiac muscle is well equipped to metabolize lactate, acetoacetate or β-hydroxybutyrate as an energy substrate (Halestrap *et al.*, 1997). Thus, in association with muscle production of lactate in either exercise or response to immune challenge, the heart (as well as the kidney and intestine) can serve as a sink for the elimination of these carbon sources, sparing glucose for use by other tissues.

There is also a dynamic relationship between muscle breakdown during infection and sepsis and the production of muscle-derived glutamine for immune cell function (Newsholme and Calder, 1997). As the immune response increases and the metabolic demands on monocytes and macrophages increase, glutamine released in muscle breakdown is utilized as a carbon source by immune cells, ensuring that they can proliferate and function while glucose is diverted to other tissues more dependent on this source of energy (Calder, 1995). The breakdown of muscle protein appears to be dependent on the direct synergistic catabolic effect of TNF-α and IL-1β (Zamir *et al.*, 1992) and disruption of normal growth hormone and IGF-I regulation by pro-inflammatory cytokines (Elsasser *et al.*, 1995; Elsasser *et al.*, 1998a).

A hierarchy of nutrient use priority also exists among tissues which are subcomponents of a major tissue type; for example, within the larger grouping of skeletal muscle, postural muscle such as the psoas major may have a different priority from a muscle primarily used for locomotor activity such as the rectus femoris. This theme can be broadened to encompass different visceral organs, regional adipose sites such as intramuscular, pelvic or renal fat, or even leukocyte and lymphocyte populations. In a recent growth trial in our laboratory, different tissues and structures in the body were affected differentially by the presence of infection stress. In addition, where growth hormone was used experimentally to determine whether the use of an anabolic hormone could decrease losses in protein gain in the body, we observed some catabolic effects additive to those of infection (Elsasser *et al.*, 1998a).

Data in Table 4.2 illustrate that average daily overall carcass protein and fat gain are significantly decreased by parasitic infection (Elsasser *et al.*, 1998a). However, GH administration does affect protein gain, but acts synergistically with infection to cause the net mobilization of adipose tissue. Similarly, there is evidence of an impact of infection on some visceral organs, but not others. Overall average daily organ gain was not statistically decreased by infection. However, while the terminal weights and rates of weight gain of heart, kidney and liver may not have been affected by infection, the growth rate and chemical composition (fat, protein, water, ash) of the intestine, rumen, abomasum, omasum and reticulum as well as some muscles was significantly altered. Interestingly, treatment of these calves with GH appeared to shift nutrient use in infected animals further away from a maintenance protein use in the intestine. Finally, carcass ash, a measure of bone content, is not affected by infection, suggesting that the skeletal axis's relative priority was conserved during stress. Thus, the processes associated with the response to infection (i) decrease the protein accretion response of some organs to the exogenous administration of GH; (ii) augment the antilipogenic aspects of GH in some adipose depots but not

Table 4.2. Effects of parasitic infection[a] and growth hormone treatment[b] on aspects of growth and composition of tissues of different organs in calves (adapted from Elsasser *et al.*, 1998a).

	−GH		+GH				
Tissue or component	Control	Infection	Control	Infection	SEM	Effect of infection $P <$	Effect of GH $P <$
Average daily carcass (ADC) gain (kg day^{-1})	0.64	0.36	0.64	0.35	0.05	0.02	0.15
ADC protein gain (g day^{-1})	129	52	122	48	12	0.005	0.12
ADC fat gain (g day^{-1})	93	11	68	−23	12	0.01	0.03
Carcass ash (%)	10.31	12.02	11.23	12.00	1.21	0.10	0.69
Average daily visceral organ gain (kg day^{-1})	0.17	0.13	0.22	0.11	0.02	0.56	0.22
Heart weight (kg)	0.83	1.01	0.83	0.98	0.07	0.26	0.24
Kidney weight (kg)	0.68	0.72	0.71	0.67	0.04	0.55	0.57
Liver weight (kg)	3.89	3.83	4.37	3.62	0.24	0.37	0.02 (GH × Infect., 0.09)
Small intestine weight (kg)	7.71	7.72	7.80	7.14	0.46	0.36	0.82
Average daily intestinal protein gain (g day^{-1})	3.19	3.04	4.13	0.21	0.10	0.05	0.24 (GH × Infect., 0.05)
Average daily intestinal fat gain (g day^{-1})	39.2	10.5	40.2	5.8	6.81	0.66	0.87
Rumen, omasum, reticulum, abomasum weight (kg)	7.48	6.28	8.5	5.9	0.29	0.05	0.08

Values represent least squares means for $n = 5–6$ per treatment.
[a] *Sarcocystis cruzi*, 250,000 oocysts per os.
[b] Pituitary-derived (USDA bGH-B1) GH, 12.5 mg per calf day^{-1} for 35 days.

others; and (iii) uncouple the normal regulation of IGF-I by GH. This illustrates how the response of cells to stress varies from location to location (and, therefore, purpose to purpose) and how stress modulates responses to effector molecules (like GH or insulin).

Not all muscles react to stress equally. Muscle growth responses to stress are an interesting and somewhat complex issue as well, as seen in Fig. 4.4 in the case of two muscles differing in location, anatomical function and biochemical make-up. In this study the impact of disease stress on the different muscles was at first masked by the inflammatory oedema response (increased muscle water content in infected calves). Protein gain, but not intramuscular fat gain, was markedly affected by infection. In rectus femoris, the decrease in protein gain associated with infection was not significant and was largely prevented by GH injection. In contrast, protein gain in psoas major was negated by infection and could not be maintained by GH even though protein gain in this muscle was increased by GH treatment of healthy animals. The significant decrease in carcass fat gain compared with the minimal effect on intramuscular fat gain suggests that the muscle fat depot is relatively refractory to catabolic effects of infection whereas other depots like back fat and subcutaneous fat may be relatively more capable of being mobilized by the endocrine and immune gradient effectors. Thus, a partitioning priority may exist not only between different tissue pools but also within pools as dictated by function and physiological purpose of the particular tissue.

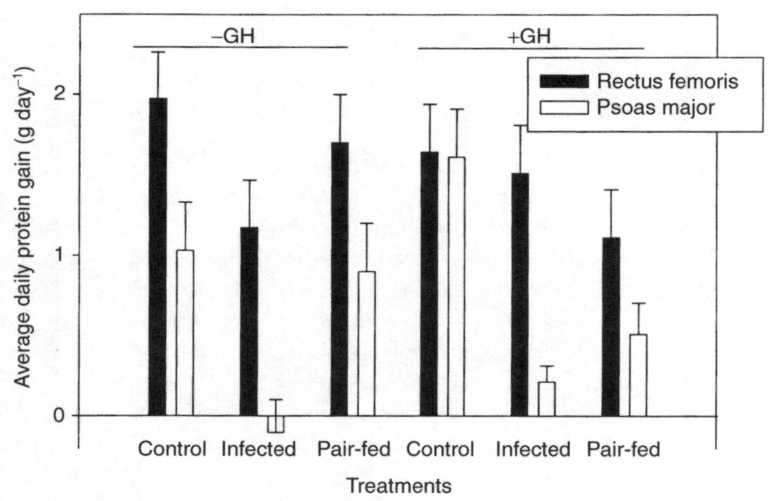

Fig. 4.4. The average daily protein gain of locomotor muscle (rectus femoris) is largely unaffected by infection with *Sarcocystis cruzi*, level of intake and the use of exogenous GH treatment. In contrast, the growth of a postural muscle (psoas major) was significantly decreased by infection and the anabolic effect of GH was abolished.

Appetite is often depressed during infectious disease challenges. The nutrient supply to support the increased metabolic and caloric demand is low and the difference is made up by catabolic processes that mobilize muscle protein and fat stores (Beisel, 1988). As stated by Beisel, the severity of the metabolic compromise and the subsequent diversion of nutrients is proportional to the severity of the stress. Growth is accomplished, or rather permitted, after basal metabolic needs are met but is often the first physiological process held in check with the onset of stress. Because stress demands on nutrient utilization supersede those of growth, tissue accretion will occur only if the metabolic needs of the immune system and the acute phase response are met.

At present, it is impossible to interpret the multiple cytokine interactions that affect muscle metabolism in states of health, let alone disease. Whereas it is apparent that the pro-inflammatory cytokines and glucocorticoids participate actively in muscle proteolysis and degradation during infection and sepsis (Douglas *et al.*, 1991; Zamir *et al.*, 1992), newly discovered cytokines such as interleukin-15 have specific anabolic effects in skeletal muscle (Quinn *et al.*, 1995) that may be the signal for muscle recovery from a disease state. Furthermore, pathologically high levels of some pro-inflammatory cytokines, such as TNF-α, can significantly perturb mitochondrial energetics, the cytochrome chain, and therefore ATP production and turnover (Lancaster *et al.*, 1989). These adverse effects of cytokines on mitochondrial energetics disrupt the energy needed to maintain the electrochemical gradients, transport processes and oxidation–reduction environment necessary for proper cell function. The culmination of prolonged or severe disruption of these mitochondrial-dependent functions is cytotoxicity, cell death and necrosis. In fact, some cytokines like TNF-α are capable of up-regulating apoptotic mechanisms in cells and effectively induce premature cell death (Natoli *et al.*, 1998).

A schematic of tissue prioritization in response to stress is shown in Fig. 4.5. The endocrine–immune gradient fine-tunes the basic priority of tissues to obtain and utilize, or donate, nutrients. Not only are there priorities between major tissue beds, but also subpriorities across different regional components of the larger tissue type. Across these different tissues, there can be greater or lesser impact of endocrine and immune effectors in shaping local metabolism, and there is also a high degree of plasticity in the capacity of tissue responses to be up-regulated or down-regulated. Many organs have a high metabolic rate and demand for nutrients, are outside the normal regulatory input of the endocrine–immune gradient, and can adapt metabolism to utilize energy resources that are not usable by other tissues. There is also a degree of relative overlap between different tissues' metabolic priority. In addition, this scheme shown in Fig. 4.5 implies that across the scope of physiological processes, the gradient of responses is affected by diet quality and food intake. Furthermore, the impact of stress on the absorption of nutrients perturbs the

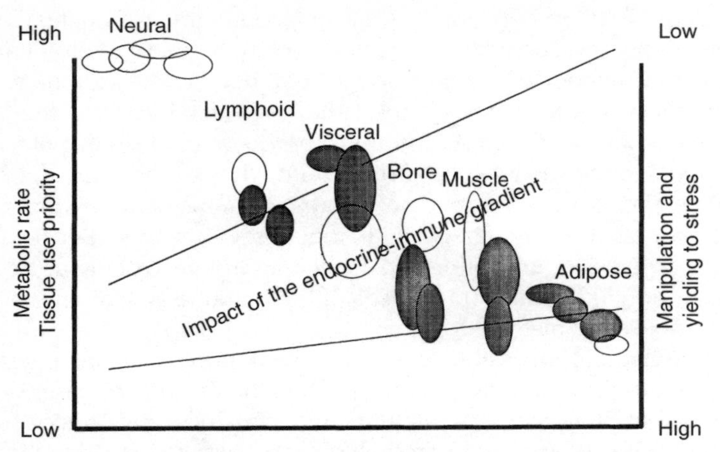

Fig. 4.5. A schematic representation of the priority of nutrient use by various tissues as mediated by a gradient of endocrine and immune response factors. Shaded ovals represent those tissues that respond to hormone and cytokine effects; unshaded ovals are tissues that minimally respond to hormones and cytokines. Within any tissue type (i.e. muscle, adipose) there may be a wide range in the priority of individual tissues or cells, and across tissues there is some degree of priority overlap.

relationship between what an animal eats and what is available to actually reach tissues.

A further adaptation of the Hammond model (Elsasser and Steele, 1992) incorporated the introduction of immune and lymphoid tissues to the prioritization scheme. The positioning of the immune cell priority just below that of neural tissue suggests not only the increased priority of this tissue when induced through disease challenge but also the relative impact on ultimate survival of the host. Within the immune system, distinct cell populations have different priorities during stress depending upon their type and state of activation. For example, neutrophils are mobilized and circulate at a higher state of activation during stress, whereas the activities of T-cytotoxic lymphocytes and macrophages are suppressed (Wilckens, 1995).

It is important to point out and clarify the relationship between prioritization of nutrient use by a tissue and the absolute use of that nutrient by a tissue. By virtue of the total mass of a given tissue it may appear that a tissue 'uses' a large percentage of available nutrients. In contrast, tissues vital to survival (brain, heart, kidney) may use much less of the total available nutrition but the use per gram of tissue and the associated oxygen consumption can be high. The significant discriminator here is the mass of the tissue in the body relative to its function. Klasing (1998) estimated that, at most, the lymphoid tissue component of the immune system (together with connective and circulating cells) accounts for less than 5% of the body's tissues. Even when faced with an immune challenge, only a portion of the immune system responds, so the body's immune system does not utilize a large amount of nutrients compared with other anabolic processes like growth. However when the immune system orchestrates a systemic acute phase response, nutritional resources are diverted to the liver for the generation of acute phase proteins and to other tissue to support the increased protein turnover needs. The biochemical mechanisms by which 'high priority' tissues like the immune system compete for nutrients are optimized in terms of affinity constants, transmembrane transport kinetics and reaction K_m (substrate concentration at half-maximal reaction rate) and V_{max} (substrate concentration at which all enzyme is reacting (saturation) and the product generation proceeds at maximum rate). In fact, some membrane nutrient transporters are up-regulated in times of infection and stress to ensure further that the substrates needed for biological response processes are there to support the needed function. For example, the burst of NO needed during the pathogen inactivation is derived from the metabolism of arginine via inducible nitric oxide synthase (iNOS; Lancaster, 1992). In order to ensure that there is sufficient arginine substrate for this purpose, some cytokines like TNF-α increase the kinetics of the transporters that mediate arginine uptake by some cells (Wu and Morris, 1998). Thus, nutritional and hormonal states (i.e. diet protein and energy content, and growth hormone administration) that increase or decrease arginase activity (Elsasser *et al.*, 1996) in tissues will also affect the availability of arginine for the iNOS pathway (Wu and Morris, 1998; Fig. 4.6).

Core temperature, fever, appetite and calories

One of the most basic responses by a host to an encounter with a pathogen is the development of fever. The relationship between infection-related fever response and metabolism is not entirely straightforward. Metabolic rate increases by approximately 10% for a given 1°C increase in body temperature (reviewed by Baracos *et al.*, 1987; Kluger, 1991). However, two caveats temper the directness of the relationship between fever and the perceived costs to metabolism in terms of caloric expenditure. First, the increases in core body temperature over 'normal' temperature customarily associated with the host fever response to disease may or may not be derived from increases in heat production from increased caloric burning. In fact, patterns of fever during illness are moderated by environmental factors. Depending on the ambient environmental temperature, fever may be due to increased generation of metabolic-derived heat or from decreased dissipation of heat (Kluger, 1991). Second, some elements of the hyperthermia of host response are more directly associated with the implementation of heat conservation mechanisms, which occur through blood shunting away from the surface and periphery, circulatory redistribution, and capillary and arteriolar

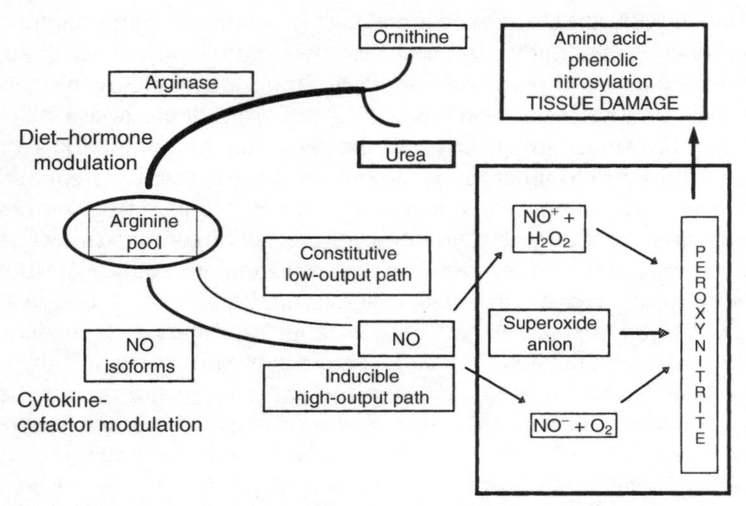

Fig. 4.6. Relationships between pathways that can utilize the common metabolic substrate arginine. Dietary supply and hormonal regulation of the K_m and V_{max} of transport systems affect arginine uptake by cells, whereas the activity of arginase and nitric oxide synthetase (NOS) is regulated by cytokines and determines the loss of arginine. Depending upon the oxidation–reduction environment in the cell, NO can decay to forms that react strongly with superoxide, form peroxynitrite and promote cytotoxic reactions in the cell.

constriction in the peripheral limbs. Hyperthermic responses involving increases in heat production (i.e. calorie burning) are more pronounced at low ambient temperatures than in warmer ones. However, some of the complications associated with the response to infection can be augmented by fever during periods of high ambient temperature, as heat load and the inability adequately to release heat affect organ (central nervous system, in particular) function, and water and ion loss from diarrhoea and dehydration compromise cooling capacity.

Metabolic changes due to stress are greatest if increased heat production and decreased feed intake occur concomitantly. In this situation, the needed calories come from the catabolism of tissues. Even when food intake is normal (Elsasser *et al.*, 1986), metabolic inefficiencies due to the acute phase response are evident and contribute to poor use of nutrients by tissues. Calves chronically infected with the protozoan parasite *Sarcocystis cruzi* gain less weight than control or non-infected pair-fed conspecifics, largely due to significantly lower nitrogen uptake from the gut and lower dietary nitrogen utilization efficiency throughout the post-infection period. In these and other studies (Elsasser *et al.*, 1988), nitrogen retention during infection is highly correlated with plasma concentrations of the anabolic hormone IGF-I and a reduced capability for GH to increase plasma concentrations of IGF-I as it does in healthy, well-fed calves.

Growth hormone may be beneficial in decreasing some aspects of tissue wasting associated with undernutrition, in quicker recovery from trauma-related tissue wasting and in an overall increase in immunocompetence resulting in a greater survival during infection. More recent experiments investigated the relationship between metabolism and the host response to parasitic infection by determining whether it was possible to modulate tissue wasting in cachectic calves with the use of GH. Young calves were infected with a moderate dose (200,000–250,000 oocysts per os, *S. cruzi*) and treated for 35 days with either bovine GH (0.1 mg kg^{-1} day^{-1}) or an excipient buffer (Elsasser *et al.*, 1998a). A variable termed the infection 'response index' was developed (average daily increase in rectal temperature for 21 days post-onset of the acute phase response) and regressed on the change (decline to nadir) in plasma concentration in IGF-I, an indicator of relative plane of nutrition and overall metabolic state (Steele and Elsasser, 1989). A substantial negative linear correlation (Fig. 4.7) was apparent, suggesting that the more severe the response to infection (the greater the change and duration in fever) the greater the impact in the decline in IGF-I. Growth hormone did not offset the magnitude of the decline in IGF-I seen due to stress.

The cost to metabolism during infection or infestation does not necessarily result from a redirection of nutrients and calories into fever and hyperthermic responses. Cole and Guillot (1987) provided interesting evidence on the relationship between metabolism and varying degrees of stress in a study in which the impact of exoparasitism on basal metabolic rate and maintenance energy requirement of cattle as a function of the percentage of body surface area infected with *Psoroptes ovis* mites was investigated. The data (Fig. 4.8) suggest that, up to a point, the increased need for nutrients to support increased metabolism from stress associated with the exoparasitism is compensated for by increased intake. However, when caloric needs cannot be met by intake, the required supplemental energy is obtained in a linear fashion by increased tissue catabolism. The increased maintenance energy is directly proportional to the surface area of infestation and thus the severity of the stress.

Subclinical infections and other undetected stressors can affect metabolism and growth. For example, chickens and pigs raised in conventional production facilities grow more slowly and less efficiently than animals kept in more sanitary environments. Some of the lost growth rate can be recovered by feeding subtherapeutic levels of antibiotics (NRC, 1998). Klasing and his colleagues suggested that dietary antibiotics act by decreasing the frequency and intensity of challenges by opportunistic bacteria and consequently preventing the stress response associated with their activation of the immune system (Roura *et al.*, 1992). The immune stress and associated metabolic diversions from the normal patterns of nutrient channelling (i.e. assimilation for growth and development) are related to patterns of low-level inflammatory

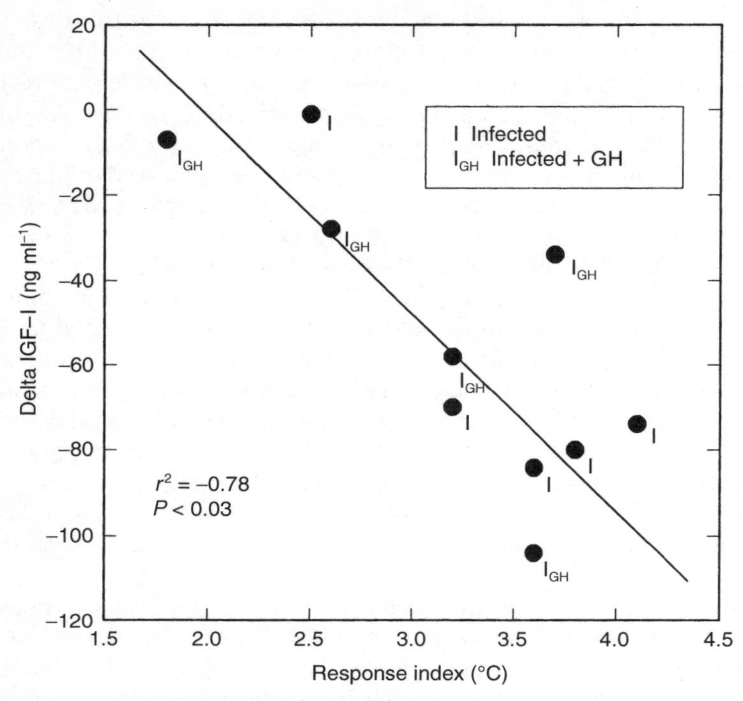

Fig. 4.7. As a measure and reflection of the relationship between the severity of an infection stress and the negative impact on growth, there is a significant negative correlation between the decrease in the anabolic hormone IGF-I and the magnitude and duration (response index) of fever on calves infected with a muscle parasite, *Sarcocystis cruzi*. A similar negative correlation was present between the response index and average daily carcass protein accretion. Protein accretion is a necessary parameter rather than weight gain because a significant contribution to weight is made from the inflammatory oedema in the muscles and tissues, and the watery oedema masks the true effect on the tissue anabolism.

cytokines such as IL-1 and TNF-α (Klasing *et al.*, 1987; Elsasser *et al.*, 1995, 1997b).

The pancreas: a chronic and acute phase response shock organ

The pituitary gland and somatotropic axis are not the only metabolically important endocrine glands targeted during disease stress. It has been recognized for years that some of the most pronounced impacts of stress and disease on animals manifest themselves in terms of carbohydrate metabolism and perturbations in pancreatic insulin and glucagon secretion. Two

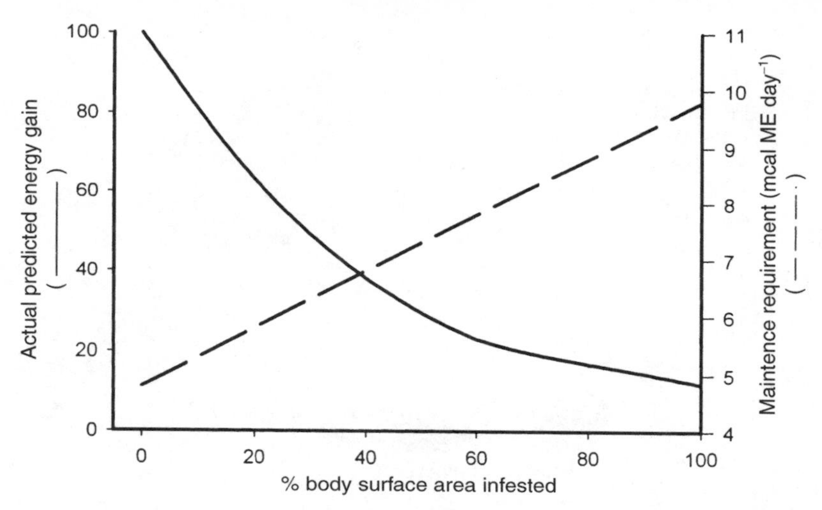

Fig. 4.8. The effect of surface area infestation on changes in basal metabolic energy requirements (adapted from Cole *et al.*, 1987).

potential impacts of stress on pancreatic function and carbohydrate utilization in muscle and adipose tissue are summarized in Fig. 4.9. In the first, the acute actions of stress effectors culminate in poor control of insulin release. Disturbances in the dynamics of insulin secretion leading to poor nutrient use by peripheral tissues were seen in experiments with parasitized calves undergoing an acute phase response, where intravenous challenge with arginine (an insulin secretagogue) was used to unmask the underlying secretion problem (Elsasser *et al.*, 1986). Secondly, immune insults in the pancreas associated with nitric oxide induction and cytotoxic damage to islet cells predispose animals to type-I diabetes-like conditions (Corbett and McDaniel, 1996) that can be further exacerbated with subsequent bouts of even low-level disease stressors (Elsasser *et al.*, 1999).

There is a temporal relationship between the origin of a disease stress and the mechanism by which the impact on metabolism is manifested. For example, when an animal first encounters the abrupt onset of the acute phase response of experimental endotoxin challenge (time course 1–6 h after challenge) the immediate biphasic hyperglycaemic–hypoglycaemic response is highly correlated over time with the peripheral release of cytokines and their effects on the waves of glucocorticoid, catecholamines, prostaglandins and reflex release of both insulin and glucagon coupled to a cytokine-induced peripheral insulin resistance (Ciraldi *et al.*, 1998). In fact, one theory suggests that TNF-α from peripheral immune cells as well as adiposites and muscle cells may link obesity, diabetes and peripheral insulin resistance (Hotamisligil and Spiegelman, 1994). TNF-α also appears to cause a differential effect on glucose transporters (GLUT-1 and GLUT-4),

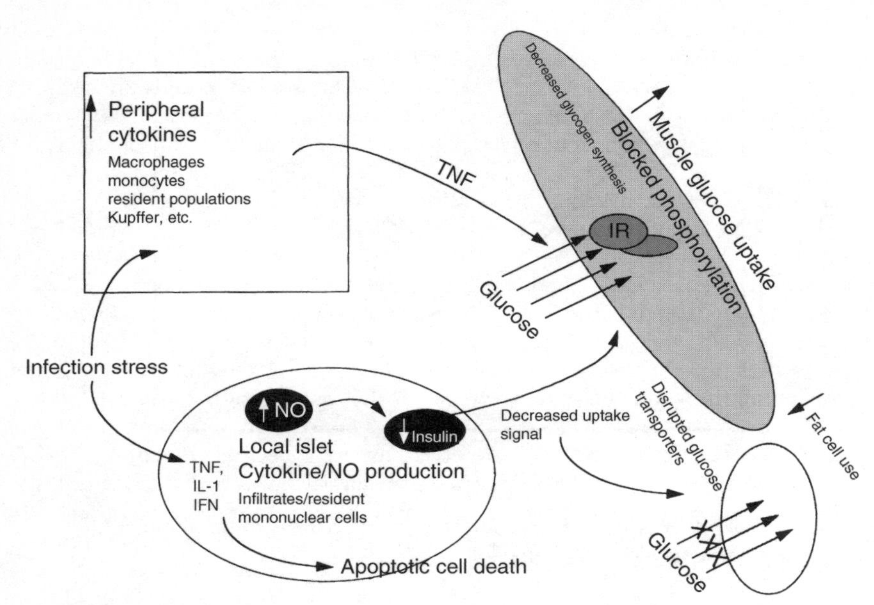

Fig. 4.9. Cytokine interactions at the level of the pancreas as well as in target tissue (muscle and adipose) are capable of yielding differential tissue responses. Where the need calls for it, the blockade of a nutrient's use by one tissue is offset by the increased ability of another tissue to utilize it, thus the 'yin and yang' of interplay between muscle and adipose depots.

insulin receptor (IR) phosphorylation and activation (Cheung *et al.*, 1998), and glycogen synthesizing and catabolizing enzymes in muscle and adipose (Ciraldi *et al.*, 1998). The ability of TNF-α to increase glucose uptake into muscle may be a compensatory mechanism to offset the insulin resistance that is apparently caused by the same cytokine (Ciraldi *et al.*, 1998). Mandrup-Poulsen *et al.* (1996) and McDaniel *et al.* (1996) suggested that the local islet production of NO under the influence of pancreatic localized cytokine production during disease stress limits the release of insulin and thus complicates hormonal regulation of sugar use by tissues. Our laboratory extended these observations by demonstrating that iNOS (the enzyme that forms NO when induced by cytokines) and the insulin secretostatic hormone (Martinez *et al.*, 1996) adrenomedullin are sharply increased in the same cells (Fig. 4.10) during mild forms of two different disease stressors, i.e. occult parasitic infection and endotoxin challenge. The pattern of unregulated iNOS was co-localized to the same cells that increased in adrenomedullin content, and the degree of up-regulation was similar between stressors. When the experimental challenge with endotoxin was applied in addition to the pre-existing occult parasitic infection, the release of adrenomedullin (Fig. 4.11) and nitrate (the stable product of NO production from arginine, Fig. 4.12) into plasma was significantly increased only in

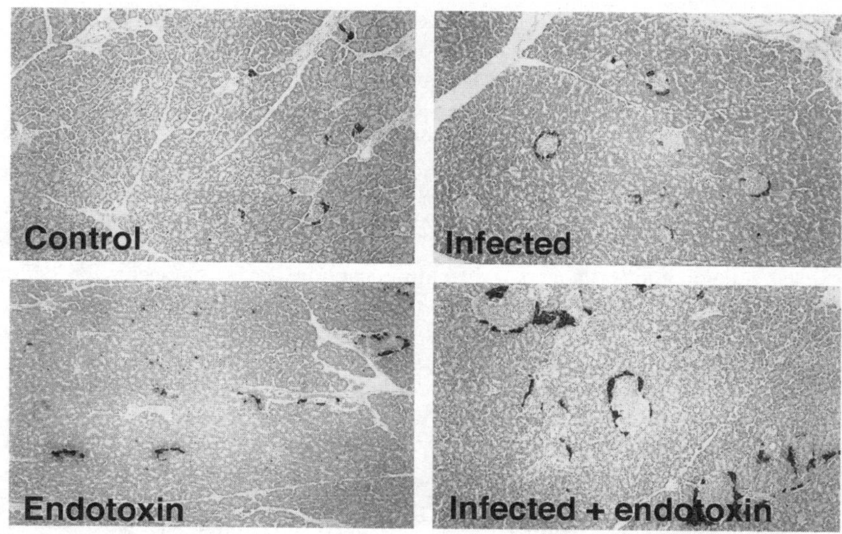

Fig. 4.10. Immunohistochemical localization of adrenomedullin in pancreatic islets of control calves and calves challenged with endotoxin, parasitized or challenged by both endotoxin and parasitism. The marked up-regulation of adrenomedullin (and parallel up-regulation of inducible nitric oxide synthase, not shown) are taken as indications that the combined stresses culminate in the local elaboration of regulatory proteins that decrease the release of insulin from the pancreas.

calves experiencing the combined infection and endotoxin challenges. We believe that these peripheral augmented cytokine and hormone responses as well as the islet-localized increases in both adrenomedullin and NO (via the increased iNOS activity) contribute to the lasting disturbances in insulin regulation and growth observed previously in similarly infected calves (Elsasser *et al.*, 1986).

Collectively, these observations suggest that stress has an impact on the pancreas to compromise the normal secretion of hormones that regulate metabolism. This results in an imbalance in the signals that are responsible for the conservation and uptake of both glucose and amino acids and potentially affect both synthesis and storage of glycogen and protein.

Breakpoint stress: where host response contributes to the pathology

Experiments discussed previously in this chapter point out that low levels of disease stress and the accompanying short-lived metabolic perturbations are relatively benign to the animal in the long run. The host easily adapts to low-level stress by processes involving such elementary activities as

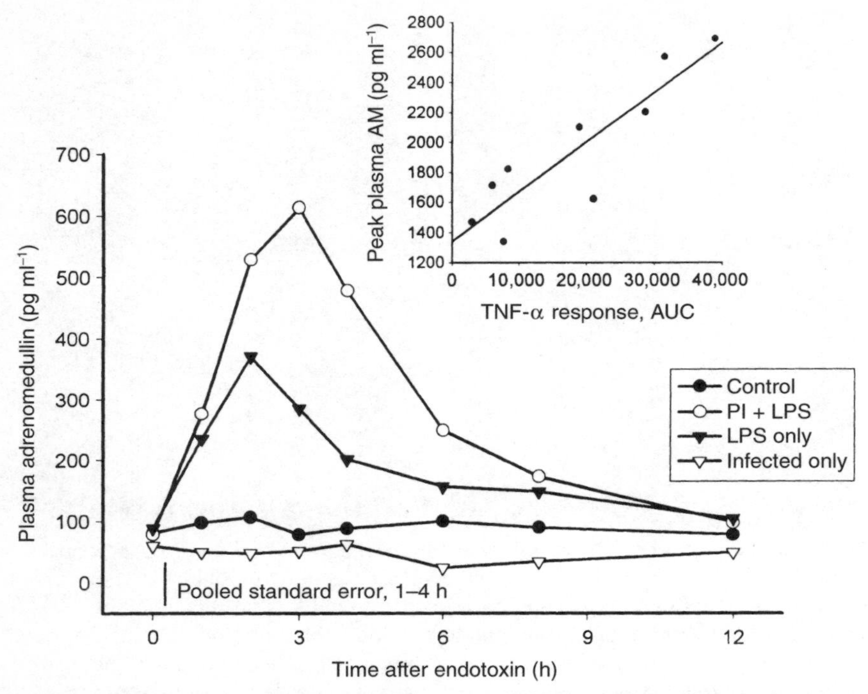

Fig. 4.11. Plasma concentrations of adrenomedullin are increased by endotoxin challenge but further synergistically augmented in calves challenged with endotoxin and harbouring an occult parasitic infection. The inset demonstrates the tight relationship between the elaboration of the TNF response to the endotoxin and the peak response in adrenomedullin.

increasing intake to offset a change in increased metabolism (Cole and Guillot, 1987). Similarly, many chronic, occult infections that decrease gain or the efficiency of nutrient use for growth are not recognized as a pathology (Elsasser *et al.*, 1999). There are limits, however, to how much an animal can adapt to and accommodate the stress. In the first study, as the level of stress increased with the greater surface area of infestation, the point was reached where the animal could no longer adapt to the challenge. In the second study, where another low-level stress was added to the first, a point was reached where the host responded to the stress with retarded growth and cytotoxic reactions. When response is such that the pro-inflammatory cytokine and free radical milieu are harmful, a 'breakpoint' is reached and further pathological consequences can be expected. Breakpoint responses can be observed under a variety of stressful situations and are largely caused by one primary biochemical format – the development of free radical-dependent chemical modification of intracellular proteins and lipids. Physical manifestations of this are apparent in cytokine-induced cell

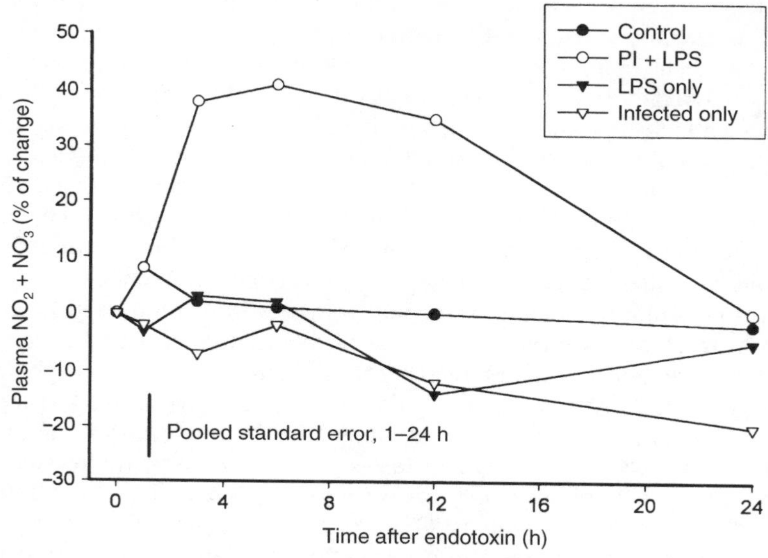

Fig. 4.12. Plasma nitrite and nitrate, the stable breakdown products of NO, increased in plasma only in conjunction with the combined parasitic and endotoxin challenges in calves.

apoptosis, general cytotoxicity and impaired signal transduction (Lancaster *et al.*, 1989; Oshima, 1990; Natoli *et al.*, 1998).

Free radicals

TNF-α and other cytokines released acutely during endotoxin challenge and septicaemia trigger the production of NO from arginine and oxygen via inducible iNOS (Lancaster, 1992; Beckman and Koppenol, 1996). Superoxide anion is also a free radical associated with tissue response to inflammatory stimulus (McCord, 1995). Depending on the intracellular oxidation/reduction state (Stamler *et al.*, 1992), NO may be further altered to several other highly chemically reactive species such as peroxynitrite ($ONOO^-$) in the reaction with superoxide anion (Beckman *et al.*, 1992; Ischiropoulos *et al.*, 1992) or nitroxyl anion (please refer to Fig. 4.6). Pathological conditions are associated with the ability of $ONOO^-$ to nitrate intracellular tyrosine residues of proteins as illustrated by the presence of $ONOO^-$ nitrated proteins in atherosclerosis (Beckman *et al.*, 1992, 1994). Furthermore, when proteins such as those in the catalytic portion of tyrosine kinases are nitrated, the ability to activate these proteins by phosphorylation is blocked (Oshima, 1990; Castro *et al.*, 1994).

Cytokines are a substantive link between how a host detects and perceives the threat of infection and how the body mobilizes a response to the threat (Sherry and Cerami, 1988). Data indicate that the local paracrine cytokine responses of tissues coordinate systemic cytokine responses to initiate, change or maintain a cascade of bioregulators including other cytokines, prostaglandin derivatives, glucocorticoids and endocrine hormones. This cascade results in altered metabolism and the decreased propensity for growth during the period of immune challenge. The pattern of cytokines is essential in directing the components of the immune system to attack invading pathogens and in diverting nutrients to support these efforts. However, there are times when the host response is unbalanced with respect to the magnitude of the challenge and the capabilities of the internal environment. The result of this imbalance in response ranges from protracted recovery times to frank additive pathology. One of the components of the immune response initiated and maintained by the pattern of cytokines elaborated during infection stress is the up-regulation of constitutive as well as inducible isoforms of NOS and the generation of NO from substrate arginine. Once formed, NO decays to other forms of reactive nitrogen intermediates (RNIs) and the path of decay dictates whether NO is beneficial or harmful to cells adjacent to those in which NO is produced. The path of decay depends in part on the internal oxidation/ reduction atmosphere inside the cell. Evidence presented here suggests that the formation of peroxynitrite from NO and superoxide radical contributes to perturbed cell function and affects growth through several mechanisms. One of the mechanisms postulated here is the nitrosylation of liver proteins that impact on the regulation of IGF-I. The magnitude of the response can be modulated by specific effects of hormones and nutrients on otherwise normal aspects of metabolism.

In a study where steers were challenged with endotoxin at 0, 0.2, 1.0 or 3 $\mu g\ kg^{-1}$ i.v. for 4 consecutive days, the change in plasma concentration of IGF-I was used as an indication of disruption of GH regulation of IGF-I control and thus an indication of metabolic regulatory perturbation (Elsasser *et al.*, 1997a). Plasma IGF-I was not affected by repeated challenge with 0.2 $\mu g\ kg^{-1}$ but was significantly decreased within 1 day at the two higher doses of endotoxin used. Plasma concentrations of IGF-I continued to decline to progressively lower levels when doses of 1.0 and 3 $\mu g\ kg^{-1}$ were repeatedly administered to calves. The greatest average decrease in plasma IGF-I concentrations was observed in calves dosed with 1.0 $\mu g\ kg^{-1}$. Only those displaying immunohistochemical presence of nitrotyrosine, a marker for $ONOO^-$ nitrated protein modification and RNI stress, displayed the prolonged perturbation in plasma IGF-I. The pattern of positive immunostaining in liver specimens suggests that the response to endotoxin resulting in production of nitrated proteins was not uniform across the liver but largely contained in periportal cell regions and hepatocytes located on the border of the hepatic central veins. The presence of positive nitrotyrosine immunostaining correlated with decreased plasma concentrations of IGF-I suggests that the formation of $ONOO^-$ contributed to tissue toxicity which affected the production of IGF-I (Elsasser *et al.*, 1998b, c). These observations are consistent with those of Billiar *et al.* (1989) who observed that arginine was required for Kupffer cell-mediated production of RNIs that were cytotoxic to hepatocytes.

Toxin–disease interactions

In the south-eastern US, fescue grass forms an important component of ruminant diets. However, most of this grass is infected with an endophyte. The endophyte infection of this grass causes a toxicosis in animals characterized by poor temperature regulation, low plasma prolactin concentrations, decreased rate of gain, and perturbations in reproduction and lactation. Evidence suggests that a significant factor causing these signs is the presence of a class of ergot-like alkaloids with dopaminergic activity (Aldrich *et al.*, 1993). Studies conducted by Filipov *et al.* (1999) examined the propensity for this underlying toxicosis to interact with a simulated additive disease stress, which was simulated by endotoxin administration. Calves grazing fescue were affected by the toxin, as evidenced by the significant decrease in prolactin levels. When challenged with endotoxin ($0.2\ \mu g\ kg^{-1}$, i.v.), plasma TNF-α and cortisol responses were significantly higher in animals grazed on toxic fescue compared with cattle grazed on an endophyte-free fescue (Fig. 4.13). In addition, the time of recovery was protracted in these animals. These data support the concept that an underlying chronic stress can make an animal more sensitive and responsive to additional stress inputs. Collectively, these stresses push animals to reach the 'breakpoint' sooner, decreasing the ability to resist disease. In these situations, management of stress is made more complicated.

Conclusion

Metabolism and the use of nutrients are affected by stress in several ways. Some mild stresses are handled simply by behaviours that allow the animal to remove itself from the stress or to adapt to the increased metabolism need by increasing feed intake. On the other hand, when stressor responses are strong or protracted enough to affect metabolism, there is a priority that dictates an order by which tissue beds are affected. An endocrine–immune gradient of hormone and cytokine interactions fine-tunes the metabolic activities of the cells. When specific endocrine glands such as the pituitary and pancreas are the targets of a stress, the perturbed secretion of hormones contributes to metabolic impairment and loss of acute regulation of cell metabolism. Similarly, while pro-inflammatory cytokine responses by immune cells are necessary to initiate signals to counter infectious

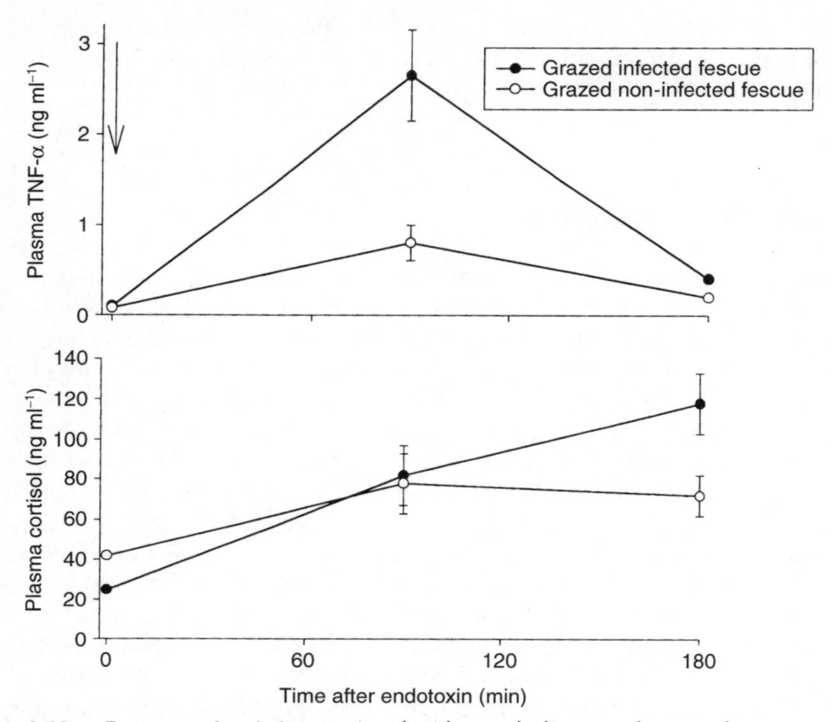

Fig. 4.13. Fescue toxicosis is associated with metabolic perturbations that decrease growth and weight gain in cattle. These data illustrate that if cattle graze a grass with this endophyte-derived toxin, the severity of metabolic responses to an infection stress increases because of increased TNF and cortisol responses to the challenge (from Filipov *et al.*, 1999).

challenges in the body, the oversecretion of these effectors can result in inappropriate increases in cytokine-driven free radical production and result in chemical modifications of cell proteins or lipids that limit the efficiency of metabolism by affected cells. Strategies that prevent stresses from reaching the 'breakpoint' where the stress response contributes to the pathology are needed. These strategies should be designed to prevent stress-induced pathology such as the accumulation of oxidatively damaged proteins while maintaining an appropriate balance between physiological processes that compete for nutrients.

References

Abebe, G., Shaw, M.K. and Eley, R. (1993) *Trypanosoma congolense* in the microvasculature of the pituitary gland of experimentally infected Boran (*Bos indicus*) cattle. *Veterinary Pathology* 30, 401–409.

Aldrich, C.G., Paterson, J.A., Tate, J.L. and Kerley, M.S. (1993) The effects of endophyte-infected tall fescue consumption on diet utilization and thermal regulation in cattle. *Journal of Animal Science* 71, 164–170.

Baracos, V.E., Whitmore, W.T. and Gale, R. (1987) The metabolic cost of fever. *Canadian Journal of Physiology and Pharmacology* 65, 1248–1253.

Beckman, J.S. and Koppenol, W.H. (1996) Nitric oxide, superoxide and peroxynitrite: the good, the bad and the ugly. *American Journal of Physiology* 271, 1424–1437.

Beckman, J.S., Ischiropoulis, H., Zhu, L., Van der Woerd, M., Smith, C., Chen, J., Harrison, J., Martin, J. and Tsai, M. (1992) Kinetics of superoxide dismutase- and iron-catalyzed nitration of phenolics by peroxynitrite. *Archives of Biochemistry and Biophysics* 298, 438–445.

Beckman, J.S., Ye, Y.Z., Anderson, P., Chen, M.A., Accavetti, M., Tarpey, M.M. and White, C.R. (1994) Extensive nitration of protein tyrosines observed in human atherosclerosis detected by immunohistochemistry. *Biological Chemistry Hoppe-Seyler* 375, 81–88.

Beermann, D.H. and Devol, D.L. (1992) Effects of somatotropin releasing hormone, and somatostatin on growth. In: Pearson, A.M. and Dutson, T.R. (eds) *Growth Regulation in Farm Animals*. Elsevier, New York, pp. 373–346.

Beisel, W. (1988) The effects of infection on growth. In: Rumsey, T.S. (ed.) *The Beltsville Symposium XII: Biomechanisms Regulating Growth and Development*. Kluwer, New York, pp. 395–408.

Bettger, W.J. and McKeehan, W.L. (1986) Mechanisms of cellular communication. *Physiological Reviews* 66, 1–35.

Beutler, B. and Cerami, A. (1989) The biology of cachectin/TNF: a primary mediator of the host response. *Annual Review of Immunology* 7, 625–655.

Billiar, T.R., Curran, R.D., West, M.A., Hoffmann, K. and Simmons, R.L. (1989) Kupffer cell cytotoxicity to hepatocytes in co-culture requires L-arginine. *Archives of Surgery* 124, 1416–1421.

Breier, B.H., Gluckman, P.D. and Bass, J.J. (1988) The somatotropic axis in young steers: influence of nutritional status and 17 beta-estradiol on hepatic high- and low-affinity somatotropic binding sites. *Journal of Endocrinology* 116, 169–177.

Calder, P.C. (1995) Fuel utilization by the immune system. *Proceedings of the Nutrition Society* 54, 65–82.

Castro, L., Rodriquez, M. and Radi, R. (1994) Aconitase is rapidly inactivated by peroxynitrite but not by its precursor, nitric oxide. *Journal of Biological Chemistry* 269, 29,409–29,415.

Cheung, A.T., Ree, D., Kolls, J.K., Fuselier, J., Coy, D.H. and Bryer-Ash, M. (1998) An *in vivo* model for elucidation of the mechanism of tumor necrosis factor-α (TNF-α)-induced insulin resistance: evidence for differential regulation of insulin signaling by TNF-α. *Endocrinology* 139, 4928–4935.

Ciraldi, T.P., Carter, L., Mudaliar, S., Kern, P.A. and Henry, R.R. (1998) Effects of tumor necrosis factor-α on glucose metabolism in cultured human muscle cells from nondiabetic and Type-2 diabetic subjects. *Endocrinology* 139, 4793–4800.

Clemmons, D.R. and Underwood, L.E. (1991) Nutritional regulation of IGF-I and IGF-binding proteins. *Annual Review of Nutrition* 11, 393–412.

Clemmons, D.R., Cascieri, M.A., Camacho-Hubner, C., McCusker, R.H. and Bayne, M.L. (1990) Discrete alterations of the insulin-like growth factor I molecule which alters its affinity for the insulin-like growth factor-binding proteins results in changes in bioactivity. *Journal of Biological Chemistry* 265(21), 12,210–12,216.

Cohick, W.S. and Clemmons, D.R. (1993) The insulin-like growth factors. *Annual Review of Physiology* 55, 131–153.

Cole, N.A. and Guillot, F.S. (1987) Influence of *Psoroptes ovis* on the energy metabolism of heifer calves. *Veterinary Parasitology* 23, 285–295.

Corbett, J.A. and McDaniel, M.L. (1996) The use of aminoguanidine, a selective iNOS inhibitor, to evaluate the role of nitric oxide in the development of autoimmune diabetes. *Methods* 10, 21–30.

Daughaday, W.H. and Trivedi, B. (1991) Clinical aspects of GH binding proteins. *Acta Endocrinologica* 124, 27–32.

Davis, S.L. (1998) Environmental modulation of the immune system via the endocrine system. *Domestic Animal Endocrinology* 15, 283–289.

Dayton, W.R. and Hathaway, M.R. (1992) Control of animal growth by glucocorticoids, thyroid hormones, autocrine and paracrine growth factors. In: Pearson, A.M. and Dutson, T.R. (eds) *Growth Regulation in Farm Animals.* Elsevier, New York, pp. 17–46.

Demling, R.H., Lalonde, C.C., Jin, L.J., Albes, J. and Fiori, N. (1986) The pulmonary and systemic response to recurrent endotoxemia in the adult sheep. *Surgery* 100, 876–883.

Douglas, R.G., Gluckman, P.D., Breier, B.H., McCall, B., Parry, B. and Shaw, J.H.F. (1991) Effects of recombinant IGF-I on protein and glucose metabolism in rTNF-infused lambs. *American Journal of Physiology* 261, 606–614.

Elsasser, T.H. (1979) TRH and prolactin: studies on prolactin secretory dynamics. PhD Thesis, Medical University of South Carolina, Charleston, USA.

Elsasser, T.H. and Steele, N.C. (1992) Growth hormone directed nutrient partitioning: Immune system-pituitary gland communication. In: Bray, G.A. and Ryan, D.H. (eds) *The Science of Food Regulation: Food Intake, Taste, Nutrient Partitioning, and Energy Expenditure.* Louisiana State University Press, Babon, Rouge, Los Angeles, pp. 164–186.

Elsasser, T.H., Rumsey, T.S., Hammond, A.C. and Fayer, R. (1986) Perturbed metabolism and hormonal profiles in calves infected with *Sarcocystis cruzi. Domestic Animal Endocrinology* 3, 277–287.

Elsasser, T.H., Rumsey, T.S., Hammond, A.C. and Fayer, R. (1988) Influences of parasitism on plasma concentrations of growth hormone, somatomedin-C and somatomedin binding proteins in calves. *Journal of Endocrinology* 116, 101–199.

Elsasser, T.H., Fayer, T., Rumsey, T.S. and Hammond, A.C. (1990) Plasma and tissue concentrations and molecular forms of somatostatin in calves infected with *Sarcocystis cruzi. Domestic Animal Endocrinology* 7, 537–548.

Elsasser, T.H., Caperna, T.J. and Fayer, R. (1991) Tumor necrosis factor-α affects pituitary hormone secretion by a direct pituitary interaction. *Proceedings of the Society for Experimental Biology and Medicine* 198, 547–553.

Elsasser, T.H., Caperna, T.J. and Rumsey, T.S. (1995) Endotoxin administration decreases plasma insulin-like growth factor-I and IGF binding protein-2 in Angus × Hereford steers independent of changes in nutritional intake. *Journal of Endocrinology* 144, 109–117.

Elsasser, T.H., Rosebrough, R., Rumsey, T.S. and Moseley, W.M. (1996) Hormonal and nutritional modulation of hepatic arginase activity in growing cattle. *Domestic Animal Endocrinology* 13, 219–228.

Elsasser, T.H., Kahl, S., Steele, N.C. and Rumsey, T.S. (1997a) Nutritional modulation of somatotropic axis–cytokine relationships. *Comparative Biochemistry and Physiology* 116, 209–221.

Elsasser, T.H., Kahl, S., Sartin, J.L. and Rumsey, T.S. (1997b) Free-radical modulation of IGF-I downregulation in endotoxin (LPS)-induced disease stress in calves. *Journal of Animal Science* 75 (Suppl. 1, abstract), 164.

Elsasser, T.H., Sartin, J.L., McMahon, C., Romo, G., Fayer, R., Kahl, S. and Blagburn, B. (1998a) Changes in somatotropic axis response and body composition during growth hormone administration in progressive cachectic parasitism. *Domestic Animal Endocrinology* 15, 239–255.

Elsasser, T.H., Kahl, S., Rumsey, T.S. and Collier, R.J. (1998b) Recombinant bovine somatotropin (bST) increases plasma NO_2 and NO_3 response to endotoxin in cattle. *Journal of Animal Science* 76 (Suppl. 2), 117.

Elsasser, T.H., Kahl, S., Rumsey, T.S. and Blum, J.W. (1998c) Cytokine-related free radical modulation of reduced growth performance in disease models. In: Blum, J., Elsasser, T. and Guilloteau, P. (eds) *Proceedings of the Symposium on Growth in Ruminants.* University of Bern, Switzerland, pp. 197–206.

Elsasser, T.H., Sartin, J.L., Martinez, A., Kahl, S., Montuenga, L.M., Pio, R., Fayer, R., Miller, M.J. and Cuttitta, F. (1999) Underlying disease stress augments plasma and tissue adrenomedullin (AM) responses to endotoxin: co-localized increases in AM and iNOS within pancreatic islets. *Endocrinology* 140, 5402–5410.

Etherton, T.D. and Bauman, D.E. (1998) Biology of somatotropin in growth and lactation of domestic animals. *Physiological Reviews* 78, 745–761.

Fan, J., Char, D., Badby, G.J., Gelato, M.C. and Lang, C.H. (1995) Regulation of insulin-like growth factor-I and IGF binding proteins by tumor necrosis factor. *American Journal of Physiology* 269, 1204–1212.

Filipov, N.M., Thompson, F.N., Stuedemann, J.A., Elsasser, T.H., Kahl, S., Young, C.R., Sharma, R.P. and Smith, C.K. (1999) Increased responsiveness to intravenous lipopolysaccharide (LPS) challenge in steers grazing endophyte-infected tall fescue compared to steers grazing endophyte-free tall fescue. *Journal of Endocrinology* 163, 213–220.

Fong, Y. and Lowry, S.F. (1990) Tumor necrosis factor in the pathophysiology of infection and sepsis. *Clinical Immunology and Immunopathology* 55, 157–170.

Griffith, O.W. and Stuehr, D.J. (1995) Nitric oxide synthesis: properties and catalytic mechanism. *Annual Review of Physiology* 57, 707–736.

Halestrap, A.P., Wang, X., Pool, R.C., Jackson, V.N. and Price, N.T. (1997) Lactate transport in heart in relation to myocardial ischemia. *American Journal of Cardiology* 80, 17a–25a.

Hammond, J. (1944) Physiological factors affecting birth weight. *Proceedings of the Nutrition Society* 2, 8–12.

Hammond, J. (1952) Physiological limits to intensive production in animals. *British Agriculture Bulletin* 4, 222–225.

Hammond, J., Jr, Bowman, J.C. and Robinson, T.J. (1983) *Hammond's Farm Animals,* 5th edn. Edward Arnold Publishing, London.

Heiman, M.L., Chen, Y. and Caro, J.F. (1998) Leptin participates in the regulation of glucocorticoid and growth hormone axes. *Journal of Nutritional Biochemistry* 9, 553–559.

Hotamisligil, G.S. and Spiegelman, B.M. (1994) Tumor necrosis factor-α: a key component of the obesity–diabetes link. *Diabetes* 43, 1271–1278.

Huntington, G.B. and Reynolds, C.K. (1987) Oxygen consumption and metabolite flux of the bovine portal-drained viscera and liver. *Journal of Nutrition* 117(6), 1167–1173.

Ischiropoulos, H., Zhu, L., Tsai, M., Martin, J.C., Smith, C.D. and Beckman, J.S. (1992) Peroxynitrite-mediated tyrosine nitration catalyzed by superoxide dismutase. *Archives of Biochemistry and Biophysics* 298, 431–437.

Jaeschke, H. (1995) Mechanisms of oxidant stress-induced acute tissue injury. *Proceedings of the Society for Experimental Biology and Medicine* 209, 104–111.

Kahl, S., Elsasser, T.H. and Blum, J.W. (1997) Nutritional regulation of plasma tumor necrosis factor-α and plasma and urinary nitrite/nitrate responses to endotoxin in cattle. *Proceedings of the Society for Experimental Biology and Medicine* 215, 370–376.

Kim, H.S., Richardson, R.L. and Hausman, G.J. (1998) The expression of insulin-like growth factor-1 during adipogenesis *in vivo*: effect of thyroxine. *General and Comparative Endocrinology* 112, 38–45.

Klasing, K.C. (1998) Nutritional modulation of resistance to infectious disease. *Poultry Science* 77, 1119–1125.

Klasing, K.C., Laurin, D.E., Peng, R.C. and Fry, D.M. (1987) Immunologically-mediated growth depression in chicks: influence of feed intake on corticosterone and IL-1. *Journal of Nutrition* 117, 1629–1637.

Kluger, M.J. (1991) Fever: role of pyrogens and cryogens. *Physiological Reviews* 71, 93–108.

Lancaster, J. (1992) Nitric oxide in cells. *American Scientist* 80, 248–257.

Lancaster, J.R., Laster, S.M. and Gooding, L.R. (1989) Inhibition of target cell mitochondrial electron transfer by tumor necrosis factor. *FEBS Letters* 248, 169–174.

Lefcourt, A.M. (1986) Usage of the term 'stress' as it applies to cattle. *Vlaams Diergeneekundig Tijdschrift* 55, 258–265.

Mandrup-Poulsen, T., Nerup, J., Reimers, I., Pociot, F., Anderson, H.U., Bjerre, U. and Bergholdt, R. (1996) Cytokines and the endocrine system. II: Roles of substrate metabolism, modulation of thyrodial and pancreatic cell functions and autoimmune endocrine diseases. *European Journal of Endocrinology* 134, 21–30.

Martinez, A., Weaver, C., Lopez, J., Bhathena, S.J., Elsasser, T.D., Miller, M.J., Moody, T.W., Unsworth, E.J. and Cuttitta, F. (1996) Regulation of insulin secretion and blood glucose metabolism by adrenomedullin. *Endocrinology* 137, 2626–2632.

McCord, J.M. (1995) Superoxide radical: controversies, contradictions, and paradoxes. *Proceedings of the Society for Experimental Biology and Medicine* 209, 112–117.

McDaniel, M.L., Kwon, G., Hill, J.R., Marshall, C.A. and Corbett, J.A. (1996) Cytokines and nitric oxide in islet inflammation and diabetes. *Proceedings of the Society for Experimental Biology and Medicine* 211, 24–32.

Morin, C.L., Gayles, E.C., Podolin, D.A., Wei, Y., Xu, M. and Pagliassoti, M.J. (1998) Adipose tissue-derived tumor necrosis factor activity correlates with fat cell size but not insulin action in aging rats. *Endocrinology* 139, 4998–5005.

Natoli, G., Costanza, A., Guido, F., Moretti, F. and Levrero, M. (1998) Apoptotic, non-apoptotic, and anti-apoptotic pathways of tumor necrosis factor signaling. *Biochemical Pharmacology* 56, 915–920.

Newsholme, E.A. and Calder, P.C. (1997) The proposed role of glutamine in some cells of the immune system and speculative consequences for the whole animal. *Nutrition* 13, 728–730.

NRC (1998) *The Use of Drugs in Food Animals: Benefits and Risks.* National Research Council, National Academy Press, Washington, DC, pp. 23–61.

Oshima, H. (1990) Nitrosylation of tyrosine kinase by tetranitromethane blocks phosphorylation in the EGF signaling cascade. *Fundamentals in Chemical Toxicology* 28, 647–652.

Powanda, M.C. (1980) Host metabolic alterations during inflammatory stress as related to nutritional status. *American Journal of Veterinary Research* 41, 1905–1911.

Quinn, L., Haugk, K.L. and Grabstein, K.H. (1995) Interleukin-15: a novel anabolic cytokine for skeletal muscle. *Endocrinology* 136, 3669–3672.

Ramsay, T.G., Chung, I.B., Czerwinski, S.M., McMurtry, J.P., Rosebrough, R.W. and Steele, N.C. (1995) Tissue IGF-I protein and mRNA responses to a single injection of somatotropin. *American Journal of Physiology* 269(part 1), E627–E635.

Rodriquez-Arnao, J., Meill, J.P. and Ross, J.M. (1993) Influence of the thyroid axis on the GH/IGF-I axis. *Trends in Endocrinology and Metabolism* 4, 169–173.

Roura, E., Homedes, J. and Klasing, K.C. (1992) Prevention of immunologic stress contributes to the growth-permitting ability of dietary antibiotics in chicks. *Journal of Nutrition* 122, 2383–2390.

Salomon, F., Cuneo, R. and Sonksen, P.H. (1991) Growth hormone and protein metabolism. *Hormone Research* 36 (Suppl. 2), 41–43.

Sherry, B. and Cerami, A. (1988) Cachectin/TNF exerts endocrine, paracrine and autocrine control of inflammatory processes. *Journal of Cell Biology* 107, 1269–1277.

Stamler, J.S. and Feelisch, M. (1996) Biochemistry of nitric oxide and redox-related species. In: Feelisch, M. and Staler, J.S. (eds) *Methods in Nitric Oxide Research.* John Wiley & Sons, New York, pp. 19–27.

Stamler, J.S., Singel, D.J. and Loscalzo, J. (1992) Biochemistry of nitric oxide and its redox-activated forms. *Science* 258, 1898–1902.

Steele, N.C. and Elsasser, T.H. (1989) Regulation of somatomedin production, release, and mechanisms of action. In: Campion, D., Hausman, G. and Martin, R. (eds) *Animal Growth Regulation.* Kluwer Academic/Plenum Publishing, Norwell, Massachusetts, pp. 295–315.

Tannenbaum, G.S. (1991) Neuroendocrine control of growth hormone secretion. *Acta Paediatrica Scandinavia* s372, 5–16.

Touchberry, R.W. (1984) Emerging technology for the production of food animals through land and aquaculture systems. In: English, B.C., Maetzold, J.A., Holding, B.R. and Heady, E.O. (eds) *Future Agricultural Technology and Resource Conservation.* Iowa State University Press, Ames, Iowa, pp. 500–504.

Tsukahara, H., Gordienko, D.V., Tonshoff, B., Gelato, M.C. and Goligorsky, M.S. (1994) Direct demonstration of insulin-like growth factor-I-induced nitric oxide production by endothelial cells. *Kidney International* 45(2), 598–604.

Wilckens, T. (1995) Glucocorticoids and immune function: physiological relevance and pathogenic potential of hormonal dysfunction. *Trends in Pharmacological Sciences* 16, 193–197.

Wu, G. and Morris, S.M. (1998) Arginine metabolism: nitric oxide and beyond. *Biochemical Journal* 336, 1–17.

Zamir, O., Hasselgren, P., Kunkle, S.L., Frederick, J., Higashiguchi, T. and Fischer, J.E. (1992) Evidence that tumor necrosis factor participates in the regulation of muscle proteolysis during sepsis. *Archives of Surgery* 127, 170–178.

Immune System Response to Stress

F. Blecha

Department of Anatomy and Physiology, College of Veterinary Medicine, 228 Veterinary Medical Sciences, Manhattan, Kansas, USA

Introduction

Organisms have evolved many strategies to defend against stressful environments and conditions. Indeed one could argue that the immune system is the result of the evolving response of the host to the stress of pathogen challenge. Exactly how the immune system responds to stress, however, is a difficult question to answer, in part because of the complexities of the host immunity and the stress response systems.

These complexities are illustrated by the myriad of choices that must be made when one attempts to unravel the influence of stress on the immune system. For example, decisions have to be made as to whether one should evaluate innate or acquired immunity (Blecha, 2000). If innate immunity is chosen, then which aspect will be studied: cellular or humoral? If cellular is chosen, then which cell type will be evaluated: neutrophils, macrophages or natural killer cells? If one evaluates acquired immunity, the same decision as to whether to evaluate cellular (T or B lymphocytes) or humoral (immunoglobulin) immunity must be made and is further complicated by the decision regarding whether to evaluate passive or active acquired immunity. Similarly, what aspect of the stress response should one evaluate in the context of immune function: the classical increased glucocorticoid response via activation of the hypothalamic–pituitary–adrenal (HPA) axis or increased sympathetic activity via activation of the hypothalamic autonomic nervous system? The danger of generalizing the influence of stress on the immune system is apparent. However, despite these difficulties, progress has been made in understanding how the immune system responds to stress.

A reflection of the activity and interest in the area of stress and immunity is illustrated by the nearly 200 review articles written on this topic in the last decade, including several reviews relative to food animals (Blecha, 1988; Griffin, 1989; Dohms and Metz, 1991; Kelley *et al.*, 1994; Minton, 1994; Davis, 1998; Griffin and Thomson, 1998). This body of literature provides evidence from the disciplines of neuroscience and immunology for the existence of bidirectional interactions between the immune and nervous systems. This crosstalk between the immune and nervous systems is mediated in large part by cytokines and is undoubtedly involved in an animal's response to stress. The effect of stress on cytokine production and the bidirectional communication between the immune and neuroendocrine systems have been the focus of several recent reviews (Faith *et al.*, 1998; Weigent and Blalock, 1998; Weizman and Bessler, 1998). This chapter will focus on research evaluating the influence of stress on immune function in cattle, pigs and sheep. I will first discuss studies that investigated the influence of management and environmental stressors on immune function. A logical extension of these studies, which I will also address, is the influence of activation of the HPA axis on host immunity. Finally, I will summarize results of laboratory studies using model stressors to determine the impact of stress on the immune system.

Management and environmental stressors and the immune system

The assumption that stress influences host immunity arises from observations of increased disease occurrence in animals exposed to extreme, stressful environments. Although it is important to remember that the link between stress and animal well-being is a continuum and is influenced by genetics, age, nutrition and experiences (Hahn and Becker, 1984), it is clear that many intensive livestock production management practices affect host immunity and disease susceptibility. One of the most frequently used examples of this relationship is bovine respiratory disease in shipped cattle.

The bovine respiratory disease complex has been studied intensely for decades, yet it continues to cause large economic losses in the cattle industry (Loan, 1984). This multifaceted disease complex involves the interaction of respiratory viruses and bacteria and stress. Several years ago we conducted a study specifically to address the impact that the stress of long-distance transport had on immune function in feeder calves (Blecha *et al.*, 1984). Calves that were shipped 700 km for 10 h developed a leukocytosis that was caused primarily by a neutrophilia. Neutrophilia is a common observation in stressed animals and is thought to be caused by a glucocorticoid-induced change in neutrophil trafficking and release from bone marrow reserves (Kelley *et al.*, 1981; Roth and Kaeberle, 1982; Roth *et al.*, 1982; Murata *et al.*, 1987). Lymphocyte proliferative responses were decreased in the shipped calves immediately after the stress of shipping (Fig. 5.1); however, concentrations of plasma cortisol did not differ between control and shipped calves. Considering the abundant evidence linking elevated glucocorticoid levels with suppressed immune responses, it is unclear why decreased lymphocyte proliferation in shipped calves was not correlated with increased concentrations of cortisol. It may be that these findings are simply a reflection of the blood sampling procedure and frequency, such that the peak cortisol concentration occurred prior to sampling at unloading.

The importance of well-controlled experiments in determining the influence of shipping stress on immune parameters was shown in experiments on pigs conducted by McGlone and colleagues (McGlone *et al.*, 1993; Hicks *et al.*, 1998). Experiments conducted in a field situation, where it was difficult to handle the pigs for blood collection, failed to show a shipping-induced increase in plasma cortisol because of the high baseline concentrations. However, in an experiment where many more experimental variables were controlled, shipping stress caused increased cortisol concentrations and neutrophilia. In addition, body weight was reduced by 5.1% in

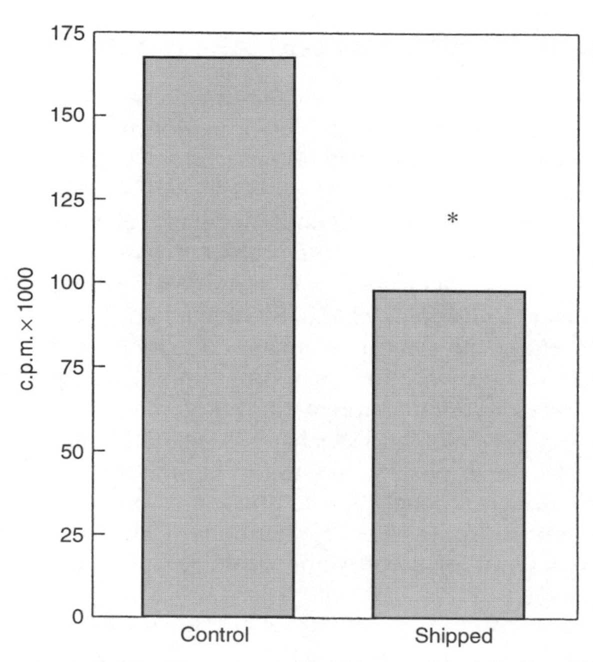

Fig. 5.1. The stress of shipping suppressed mitogen-stimulated lymphocyte proliferation in 9-month-old feeder calves. Forty calves were allotted by weight and breed to control or shipped groups. Mitogen-stimulated lymphocyte proliferation was evaluated prior to transport, at unloading and 1 week after transport. After 10 h of transport, concanavalin A-stimulated lymphocyte proliferation was lower in shipped calves; $*P < 0.05$. Adapted from Blecha *et al.* (1984).

shipped pigs compared with non-shipped controls. Increased cortisol concentrations in shipped pigs did not translate into lower immune function as measured by natural killer (NK) cell cytotoxicity. In fact, just the opposite was observed: pigs with higher plasma concentrations of cortisol had increased NK cell cytotoxicity (McGlone *et al.*, 1993). An interesting and important aspect of these studies was that a pig's social status interacted with the stress of shipping. Among the shipped pigs, submissive pigs and those of intermediate social status had decreased NK cell cytotoxic responses as compared with the socially dominant pigs. Other studies have also failed to show a direct correlation between concentrations of plasma cortisol induced by various management stressors and immune function (Blecha *et al.*, 1985; Lund *et al.*, 1998).

Collectively, studies of the effect of shipping stress on immune function in cattle and pigs suggest that other factors, in addition to cortisol, are involved in shipping-induced immune alterations. Furthermore, they highlight the importance of considering the influence of social status when evaluating the effect of stress on the immune system.

Activation of the HPA axis and the immune system

Although acute increases in glucocorticoids are considered a classic characteristic of the stress response, as indicated previously it is often difficult to establish a relationship between stress-induced activation of the HPA axis and alterations in immune function. It is well established that incubation of cattle and swine immune cells with physiologically relevant high concentrations of cortisol decreases lymphocyte proliferative responses, interleukin-2 (IL-2) production and neutrophil functions (Westley and Kelley, 1984; Blecha and Baker, 1986; Salak *et al.*, 1993). Synthetic glucocorticoids have been used to study the effects of increased adrenal glucocorticoids on immune function; however, they may not represent an acceptable model for studying stress and immune interactions (Roth and Kaeberle, 1985; Roth and Flaming, 1990; Minton and Blecha, 1991; Saulnier *et al.*, 1991). These findings have led to *in vivo* studies using the anterior pituitary hormone, adrenocorticotropic hormone (ACTH), and more recent experiments using hypothalamic corticotropin-releasing hormone (CRH), to increase physiological concentrations of cortisol and evaluate subsequent alterations in immune capacity.

Feeder calves injected with ACTH twice daily for 2 days had concentrations of plasma cortisol three times higher than saline-injected control calves (Blecha and Baker, 1986; Fig. 5.2). This ACTH-induced increase in cortisol caused a leukocytosis, primarily of mature and immature neutrophils. These findings are similar to changes in cortisol concentrations and leukocyte profiles often found in acutely stressed animals. Importantly, this experimental paradigm also decreases mitogen-induced lymphocyte proliferation

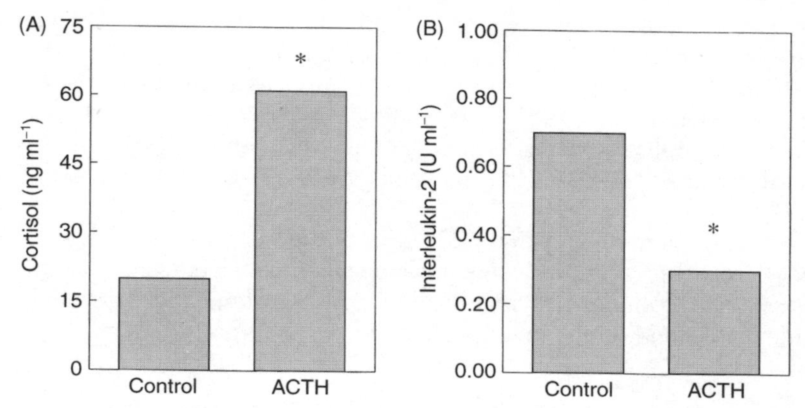

Fig. 5.2. Administration of ACTH to 9-month-old feeder calves increased plasma concentrations of cortisol and decreased interleukin-2 production. Calves were injected with ACTH (1 IU kg^{-1}; $n = 6$) or saline ($n = 6$) twice daily for 2 days. Plasma cortisol concentrations (A) and lymphocyte production of interleukin-2 (B) after 2 days of ACTH injections were different from saline-injected calves; $*P < 0.05$. Adapted from Blecha and Baker (1986).

and IL-2 production in cattle (Fig. 5.2) and pigs (Blecha and Baker, 1986; Wallgren *et al.*, 1994).

Activation of the HPA axis, administration of exogenous cortisol and blockage of cortisol synthesis have been used to investigate the *in vivo* effects of glucocorticoids on NK cell activity in pigs (Salak-Johnson *et al.*, 1996). A single intravenous bolus of ACTH causes plasma cortisol concentrations to double within 2 h of injection in 7- to 9-week-old gilts. However, NK cell cytotoxicity measured within this same time frame is not influenced by ACTH treatment. In an earlier study, ACTH administration increased porcine NK cell activity (McGlone *et al.*, 1991). A pharmacologically induced threefold increase in plasma cortisol concentrations does not alter porcine NK cell activity at 1 or 2 h after injection (Salak-Johnson *et al.*, 1996). However, infusion of 400 μg of cortisol, which causes a sevenfold increase in plasma cortisol concentrations, does decrease NK cell cytotoxicity at 1 h post-infusion, but not at 2 h after cortisol injection. If increased concentrations of cortisol cause a decrease in NK cell activity, then blocking the synthesis of cortisol should influence, and perhaps increase, this response. Pigs fed metyrapone, which blocks the conversion of deoxycortisol to cortisol, had plasma concentrations of cortisol 10- to 30-fold lower than controls (Salak-Johnson *et al.*, 1996). However, pigs with these low concentrations of plasma cortisol also had NK cell cytotoxic responses that were significantly lower than those of control pigs. The authors speculated that the low cortisol concentrations caused a loss of negative feedback, which probably increased concentrations of central CRH, ACTH, β-endorphin and catecholamines. Increased central CRH and subsequent release of catecholamines decreases NK cell activity (Irwin *et al.*, 1987, 1988, 1990). However, in pigs, intracerebroventricular administration of porcine CRH only marginally decreases NK cell activity but significantly impairs neutrophil chemotaxis (Salak-Johnson *et al.*, 1997). Depending on the mitogen used, porcine lymphocyte proliferative responses are either decreased or not influenced by central administration of CRH (Johnson *et al.*, 1994; Salak-Johnson *et al.*, 1997).

These studies suggest that activation of the HPA axis stimulates some aspects of acute stress-induced immune alterations. However, they also illustrate the complexities involved in pharmacologically modelling the influence of stress on the immune system.

Model stressors used to study the immune system

Because of the difficulties in conducting adequately controlled *in vivo* experiments that mimic stressful field conditions and because of the inherent experimental bias associated with pharmacologically influencing the HPA axis, researchers have sought to develop laboratory stress models to study the influence of stress on the immune system (Roth and Flaming, 1990; Minton, 1994). Laboratory stress models used in large domestic animals include acute physical exertion, restraint, and restraint and isolation paradigms.

An early laboratory stress model used to simulate stress-induced influences on bovine respiratory disease is acute physical exertion (Blecha and Minocha, 1983). This model involves running a calf for 5 min on a large-animal treadmill, inclined at 3°, at speeds of 1.8–2.2 m s^{-1}. Hereford steers exposed to this stress paradigm had lower lymphocyte proliferative responses, and serum from stressed calves enhanced the growth of the bovine respiratory disease virus bovine herpesvirus-1. Although this laboratory model established that acute physical exertion is a stressor that can be used to study stress-induced immune and viral alterations in calves, it is an expensive and difficult stress model to use in large animals. In addition, it is not clear how closely acute physical exertion mimics the stress of long-distance transport of calves. For these reasons, this stress model has not been widely used in domestic farm animals, and other stress paradigms, like restraint, have become more common.

Physical restraint has been used for many years as a model to study stress and immune function interactions in laboratory animals (Blecha *et al.*, 1982a, b; Blecha and Topliff, 1984; Morrow-Tesch *et al.*, 1993). Over the last few years this same stressor has been used to evaluate the influence of acute stress on immune function in pigs and sheep (Klemcke *et al.*, 1990; Minton and Blecha, 1990; Coppinger *et al.*, 1991; Minton *et al.*, 1992; Brown-Borg *et al.*, 1993). Although restraint increases ACTH and cortisol concentrations in pigs and lambs and alters some immune parameters, including lymphocyte proliferative responses, IL-2 production and expression of some leukocyte differentiation antigens, the responses have been variable. Because of this, we conducted a study to determine the contribution of cortisol to the suppression in lymphocyte functions in restrained and isolated lambs (Minton *et al.*, 1995). Cortisol was infused intravenously into lambs to approximate the concentration and time domains that were observed in lambs exposed to restraint and isolation stress. Lambs that were subjected to a restraint and isolation stressor had decreased mitogen-induced lymphocyte proliferative responses (Fig. 5.3); however, cortisol infusion did not alter lymphocyte function. These findings show that factors other than cortisol were involved in the stress-induced immunosuppression observed in restrained and isolated lambs. Furthermore, they illustrate the importance of using appropriate laboratory models to investigate the influence of stress on the immune system.

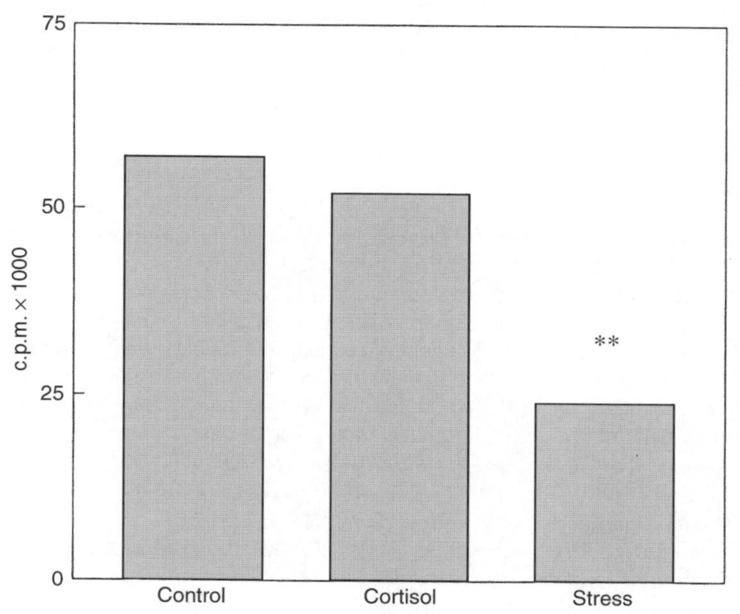

Fig. 5.3. Increased concentrations of cortisol do not account for decreased lymphocyte proliferative responses induced by restraint and isolation stress. Cortisol-infused lambs received cortisol intravenously to attain stress-induced concentrations of plasma cortisol. Restraint and isolation (stress) treated lambs were placed in right lateral recumbency with their legs bound with non-adhesive wrap and isolated from visual and tactile contact with other lambs for 6 h on 3 consecutive days. Control lambs remained undisturbed in their home stanchions. Restraint and isolation decreased concanavalin A-induced lymphocyte proliferation; $^{**}P < 0.01$. Adapted from Minton *et al.* (1995).

Summary

Extreme environmental conditions and stressful management practices influence the health and well-being of domestic farm animals. Converging evidence suggests that these stressors directly alter host immune function. These alterations have been linked to disturbances in the HPA axis and the hypothalamic autonomic nervous system. However, through the use of carefully controlled laboratory stress models, it is clear that many physiological and behavioural factors interact to influence host immunity. Only through the use of carefully conducted, well-controlled studies will our understanding of the complex interactions between various stressors and the immune system continue to be enhanced.

Acknowledgements

The author thanks his graduate students, postdoctoral fellows and technicians who made this work possible. Supported by USDA National Research Initiative Competitive Grants 93-37206-9351, 95-37204-2141 and 98-02531-1941.

References

Blecha, F. (1988) Immunomodulation: a means of disease prevention in stressed livestock. *Journal of Animal Science* 66, 2084–2090.

Blecha, F. (2000) Immunology. In: Mersmann, H.J. and Pond, W.G. (eds) *Biology of the Domestic Pig*. Cornell University Press, Ithaca, New York (in press).

Blecha, F. and Baker, P.E. (1986) Effect of cortisol *in vitro* and *in vivo* on production of bovine interleukin-2. *American Journal of Veterinary Research* 47, 841–845.

Blecha, F. and Minocha, H.C. (1983) Suppressed lymphocyte blastogenic responses and enhanced *in vitro* growth of infectious bovine rhinotracheitis virus in stressed feeder calves. *American Journal of Veterinary Research* 44, 2145–2148.

Blecha, F. and Topliff, D. (1984) Lung delayed-type hypersensitivity in stressed mice. *Canadian Journal of Comparative Medicine* 48, 211–214.

Blecha, F., Barry, R.A. and Kelley, K.W. (1982a) Stress-induced alterations in delayed-type hypersensitivity to SRBC and contact sensitivity to DNFB in mice. *Proceedings of the Society for Experimental Biology and Medicine* 169, 239–246.

Blecha, F., Kelley, K.W. and Satterlee, D.G. (1982b) Adrenal involvement in the expression of delayed-type hypersensitivity to SRBC and contact sensitivity to DNFB in stressed mice. *Proceedings of the Society for Experimental Biology and Medicine* 169, 247–252.

Blecha, F., Boyles, S.L. and Riley, J.G. (1984) Shipping suppresses lymphocyte blastogenic responses in Angus and Brahman × Angus feeder calves. *Journal of Animal Science* 59, 576–583.

Blecha, F., Pollmann, D.S. and Nichols, D.A. (1985) Immunologic reactions of pigs regrouped near or at weaning. *American Journal of Veterinary Research* 46, 1934–1937.

Brown-Borg, H.M., Klemcke, H.G. and Blecha, F. (1993) Lymphocyte proliferative responses in neonatal pigs with high or low plasma cortisol concentration after stress induced by restraint. *American Journal of Veterinary Research* 54, 2015–2020.

Coppinger, T.R., Minton, J.E., Reddy, P.G. and Blecha, F. (1991) Repeated restraint and isolation stress in lambs increases pituitary–adrenal secretions and reduces cell-mediated immunity. *Journal of Animal Science* 69, 2808–2814.

Davis, S.L. (1998) Environmental modulation of the immune system via the endocrine system. *Domestic Animal Endocrinology* 15, 283–289.

Dohms, J.E. and Metz, A. (1991) Stress – mechanisms of immunosuppression. *Veterinary Immunology and Immunopathology* 30, 89–109.

Faith, R.E., Plotnikoff, N.P. and Murgo, A.J. (1998) Cytokines, stress hormones, and immune function. In: Plotnikoff, N.P., Faith, R.E., Murgo, A.J. and Good, R.A. (eds) *Cytokines: Stress and Immunity*. CRC Press, Boca Raton, Florida, pp. 161–172.

Griffin, J.F. (1989) Stress and immunity: a unifying concept. *Veterinary Immunology and Immunopathology* 20, 263–312.

Griffin, J.F.T. and Thomson, A.J. (1998) Farmed deer: a large animal model for stress. *Domestic Animal Endocrinology* 15, 445–456.

Hahn, G.L. and Becker, B.A. (1984) Assessing livestock stress. *Agricultural Engineering* 65, 15–17.

Hicks, T.A., McGlone, J.J., Whisnant, C.S., Kattesh, H.G. and Norman, R.L. (1998) Behavioral, endocrine, immune, and performance measures for pigs exposed to acute stress. *Journal of Animal Science* 76, 474–483.

Irwin, M.R., Vale, W. and Britton, K.T. (1987) Central corticotropin-releasing factor suppresses natural killer cell cytotoxicity. *Brain, Behavior, and Immunity* 1, 81–87.

Irwin, M.R., Hauger, L., Brown, M. and Britton, K.T. (1988) CRF activates autonomic nervous system and reduces natural killer cell cytotoxicity. *American Journal of Physiology* 255, R744-R747.

Irwin, M., Vale, W. and Rivier, C. (1990) Central corticotropin-releasing factor mediates the suppressive effect of stress on natural killer cell cytotoxicity. *Endocrinology* 126, 2837–2844.

Johnson, R.W., von Borell, E.H., Anderson, L.L., Kojic, L.D. and Cunnick, J.E. (1994) Intracerebroventricular injection of corticotropin-releasing hormone in the pig: acute effects on behavior, adrenocorticotropin secretion and immune suppression. *Endocrinology* 135, 642–648.

Kelley, K.W., Osborne, C.A., Evermann, J.F., Parish, S.M. and Hinrichs, D.J. (1981) Whole blood leukocyte vs. separated mononuclear cell blastogenesis in calves: time-dependent changes after shipping. *Canadian Journal of Comparative Medicine* 45, 249–258.

Kelley, K.W., Johnson, R.W. and Dantzer, R. (1994) Immunology discovers physiology. *Veterinary Immunology and Immunopathology* 43, 157–165.

Klemcke, H.G., Blecha, F. and Nienaber, J.A. (1990) Pituitary-adrenocortical and immune responses to bromocriptine-induced hypoprolactinemia, adrenocorticotropic hormone and restraint in swine. *Proceedings of the Society for Experimental Biology and Medicine* 195, 100–108.

Loan, R.W. (1984) *Bovine Respiratory Disease: a Symposium.* Texas A&M University Press, College Station, Texas.

Lund, A., Wallgren, P., Rundgren, M., Artursson, K., Thomke, S. and Forsum, C. (1998) Performance, behaviour and immune capacity of domestic pigs reared for slaughter as siblings or transported and reared in mixed groups. *Acta Agricultura Scandinavica. Section A, Animal Science* 48, 103–112.

McGlone, J.J., Lumpkin, E.A. and Norman, R.L. (1991) Adrenocorticotropin stimulates natural killer cell activity. *Endocrinology* 129, 1653–1658.

McGlone, J.J., Salak, J.L., Lumpkin, E.A., Nicholson, R.I., Gibson, M. and Norman, R.L. (1993) Shipping stress and social status effects on pig performance, plasma cortisol, natural killer cell activity, and leukocyte numbers. *Journal of Animal Science* 71, 888–896.

Minton, J.E. (1994) Function of the hypothalamic–pituitary–adrenal axis and the sympathetic nervous system in models of acute stress in domestic farm animals. *Journal of Animal Science* 72, 1891–1898.

Minton, J.E. and Blecha, F. (1990) Effect of acute stressors on endocrinological and immunological functions in lambs. *Journal of Animal Science* 68, 3145–3151.

Minton, J.E. and Blecha, F. (1991) Cell-mediated immune function in lambs chronically treated with dexamethasone. *Journal of Animal Science* 69, 3225–3229.

Minton, J.E., Coppinger, T.R., Reddy, P.G., Davis, W.C. and Blecha, F. (1992) Repeated restraint and isolation stress alters adrenal and lymphocyte functions and some leukocyte differentiation antigens in lambs. *Journal of Animal Science* 70, 1126–1132.

Minton, J.E., Apple, J.K., Parsons, K.M. and Blecha, F. (1995) Stressor-associated concentrations of plasma cortisol cannot account for reduced lymphocyte function and changes in serum enzymes in lambs exposed to restraint and isolation stress. *Journal of Animal Science* 73, 812–817.

Morrow-Tesch, J.L., McGlone, J.J. and Norman, R.L. (1993) Consequences of restraint stress on natural killer cell activity, behavior, and hormone levels in rhesus macaques (*Macaca mulatta*). *Psychoneuroendocrinology* 18, 383–395.

Murata, H., Takahashi, H. and Matsumoto, H. (1987) The effects of road transportation on peripheral blood lymphocyte subpopulations, lymphocyte blastogenesis and neutrophil function in calves. *British Veterinary Journal* 143, 166–174.

Roth, J.A. and Flaming, K.P. (1990) Model systems to study immunomodulation in domestic food animals. In: Blecha, F. and Charley, B. (eds) *Immunomodulation in Domestic Food Animals.* Academic Press, Inc., San Diego, California, pp. 21–41.

Roth, J.A. and Kaeberle, M.L. (1982) Effect of glucocorticoids on the bovine immune system. *Journal of the American Veterinary Medical Association* 180, 894–901.

Roth, J.A. and Kaeberle, M.L. (1985) Enhancement of lymphocyte blastogenesis and neutrophil function by avridine in dexamethasone-treated and nontreated cattle. *American Journal of Veterinary Research* 46, 53–57.

Roth, J.A., Kaeberle, M.L. and Hsu, W.H. (1982) Effects of ACTH administration on bovine polymorphonuclear leukocyte function and lymphocyte blastogenesis. *American Journal of Veterinary Research* 43, 412–416.

Salak, J.L., McGlone, J.J. and Lyte, M. (1993) Effects of *in vitro* adrenocorticotrophic hormone, cortisol and human recombinant interleukin-2 on porcine neutrophil migration and luminol-dependent chemiluminescence. *Veterinary Immunology and Immunopathology* 39, 327–337.

Salak-Johnson, J.L., McGlone, J.J. and Norman, R.L. (1996) *In vivo* glucocorticoid effects on porcine natural killer cell activity and circulating leukocytes. *Journal of Animal Science* 74, 584–592.

Salak-Johnson, J.L., McGlone, J.J., Whisnant, C.S., Norman, R.L. and Kraeling, R.R. (1997) Intracerebroventricular porcine corticotropin-releasing hormone and cortisol effects on pig immune measures and behavior. *Physiology and Behavior* 61, 15–23.

Saulnier, D., Martinod, S. and Charley, B. (1991) Immunomodulatory effects *in vivo* of recombinant porcine interferon gamma on leukocyte functions of immunosuppressed pigs. *Annales de Recherches Veterinaires* 22, 1–9.

Wallgren, P., Wilén, I.-L. and Fossum, C. (1994) Influence of experimentally induced endogenous production of cortisol on the immune capacity in swine. *Veterinary Immunology and Immunopathology* 42, 301–316.

Weigent, D.A. and Blalock, J.E. (1998) Bidirectional communication between the immune and neuroendocrine systems. In: Plotnikoff, N.P., Faith, R.E., Murgo, A.J. and Good, R.A. (eds) *Cytokines: Stress and Immunity*. CRC Press, Boca Raton, Florida, pp. 173–186.

Weizman, R. and Bessler, H. (1998) Cytokines: stress and immunity – an overview. In: Plotnikoff, N.P., Faith, R.E., Murgo, A.J. and Good, R.A. (eds) *Cytokines: Stress and Immunity*. CRC Press, Boca Raton, Florida, pp. 1–15.

Westley, H.J. and Kelley, K.W. (1984) Physiologic concentrations of cortisol suppress cell-mediated immune events in the domestic pig. *Proceedings of the Society for Experimental Biology and Medicine* 177, 156–164.

Hands-on and Hands-off Measurement of Stress

6

C.J. Cook,[1] D.J. Mellor,[2] P.J. Harris,[1]
J.R. Ingram[1] and L.R. Matthews[3]

[1]Technology Development Group, Horticultural and Food Research Institute of New Zealand, Hamilton, New Zealand; [2]Animal Welfare Science and Bioethics Center, Institute of Food, Nutrition and Human Health, Massey University, Palmerston North, New Zealand; [3]Animal Behaviour and Welfare Research Center, AgResearch, Ruakura Research Center, Hamilton, New Zealand

Introduction

In order to survive, an animal must be able to adapt to environmental challenges. These adaptations are often referred to as the 'stress response', an umbrella term that in fact includes a disparate set of body-wide effects detectable as behavioural and physiological changes. These include changes within the central nervous system in sensory processing, cognition, hypothalamic and pituitary function, and neurosteroids and neurotransmitters, changes within limbic structures such as the amygdala and hippocampus, and changes in hypothalamic–pituitary–adrenal (HPA) activity. They also include changes in the levels of circulating metabolites. All of these are integrated and provide the physiological foundations for behavioural responses (for a review, see Sapolsky, 1992). The complexity and suite of changes can differ markedly from species to species, individual to individual and stressor to stressor, and can vary according to prior experience and hormonal status (Cook, 1996, 1998a).

It is well recognized that in order to understand the complexity and integration of physiological and behavioural responses to challenge it is necessary to measure at least several variables. Recent advances in biomedical technology provide the opportunity for sophisticated exploration of the functional complexity of stress responses using new animal models. Use of this technology has already led to fresh insights, and more may be expected. Nevertheless, our capacity to produce sophisticated animal models must be tempered by a humane commitment to minimizing the invasiveness of the procedure for the animal's sake, the scientific need to avoid artefacts linked to invasive procedures, and the benefits of studying animals in their 'natural' situation. A comprehensive account of all recent advances in stress measurement is not possible in the space available. We therefore provide some specific examples, derived from the work of our different groups, where progress with hands-off studies has been made. We have outlined the strengths and weaknesses of hands-on studies of hormonal and behavioural responses to painful husbandry practices elsewhere in this volume (Mellor *et al.*, Chapter 9, this volume).

Hands-on and hands-off studies

With few exceptions, the act of studying an animal affects what is being studied and can thereby confound interpretation of the results. Accordingly there is a strong need to minimize study-induced disturbances. This can be achieved by both reducing the extent of contact and the invasiveness of hands-on studies and, where practicable, engaging in hands-off studies.

Hands-on studies

Hands-on studies may be favoured in several situations. Where no hands-off study methods are available there is a trade-off between getting no data at all or acquiring some data that are possibly compromised by the need to use hands-on approaches (for example castration of young ram lambs using different analgesic regimes). Thus, hands-on monitoring of behavioural and/or physiological responses may be needed to allow repeated blood sampling, invasive preparation of animals for intensive monitoring (e.g. electrocorticographic (EcoG) recording from surgically implanted electrodes used, for example, to monitor the efficiency of preslaughter stunning), or to allow intensive behavioural observation and/or clinical care of study animals. For these types of measures it may be necessary to remove animals from their usual habitat and relocate them (e.g. indoors or to the laboratory). Also, duplication of hands-on farm or other procedures may be required to test animal responses to normal non-invasive hands-on activities (e.g. mustering, yarding, shearing, dipping, drenching, etc.), or to study animal responses to invasive hands-on procedures (e.g. branding, castration, dehorning, tail docking, clinical or experimental surgery or other procedures).

Hands-on studies can be informative within certain limits (Mellor *et al.*, Chapter 9, this volume). Caution is required when extrapolating from a hands-on experiment to the more natural hands-off situation if only the former has been studied. Familiarizing animals with hands-on procedures before the study may be helpful; however, such training is problematical as it may alter the animals' responses during the study; yet some untrained animals (individuals or species) may otherwise be too intractable for meaningful hands-on study. The use of appropriate control animals is imperative, but when control handling has measurable effects any associated interpretative limitations must be clarified. Technical difficulties associated with handling, including contamination of signals caused, for example, by inadvertent physical displacement of monitoring electrodes, must also be minimized. However informative hands-on studies may be, there would usually be value in conducting similar hands-off studies when the appropriate methods are available.

Hands-off studies

Hands-off studies may be favoured when the available hands-off methods allow observation of the required parameters. For instance, responses to usual activities and/or events that involve little or no direct hands-on contact with people may be studied without handling in the animals' usual environment. Intractable animals (individuals or species) may be studied without complications of extreme responses to the hands-on study methods themselves. Short-term and/or long-term responses to a variety of challenges may be studied in free-range and freely behaving animals.

Such hands-off studies are enhanced by the following safeguards. Monitoring equipment needs to be weather-proof (for outdoor use) and robust enough to operate effectively for the required periods on freely ranging animals. Signals from equipment on different animals must not interfere with each other, and equipment must not impede the animals' expression of their normal behaviours. Baseline values and the time required for them to be reached after attaching monitoring equipment need to be ascertained; generally non-invasive is preferred to invasive equipment to minimize post-attachment disturbances. Finally, special attention needs to be given to monitoring uncontrolled and/or unexpected phenomena that can affect free-range animals, either as individuals or as groups, and that can evoke responses which otherwise would hinder interpretation of results.

While this list is by no means complete, it does allow some guiding principles that can be used as a framework for the methodological examinations below. In many studies experimentation will begin with a hands-on study. As we then come to understand the usefulness of the measures chosen, it will advance to the comparative state of a hands-off study, the data from which in time will come to supplant and exceed the data obtained using hands-on methods. The first methodological approach presented below illustrates this.

Stress-free assessment of the HPA axis: an example of moving from hands-on to hands-off assessment through the development of remote blood sampling technologies

Changes in the activity and functioning of the HPA axis are routinely used to quantify an animal's response to stress. However, the stress associated with handling and restraint of animals with traditional blood sampling techniques can in itself cause activation of the HPA axis, thereby confounding such measurements (Seal *et al.*, 1972; Hattingh *et al.*, 1988). This effect is further exacerbated in wild or semi-domesticated animals or in species with flighty natures such as red deer (Fig. 6.1).

A variety of techniques have been developed in an effort to minimize the stress imposed on the animal during sample collection. These include rendering the animal unresponsive to the effects of handling by shooting (Hattingh *et al.*, 1984; Smith and Dobson, 1990) or sedation (Monfort *et al.*, 1993), habituation to the blood sampling procedure (Bubenik *et al.*, 1983) and the use of techniques which reduce or eliminate the degree of handling. These include measurement of corticosteroids in substances other than blood (e.g. saliva, Cook and Jacobson (1995); milk, Verkerk *et al.* (1998); urine and faeces, Palme *et al.* (1996)), and the use of indoor remote catheter systems (Ladewig and Stribrny, 1988; Monfort *et al.*, 1993) or remote portable blood sampling devices (Farrell *et al.*, 1970; Bubenik and Bubenik, 1979; Hattingh *et al.*, 1988; Mayes *et al.*, 1988; Stephan and Cybik, 1989; Goddard *et al.*, 1994; Ingram *et al.*, 1994). These existing procedures,

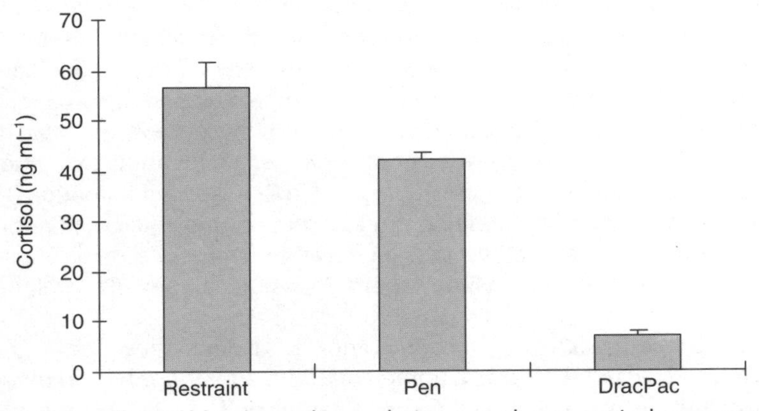

Fig. 6.1. The effect of blood sampling technique on plasma cortisol concentrations in red deer (*Cervus elaphus*). The restraint group consists of animals sampled via venepuncture in a pneumatic crush (Ingram *et al.*, 1994). The pen group consists of animals sampled via a catheter while free standing in pens (Ingram *et al.*, unpublished data). The DracPac group consists of animals sampled by a remote blood sampling device while undisturbed at pasture.

however, have limitations when it comes to assessing the activity and functioning of the HPA axis. Shooting animals affords only limited information on cortisol concentrations at the time of death, with no opportunity to follow changes over time in the same individual. Measurements obtained from sedated animals may vary due to the difficulty in maintaining consistent levels of sedation over time and between individuals (Monfort *et al.*, 1993) while some sedatives (e.g. xylazine hydrochloride, Chao *et al.*, 1984) may affect the HPA axis directly.

The process of taming and habituating an animal to the blood sampling procedure may result in animals showing atypical responses when compared with wild or semi-domesticated conspecifics. This is of particular importance when handling is an integral component of the stressor being assessed (e.g. shearing or transport of animals).

Collection of saliva or milk samples from lactating females, though less invasive than blood sampling, still requires some degree of handling. Though collection of urine or faeces is achieved without handling, the integrated estimates of HPA axis activity obtained with this technique lack the temporal precision to detect fluctuations in circulating corticosteroids, which are important in assessing responses to acute stress, or in tests of HPA axis function.

The use of static remote catheter systems for obtaining blood from undisturbed animals also involves an extended process of habituation to indoor housing and in some cases restraint of the animal, which can place limitations on the type of animal and the stressor that can be evaluated. The use of remote blood sampling systems can overcome many of these limitations. Such a technique can be applied in the assessment of a wide range of stressors and removes the requirement for repeated handling and the need for habituation to indoor housing, though some degree of habituation to carrying the sampling equipment may be required. The relatively large size of remote blood sampling systems and the requirement in most systems to store the blood sample on the animal until retrieval does, however, impose limitations upon the size of animal that can be remotely sampled as well as restricting any measurement to compounds that are relatively stable over time in whole blood.

Remote blood sampling systems do, however, lend themselves to the monitoring of glucocorticoid concentrations, which are relatively stable in whole blood for several days (Reimers *et al.* 1983), and to the sampling of large, flighty, free-ranging animals (such as red deer). We used such a system to determine both HPA axis activity and function in undisturbed free-ranging red deer and cattle. To quantify ultradian and circadian rhythms in HPA axis basal activity accurately, blood samples need to be collected over 24 h at a frequency that can accurately measure corticosteroid pulsatility (sampling at least every 20 min (Monfort *et al.*, 1993)). To measure changes in HPA axis function remotely, a device capable of administering, by remote bolus infusion, adrenocorticotropic hormone (ACTH) or corticotropin-releasing hormone (CRH) challenges and monitoring the subsequent adrenocortical response would be ideal. The DracPac remote blood sampling device was therefore developed with these specifications in mind.

The DracPac device (Fig. 6.2) consists of a pump unit containing two small peristaltic pumps for pumping heparin and drawing blood (length 150 mm, diameter 50 mm, weight 395 g), two 38-position rotary switching valves (length 200 mm, diameter 60 mm, weight 950 g), a 6 volt battery supply (Duracel, MN908) and a control box (length 160 mm, width 80 mm, height 60 mm, weight 580 g) containing a programmable microprocessor that allows programming of the start time, duration and rate of sample

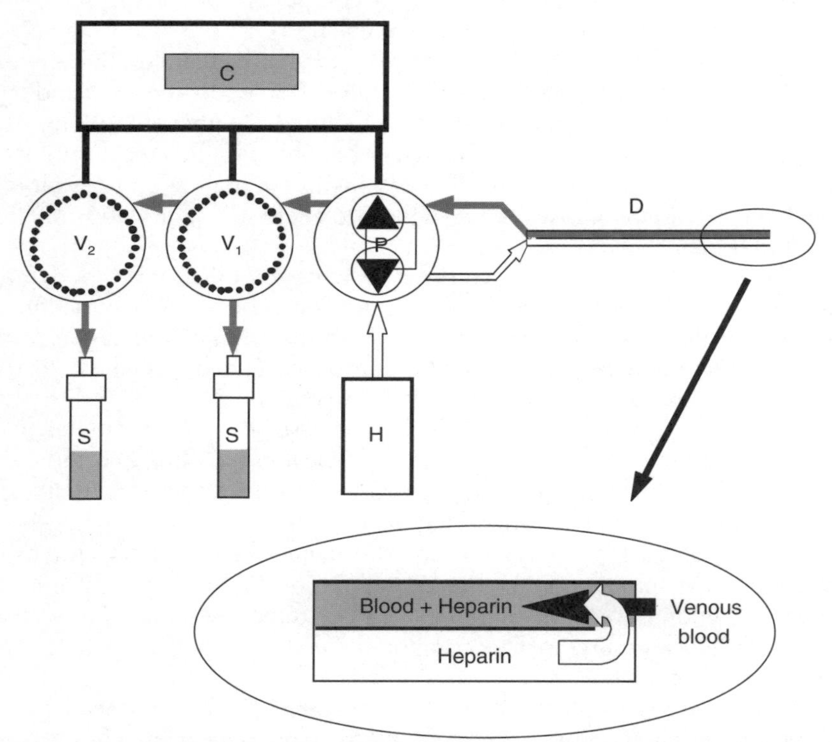

Fig. 6.2. Schematic representation of the DracPac remote blood sampling system used for automatic continuous sampling of blood from free-ranging animals. The device comprises a microprocessor control box (C), a peristaltic pump unit (P), a plastic bag containing concentrated heparin (H), two 38-port rotary switching valves (V_{1-2}), 74 separate blood collection tubes (S) (4.5 ml monovette syringe, Sarsdetd Ltd, Numbrecht, Germany) and a double lumen catheter (D)(Cavafix Duo16/18G 32 cm, Braun, Germany) which is modified by shortening the end of the catheter by ~5 cm and removing 1–2 mm of the wall between the two lumina at the tip to facilitate optimal mixing of the outflowing heparin and inflowing blood (from Ladewig and Stribrny, 1988).

collection. The device samples by pumping heparinized saline (5000 IU ml^{-1}) down one side of a double lumen catheter to the tip where it mixes with jugular blood being continuously drawn up the second lumen by the action of the second pump. The heparinized blood passes through one or both rotary switching valves and is collected into one of 74 separate blood collection tubes (4.5 ml monovette syringe, Sarsdetd Ltd, Numbrecht, Germany). Bolus infusions of substances such as ACTH and CRH can also be administered to the animal remotely by aligning a valve to a specific port containing the substance and reversing the direction of the blood pump. This step can be repeated a number of times allowing either multiple infusions of the same substance or a battery of substances to be delivered over time. The equipment and samples are kept on the animal in a backpack harness (Fig. 6.3) until sampling is completed.

Using this device, detailed 24 h profiles of plasma cortisol concentrations in red deer stags at pasture have been obtained (Fig. 6.4). These profiles are characterized by low mean plasma cortisol concentrations (6.7 ± 1.0 ng ml^{-1}) and the presence of ultradian, circadian and seasonal rhythms (Ingram *et al.*, 1999). Mean basal cortisol concentrations are similar to those reported for hand-reared Eld's deer sampled via a remote catheter system (5.4–14.5 ng ml^{-1}; Monfort *et al.*, 1993) and those reported for red deer stags blood sampled immediately after being shot dead while undisturbed at pasture (5.7 ± 3.7 ng ml^{-1}; Smith and Dobson, 1990).

The adrenocortical response to a variety of acute stressors including handling (Ingram *et al.*, 1994, 1997, 1999; Carragher *et al.*, 1997), transport (Waas *et al.*, 1997) and velveting (Matthews *et al.*, 1994) has also been determined in deer using this sampling technique (Fig. 6.5). Remote bolus

Fig. 6.3. Farmed red deer (*Cervus elaphus*) stags equipped with a remote blood sampling device and harness.

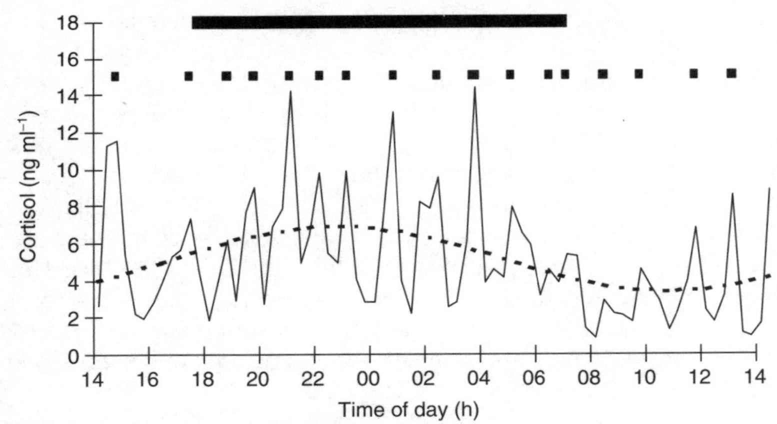

Fig. 6.4. A 24 h profile (72 × 20 min samples) of plasma cortisol concentrations collected from a red deer stag (*Cervus elaphus*). The occurrence of cortisol pulses as detected by cluster analysis is shown as points at the top of each graph. The circadian rhythm is represented by the best fitting sine curve (dashed line) for the data series derived from COSINOR analysis. The black bar at the top of the graph represents the period of darkness.

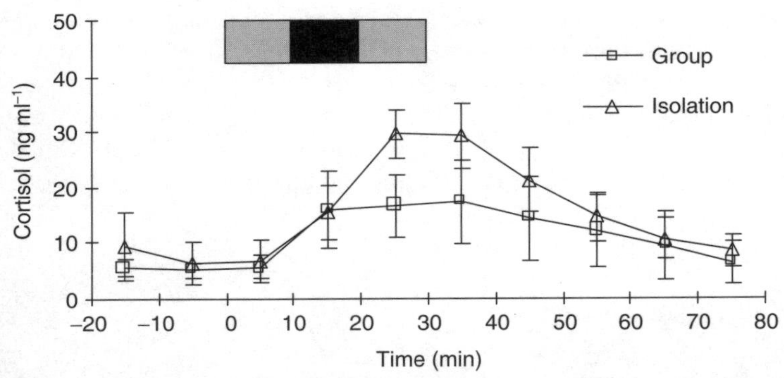

Fig. 6.5. The adrenocortical response of free-ranging red deer stags ($n = 5$) to yarding and holding in a pen for 10 min either as a group or in isolation, and subsequent recovery after release to pasture. Grey blocks indicate time to and from yards, while the black block indicates time spent in yards.

infusions of ACTH or CRH have also been used to quantify seasonal (Ingram *et al.*, 1999) and social stress-induced changes in HPA axis function in red deer. A marked reduction in adrenal responsiveness to ACTH was observed during the winter breeding season (Ingram *et al.*, 1999) while chronic social stress resulted in reductions in responsiveness to both ACTH and CRH (Fig. 6.6).

These data demonstrate the versatility of remote blood sampling as a technique for monitoring changes in both the activity and functioning of

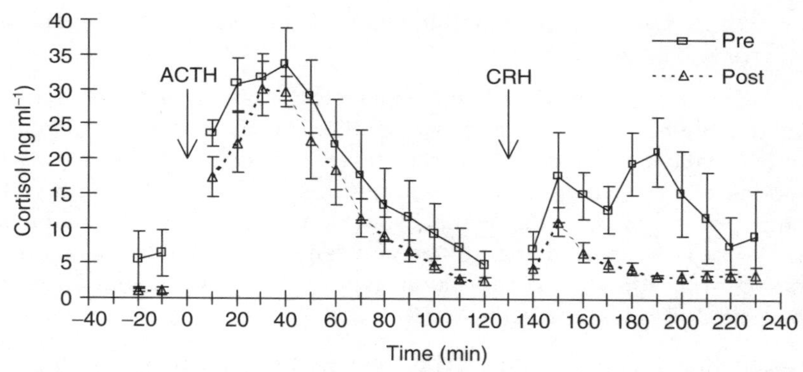

Fig. 6.6. The effect of a chronic social stressor (repeated mixing into novel groups over 12 days) on plasma cortisol responses to remote bolus infusions of ACTH (0 min) and CRH (130 min) prior to (Pre) and following (Post) the imposition of the stressor in free-ranging red deer stags.

the HPA axis in undisturbed animals as well as in response to a variety of stressors. The DracPac system can mitigate the confounding effects of blood sampling stress associated with manual methods of blood collection, thereby allowing the true effects of stress to be revealed. It is envisaged that the remote blood sampling technique will find application in a wide range of studies quantifying physiological, haematological and biochemical parameters influenced by stress. Such a development provides us with a measurement technology applicable to at least larger sized animals that, as seen in the next section, can be combined with other remote technologies to enable hands-off assessments.

Development of free-range physiological monitors: the use of telemetry and logging to obtain physiological data from an animal in the field

Stress is difficult to measure directly, and is often quantified by measuring its effects. However, if one identifies a measure that changes in a consistent and repeatable manner in response to a treatment that is generally recognized as a stressor (for example, bringing a dog near sheep), one can argue for the identification of a reliable indicator of a particular stress. Clearly this is the first step in developing a useful suite of measures to allow quantification of stress. The difficulties of this are that:

- it is generally difficult to prove that the observed response is indeed caused by stress;
- the measurement may be confounded by other phenomena, e.g. increased activity;

- the degree of response to any stressor may be reduced by habituation or increased by facilitation.

Because of these and other physiological factors, estimates of stress that consider only physiological measurements and behaviour have faced difficulties of interpretation. Stress studies need to address the complete axis from brain functions of perceiving and responding to a stressor to the outputs of this central processing and the feedback from these actions. If we have a measure of how an animal is processing sensory information while freely behaving in the field, and can measure how this changes with behaviour and applied stress, we may achieve four goals:

1. Identification of stress (in an objective, gradable manner);
2. Determination of the relativity of different stressors (how much is normal sensory processing changed?) and adaptation rates to new stressors;
3. An understanding of central nervous system involvement in stress mechanisms; and hence
4. Ways of controlling it.

Technology that allows remote monitoring of freely behaving animals in the field can remove many of the problems imposed by stress associated with experimenter intervention. However, the acquisition of data from freely behaving large animals poses a considerable challenge.

Many research groups worldwide have been using telemetric recording systems on freely behaving animals for many years. These telemetric systems are often primarily used for tracking of the animal to which they are attached, although in some cases they allow for the recording of a single (e.g. temperature), or perhaps two (e.g. heart rate and temperature), physiological parameters.

The free-range physiological monitor

To avoid the problems caused by experimenter interaction and to enable us to take multiple physiological measurements from freely behaving animals, we (HortResearch) undertook to design and build a custom ambulatory monitor for our own use. We required a system that allowed the measurement of many parameters on a freely behaving animal over long periods of time with the ability to integrate new and novel measurement technologies. This monitor needed to be able to record and analyse both the high data volumes encountered with electrocardiogram (ECG) and electroencephalogram (EEG) recordings as well as slow periodic measurements such as body temperature. We also had the requirements to allow multiple animals to be monitored simultaneously and the data downloaded or new experimental parameters uploaded to the animal from a base station without any physical intervention.

Together, these requirements led to the free-range physiological monitor (FRPM) shown in Fig. 6.7. This ambulatory physiological monitor incorporates powerful data acquisition and computing hardware, is solar powered, and includes a high-speed radio link for control and acquisition of data from multiple units by a base computer. If there is insufficient sunlight to run the FRPM, backup batteries supply power to the unit for up to 7 days. The FRPM is of robust and waterproof construction and was designed to withstand both the weather extremes and physical abuse often encountered during long-term field trials. High-density surface mount components were used throughout the design to minimize the size and weight of the unit. The unit measures $20 \times 120 \times 250$ mm and weighs 700 g (most of the weight being the aluminium case).

A fundamental justification for the development of the FRPM lies in the high data volumes inherent in the signals we wish to use in assessing stress in unconstrained animals. Raw EEG, ECG and electromyogram (EMG) signals must be sampled at hundreds or thousands of Hz, creating data rates of many kbits s^{-1}, from each of a number of subject animals simultaneously. These data rates pose a number of technical difficulties: the capacity of a digital radio data channel is soon exceeded, constant transmitter operation

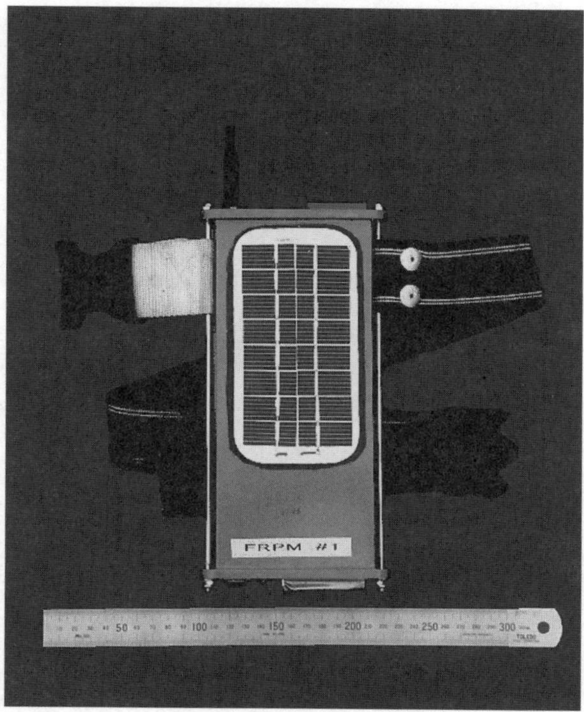

Fig. 6.7. The FRPM ambulatory monitor.

soon drains battery power, and the volume and subsequent processing of raw data at a base station rapidly become formidable. The FRPM solves these problems by processing the raw data on the animals and reducing the high volume data stream to a few summary statistics.

Initial development of the FRPM was carried out to perform the computation-intensive processes needed to obtain auditory-evoked response (AER) recordings along with heart rate and temperature measurements. It is a natural extension of the FRPM's capabilities to add measurement of other physiological parameters related to welfare and stress, some of which require data compression (such as sleep state from EEG). Other parameters to be added include body and environmental temperatures (related to environmental stressors), ruminal activity as indicated by microphonics, biosensors for cortisol and nitric oxide, EMG, and global positioning system (GPS) location measurement (to follow feeding/sleep patterns and animal interactions). Currently the FRPM is being used in animal welfare trials to determine if the AER can be used as an indicator of anxiety in sheep. Figure 6.8 shows the FRPM on a sheep for this purpose.

The FRPM has also been used to assess a potential welfare problem in the use of washing at stockyards prior to slaughter. This washing procedure, particularly if it involves sheep swimming as is common practice, may be stressful to animals. An alternative is to use a race with a high-pressure water jet. Obviously during this procedure it would be both awkward and confounding to obtain hands-on measures. To circumvent this problem, we used the FRPM to record ECG signals from sheep undergoing washing by

Fig. 6.8. A sheep during an AER isolation experiment.

either high-pressure water jet or standard sheep swim methods to monitor animal welfare aspects of the procedure (data being processed).

The auditory-evoked response as a measure of anxiety

The monitoring of central nervous system function by relatively non-invasive means in freely behaving large animals poses a considerable challenge. One approach to this is to measure neural-evoked responses in animals. Evoked responses (Cooper *et al.*, 1980; Chiappa, 1990) are an electrical manifestation of the central nervous system's reception of and response to an external stimulus such as a noise, a light or a touch. These responses can be recorded from small electrodes positioned on the head of an animal. They are extremely low in amplitude and, because of their admixture with normal background brain electrical activity and various artefacts, need to be separated out by signal processing techniques.

The evoked response consists of a series of electrical peaks that occur at consistent time intervals after stimulation. These peaks are usually grouped into three divisions according to their latency range (time at which they occur following the stimulus). These divisions are short latency (occurring up to 30 ms after the stimulus), mid-latency (30–75 ms after the stimulus) and late latency (> 75 ms) wave peaks. The short and mid-latency components generally represent the conduction of neural information from the stimulus to the cortex of the brain. The late latency components appear indicative of stimulus processing at higher cortical levels of the brain and are very much influenced by psychological state (Knott and Irwin, 1973; Wilson *et al.*, 1994).

The three most common measured evoked responses are the visual-evoked response (VER), somatosensory-evoked response (SER) and the AER. While the overall series of peaks gives information concerning the pathway of sensory processing and time of conduction, each peak also contains information concerning the function of the nervous system. As the peaks result from electrical activity generated by groups of neurons in response to stimuli, the peak interval gives information as to how long processing of a stimulus takes place at any sensory way station. The area under the peak gives an indication of how many neurons are involved. At the higher cortical levels, late latency peaks can give us important information as to how much time is spent 'working' on a particular stimulus and how much of the brain is devoted to it. This can be an indicator of how important information derived from that stimulus is to the animal. With stresses of various kinds, the relative importance of stimuli can change dramatically as will their processing within the central nervous system.

There have been many studies that indicate that the evoked response may be used as a measure of stress. The earliest indication of this was a study (Hernandez-Peon *et al.*, 1956) that showed auditory responses from the cochlear nucleus of a cat were greatly reduced when a cat was presented with live mice but returned to normal when the mice were removed. In human subjects (Bond *et al.*, 1971; Drake *et al.*, 1991), peak latencies in the recorded brainstem AER of subjects suffering from chronic anxiety are significantly shorter than those of normal subjects. The evoked response has also been used successfully as a measure of cognitive processing, or workload, during such activities as display monitoring (Isreal *et al.*, 1980) and air-to-ground attack mission training (Wilson *et al.*, 1994). These studies show that the evoked response also varies with short-term stress, in this case workload.

Based on these studies, it was decided to investigate the potential of using the AER in free-ranging animals as a potential indicator of stress. To test the hypothesis that the AER could be used as an indicator of stress in free-ranging animals, we simultaneously recorded both AER and ECG signals from freely behaving sheep (Pierson *et al.*, 1995; Griffiths *et al.*, 1996) using the FRPM physiological monitor. We used the AER because the stimulus (clicks) was easy both to generate and to present to a freely behaving animal via a small speaker in one of the animal's ears. The sheep were exposed to four 1 h experiments, each comprising three 20 min periods: a control, a stressor and a recovery period. The four stressors used comprised a control stressor (no stressor), distraction (a moving water sculpture), isolation of an individual, and the nearby presence of a dog. AER was recorded every 5 min while ECG was recorded every 5 min for 30 s. Behaviour observations were also made. Figure 6.7 shows a subject sheep during the isolation part of the experiment.

Under control conditions, the AER was repeatable. However, when a stressor such as isolation was applied to a single animal, or the flock was exposed to a predator (dog), the response could change significantly. Figure 6.9 shows a typical set of results obtained using the dog as the stressor. The control period shows a repeatable baseline waveform. During the stressor period, a considerable change in the waveform shape can be seen while the dog is present. The recovery period shows the waveform slowly reverting to its original control shape. Figure 6.10 shows heart rate measurements made during the experiment. These data show that while brain activity significantly changed while the dog was present, the dog had little effect on the heart rate of the sheep.

The results show that different subjects can exhibit quite different responses to the same stressor, making the creation of a general model for changes in the AER as a quantifier of anxiety unlikely. However, it may be a useful individual measure. The results also suggest that although increased heart rate, a conventional indicator of stress, may detect the onset of a stressor, it does not persist during a sustained level of stress in the subject. In contrast, the changes that occur in the AER waveform remain as long as the stressor is present.

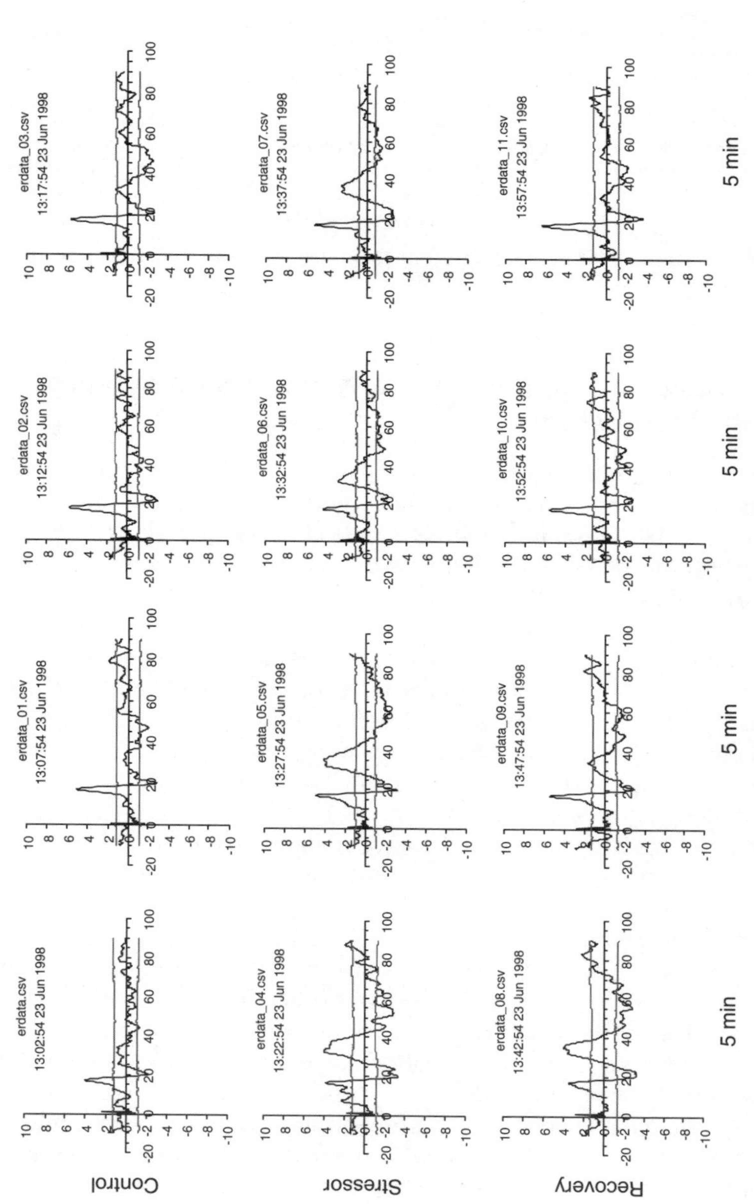

Fig. 6.9. Typical AER experiment recording from sheep before (control), during (stressor) and after (recovery) exposure to a potential predator (dog).

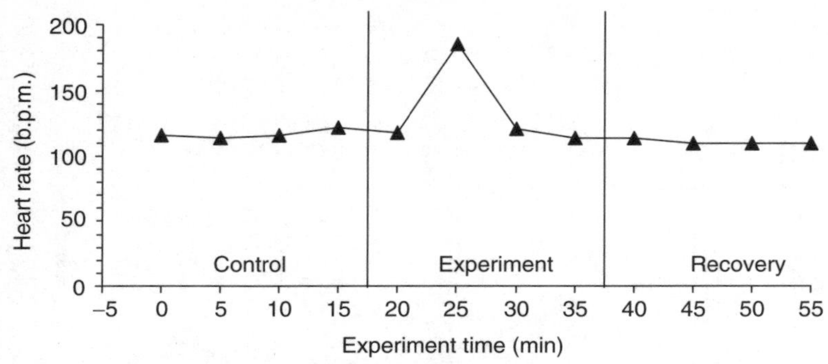

Fig. 6.10. Typical heart rate results from sheep using a dog as a stressor as in Fig. 6.9.

Biosensors and monitoring central nervous system neurohumoral states

Recent research has highlighted the extent and importance of central nervous system contributions to stress perception and response outside of the classic HPA axis (Gardner *et al.*, 1998; Kaufer *et al.*, 1998; Koepp *et al.*, 1998; Morris *et al.*, 1998). However, even at the level of the hypothalamus, stress often seems to have a differential effect that merits sophisticated tracking.

Monitoring the central nervous system requires approaches that are minimally invasive and can be adapted to conscious animal and field use. In humans, recent studies (Koepp *et al.*, 1998) using neuroimaging combined with specific radiochemistry have proved revealing in understanding the link between chemical activity and mental processing. Recent work using radiolabelled raclopride displacement to measure extracellular levels of endogenous dopamine in the nucleus accumbens during a learning situation has added a dimension of understanding to the role of reward in learning.

Unfortunately, however, not only does neuroimaging require a relatively intensive hands-on approach, it is also prohibitively expensive for most of us in the animal biology world! For some time, microdialysis techniques have allowed measurement of extracellular levels of neurotransmitters in different brain areas of conscious behaving animals (Beneviste, 1989). The advantage of using a dialysis membrane and osmotic equilibration sampling of an analyte is that samples need not be drawn off, unlike the case of standard catheterization, thus not limiting sample time and intensity. Microdialysis techniques have been somewhat limited by the time needed to sample and the need to analyse the sample offline, often necessitating hands-on and restraint of the animal, and thus considerable time delay in retrieving the sample. Within these limitations, however, microdialysis probes as small as 50 μm in diameter can be made and have been used successfully over several weeks to monitor hypothalamic function in conscious animals during stress states (Cook, 1997a, 1998b).

Recent advances in the design of these types of probes (Obrenovitch *et al.*, 1994; Cook, 1997b, c; Cook and Devine, 1998) have allowed for both real time and intensive sampling combined with electrophysiology and pharmacological manipulation capabilities. The measurement can be made *in situ* within the probe, allowing both rapid and sensitive discrimination of the analyte. These techniques are of course equally applicable to other body tissues and to blood sampling, where they can allow continuous and long-term sample measurements without the problems associated with standard catheterization. The technique also allows samples to be drawn off for offline measurement, useful to validate the real time results. The advent of polymer technologies for recognition of different analytes of interest (Ramström *et al.*, 1996) will advance this technology in both application and price. When combined with sophisticated behavioural models, new information on stress processing and individual differences will no doubt arise. The microprobe technology is already being developed to allow interface with the FRPM, which will enable free-ranging collections to be made. A recent example of microprobe use is presented in Fig. 6.11.

Increasing complexity in behavioural assessment of stress

It can be revealing to use behavioural measures in conjunction with physiological data to determine the degree of challenge to which animals are exposed. For example, in a recent study by Munksgaard *et al.* (1999), bulls were given a limited period daily during which they could lie down. This procedure continued for 49 days. The stressfulness of the procedure was determined by measuring ACTH and cortisol responses to a CRH challenge on days –32, 3, 22 and 49 relative to treatment and by determining the frequency of transition between behavioural states. The ACTH and cortisol concentrations to CRH challenge were greatly reduced 3 days after the initiation of lying deprivation, but these responses to challenge reduced over time such that at day 49 they were not different from control (non-deprived) animals or pretreatment values. On the other hand, lying deprivation resulted in an increased frequency of transitions between activities that persisted for the duration of the study. Thus, the physiological measures indicated that the bulls had adapted to the procedure whereas the behavioural data indicated that they had not.

Behavioural measures obtained from simple observation can be used as an index of disturbance for various manipulations/challenges (reviewed in Fraser and Broom, 1990; Broom and Johnson, 1993). Such measures include the intensity, duration and frequency of startle responses and defensive or flight reactions; the time required to resume normal behavioural activity

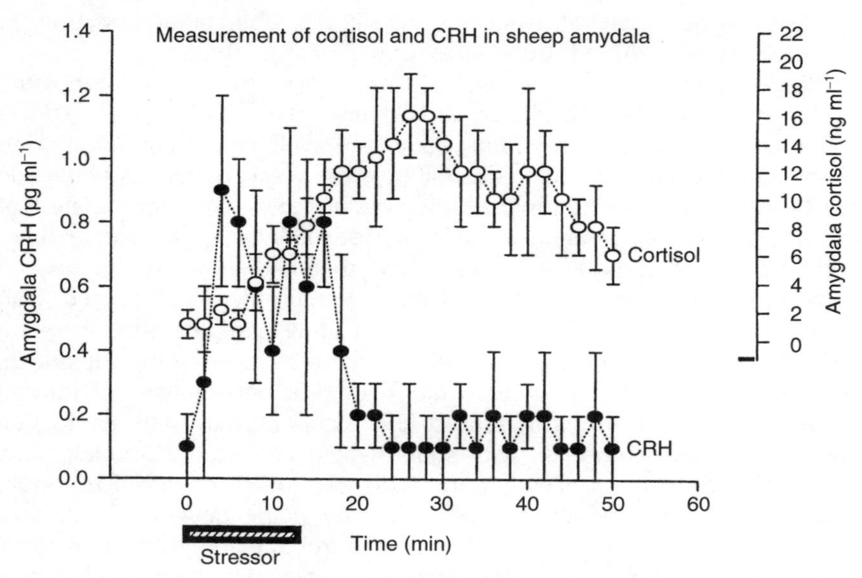

Fig. 6.11. The use of two microprobes located adjacently, 80 µm apart within the amygdala (a limbic brain region) in conscious sheep. The sheep were exposed to a mild stressor (restraint for 12 min). Measurements of both cortisol (one probe) and CRH (other probe) were taken every 2 min. CRH showed a rapid increase immediately following the start of the stressor then returned to normal quickly following the cessation of the stressor. Cortisol showed both a slower rise time and return, and mirrored (Cook, unpublished data) the pattern seen in systemic circulation (not shown).

after stress; and increases in the frequencies of aggression, stereotypies and apathetic or unresponsive behaviours. Features of animal group dynamics such as spatial relationships (inter-individual distances, Diverio *et al.*, 1993) or temporal synchronization of behaviour may also exhibit stress-induced changes. Complexity of behavioural patterns is also reduced under situations of stress (Alados *et al.*, 1996). Complex behavioural patterns are considered more energetically demanding and, as a result of the increased metabolic response to stress, a trade-off is observed in a reduction in complexity of behaviour.

However, caution must be used in interpreting behavioural changes to stress as they, like physiological measures, may simply reflect a successful adaptive response to a particular challenge. A more direct way to access information about the way an animal is responding to challenges is to ask it in preference tests (e.g. Grandin *et al.*, 1986; Fraser and Matthews, 1997). Similar information on an animal's perception of stressors can be gauged by recording time taken by an animal to approach a situation or location where an aversive treatment has taken place (Rushen, 1990).

Such simple tests of preference or aversion, however, do not readily provide a quantitative measure of the degree of aversiveness or pleasantness of environmental events or challenges: more sophisticated operant techniques based on behavioural demand functions (Dawkins, 1990) are required in order to achieve this. With this procedure, animals pay a price to gain access to or avoid environmental stimuli, and the demand (consumption) is measured as the slope of the function relating changes in consumption to changes in price. For events that the animal has a strong need to avoid, or gain access to, the slope of the function (demand elasticity) approaches zero.

We have shown (Matthews *et al.*, 1995) using this technique that hens kept in wire cages display a strong demand for a litter substrate like sand, but a less strong demand for wood shavings (Fig. 6.12). Sand is also more efficient than wood shavings at removing feather lipids (Van Liere *et al.*, 1990), one of the primary functions of dust bathing and crucial in maintaining feather condition. Thus, the psychological value of the substrate as measured by demand functions corresponds to the effectiveness of the substrate in maintaining feather condition. Thus, we can see that this behavioural technique is useful in quantifying an animal's perceptions of external events.

A logical progression of this work would be to combine this type of behavioural approach with sophisticated physiological measurements. Dopamine in the nucleus accumbens has been linked to sensations of reward and pleasure and to a drive to obtain physical elements that will increase its release (Wise, 1980). As an example of combining approaches, it would be interesting to see if choice is related to levels of dopamine in the nucleus accumbens measured using microdialysis techniques or to

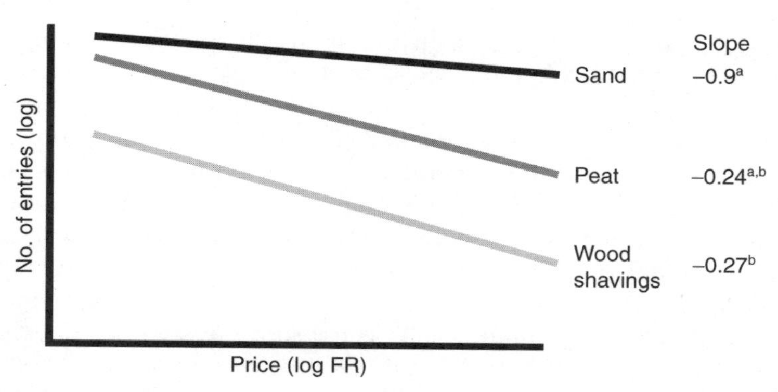

Fig. 6.12. Demand for different substrates (sand, peat and wood shavings) for use in dust bathing behaviour in hens. Superscripts (a,b) indicate significantly different slopes ($P < 0.05$). Figure adapted from Matthews *et al.* (1995).

subsequent cardiac reactivity to an imposed stressor. Thus, by using a combination of sophisticated behavioural and physiological techniques, we should be able to develop a better understanding of the adaptive capability of animals and the impact of adverse events on their health and welfare.

Summary

Stress not only refers to an animal's state when extended beyond its physiological and behavioural ability to adapt but also to the lesser responses involved in 'normal' day-to-day environmental adaptation. The disparateness of the stress response can make it difficult to quantify between this normal day-to-day adaptation and ability or inability to adapt at the more extreme levels. In particular one measure does not of necessity make a distress response (a high cortisol level in the absence of other measures may not indicate a distress), and so there is an onus on us to take multiple measures from an animal. Such multiple measures are likely to include sophisticated reference to the central nervous system, as many of the subtleties of stress responsiveness and effect are set at this level. This obviously necessitates a sophisticated approach – one of the key methodological topics discussed above. Tempered with this sophistication is the need to reduce the invasive load on an animal, the complications of interpretation that come with this load, and the need to take measurements while the animal is in its 'natural' environment with the parallel requirement of minimal experimenter intervention.

In isolation, the techniques that we have presented may seem powerful but somewhat disparate. The aims of our combined groups are to utilize this range of techniques and models simultaneously to enable assessment of the animal from the central nervous system through to the systemic response using sophisticated and relevant behavioural models. We are of the opinion that such an approach may revolutionize our understanding of both animal stress and its welfare consequences. In terms of welfare implications, the use and development of this equipment provides a less invasive way to obtain sophisticated data during potentially stressful situations, and the opportunity to develop hands-off procedures. It also offers the ability to monitor animals in the long term during normal field situations and therefore collect much-needed baseline data on the physiological function of animals. There are increased opportunities to collect data during procedures where direct hands-on intervention is both difficult and confounding (such as transport, washing or slaughter of animals). Lastly, but not exclusively, there is the opportunity to collect from an individual animal, without intensive invasive procedures, a wide variety of measures (both physiological and behavioural) to correlate with potential stress.

Acknowledgements

We acknowledge funding from the Foundation for Research, Science and Technology of New Zealand; the New Zealand Lottery Grants Commission, the New Zealand Dairy Board, Ministry of Agriculture and Fisheries Policy, and from our respective institutes.

References

Alados, C.L., Escos, J.M. and Emlen, J.M. (1996) Fractal structure of sequential behaviour patterns: an indicator of stress. *Animal Behaviour* 51, 437–443.

Beneviste, H. (1989) Brain microdialysis. *Journal of Neurochemistry* 52, 1667–1679.

Bond, A.J., James, D.C. and Ladar, M.H. (1971) Physiological and psychological measures in anxious patients. *Psychological Medicine* 4, 364–373.

Broom, D.M. and Johnson, K.G. (1993) *Stress and Animal Welfare.* Chapman and Hall, London.

Bubenik, G.A. and Bubenik, A.B. (1979) Remote-controlled, self-contained blood sampler. *CALAS/ACTAL Proceedings* 5, 51–53.

Bubenik, G.A., Bubenik, A.B., Schams, D. and Leatherland, J.F. (1983) Circadian and circannual rhythms of LH, FSH, testosterone (T), prolactin, cortisol, T_3 and T_4 in plasma of mature, male white-tailed deer. *Comparative Biochemistry and Physiology* 76A, 37–45.

Carragher, J.F., Ingram, J.R. and Matthews, L.R. (1997) Effects of yarding and handling procedures on stress responses of free-ranging red deer (*Cervus elaphus*). *Applied Animal Behaviour Science* 51, 143–158.

Chao, C.C., Brown, R.D. and Deftos, L.J. (1984) Effects of xylazine immobilization on biochemical and endocrine values in white-tailed deer. *Journal of Wildlife Diseases* 20, 328–332.

Chiappa, K.H. (1990) *Evoked Potentials in Clinical Medicine,* 2nd edn. Raven Press, New York.

Cook, C.J. (1996) Basal and stress response cortisol levels and stress avoidance learning in sheep. *New Zealand Veterinary Journal* 44, 162–163.

Cook, C.J. (1997a) Oxytocin and prolactin suppress cortisol responses to acute stress in both lactating and non lactating sheep. *Journal of Dairy Research* 64, 327–339.

Cook, C.J. (1997b) Real time extracellular measurement of neurotransmitters in conscious sheep. *Journal of Neuroscience Methods* 72, 161–166.

Cook, C.J. (1997c) Real time measurements of corticosteroids in conscious animals using an antibody based electrode. *Nature Biotechnology* 15, 467–472.

Cook, C.J. (1998a) Steroidal hormones determine sex related differences in opioid induced elevation of nociceptive threshold in sheep. *New Zealand Veterinary Journal* 46, 68–71.

Cook, C.J. (1998b) Monitoring on-line of extracellular gamma-amino-4-butyric acid using microdialysis coupled to immunosensor analysis. *Journal of Neuroscience Methods* 82, 145–150.

Cook, C.J. and Devine, C.E. (1998) Antibody based electrodes for hormonal and neurotransmitter measurements *in vivo. Electroanalysis* 10(16), 1108–1111.

Cook, C.J. and Jacobson, L.H. (1995) Salivary cortisol as an indicator of stress in sheep. *New Zealand Veterinary Journal* 43, 248.

Cooper, R., Osselton, J.W. and Shaw, J.C. (1980) *EEG Technology*, 3rd edn. Butterworths, London.

Dawkins, M.S. (1990) From an animal's point of view: motivation, fitness, and animal welfare. *Behavioral and Brain Sciences* 13, 1–9, 54–61.

Diverio, S., Goddard, P.J., Gordan, I.J. and Elston, D.A. (1993) The effect of management practices on stress in farmed red deer (*Cervus elaphus*) and its modulation by long-acting neuroleptics: behavioural responses. *Applied Animal Behaviour Science* 36, 363–376.

Drake, M.E., Pakalnis, A., Phillips, B., Padaman, H. and Hietter, S.A. (1991) Auditory evoked potentials in anxiety disorder. *Clinical Electroencephalography* 22, 97–101.

Farrell, D.J., Corbett, J.L. and Leng, R.A. (1970) Automatic sampling of blood and ruminal fluid of grazing sheep. *Research in Veterinary Science* 11, 217–220.

Fraser, A.F. and Broom, D.M. (1990) *Farm Animal Behaviour and Welfare*. Bailliere Tindall, London.

Fraser, D. and Matthews, L.R. (1997) Preference and motivation testing. In: Appleby, M.C. and Hughes, B.O. (eds) *Animal Welfare*. CAB International, Wallingford, UK, pp. 159–175.

Gardner, J., Rothwell, N. and Luheshi, G. (1998) Leptin affects food intake via CRF receptor mediated pathways. *Nature Neuroscience* 1(2), 103–109.

Goddard, P.J., Rhind, S.M., Hamilton, W.J., MacDonald, A.J., Fawcett, A.R., Soanes, C. and McMillen, S.R. (1994) The adrenocorticotrophic hormone stimulation test: its potential use and limitations in red deer (*Cervus elaphus*). *Canadian Journal of Zoology* 72, 1826–1830.

Grandin, T., Curtis, S.E., Widowski, T.M. and Thurmon, J.C. (1986) Electro-immobilisation versus mechanical restraint in an avoid–avoid chronic test for ewes. *Journal of Animal Science* 62, 1469–1480.

Griffiths, S.K., Pierson, L.L., Gerhardt, K.J., Abrams, R.M. and Peters, A.J.M. (1996) Auditory brainstem response in sheep. Part II: postnatal development. *Developmental Psychobiology* 29, 53–68.

Hattingh, J., Wright, P.G., de Vos, V., McNairn, I.S., Ganhao, M.F., Silove, M., Wolverson, G. and Cornelius, S.T. (1984) Blood composition in culled elephants and buffaloes. *Journal of the South African Veterinary Association* 55, 157–164.

Hattingh, J., Ganhao, M.F., Kruger, F.J.N., De Vos, V. and Kay, G.W. (1988) Remote controlled sampling of cattle and buffalo blood. *Comparative Biochemistry and Physiology* 89A, 231–235.

Hernandez-Peon, R., Scherrer, H. and Jouvet, M. (1956) Modification of electric activity in cochlear nucleus during attention in unanesthetized cats. *Science* 123, 331–332.

Ingram, J.R., Matthews, L.R. and McDonald, R.M. (1994) A stress free blood sampling technique for free ranging animals. *Proceedings of the New Zealand Society of Animal Production* 54, 39–42.

Ingram, J.R., Matthews, L.R., Carragher, J.F. and Schaare, P.R. (1997) Plasma cortisol responses to remote adrenocorticotropic hormone (ACTH) infusion in free-ranging red deer (*Cervus elaphus*). *Domestic Animal Endocrinology* 14, 63–71.

Ingram, J.R., Crockford, J.N. and Matthews, L.R. (1999) Ultradian, circadian and seasonal rhythms in cortisol secretion and adrenal responsiveness to ACTH in unrestrained red deer stags (*Cervus elaphus*). *Journal of Endocrinology* 162, 289–300.

Isreal, J.B., Wickens, C.C., Chesney, G.L. and Donchin, E. (1980) The event-related brain potential as an index of display-monitoring workload. *Human Factors* 22, 211–224.

Kaufer, D., Friedman, A., Seidman, S. and Soreq, H. (1998) Acute stress facilitates long lasting changes in cholinergic gene expression. *Nature* 393, 373–377.

Knott, J.R. and Irwin, D.A. (1973) Anxiety, stress and the contingent negative variation. *Archives of General Psychiatry* 29, 538–541.

Koepp, M.J., Gunn, R.N., Lawrence, A.D., Cunningham, V.J., Dagher, A., Jones, T., Brooks, D.J., Bench, C.J. and Grasby, P.M. (1998) Evidence for striatal dopamine release during a video game. *Nature* 393, 266–268.

Ladewig, J. and Stribrny, K. (1988) A simplified method for the stress free continuous blood collection in large animals. *Laboratory Animal Science* 38, 333–334.

Matthews, L.R., Carragher, J.F. and Ingram, J.R. (1994) Post-velveting stress in free-ranging red deer. *Proceedings of a Deer Course for Veterinarians, New Zealand Veterinary Association, Deer Branch* 11, 138–146.

Matthews, L.R., Temple, W., Foster, T.M., Walker, J. and McAdie, T.M. (1995) Comparison of the demand for dustbathing substrates by layer hens. In: Rutter, S.M., Rushen, J., Randle, H.D. and Eddison, J.C. (eds) *Proceedings of the 29th International Congress of the International Society for Applied Ethology*. Universities Federation for Animal Welfare, Herts, UK, pp 11–12.

Mayes, R.W., Lamb, C.S. and Colgrove, P.M. (1988) Equipment for estimating carbon dioxide turnover rate in undisturbed grazing sheep. *Proceedings of the Nutritional Society* 47, 136A.

Monfort, S.L., Brown, J.L. and Wildt, D.E. (1993) Episodic and seasonal rhythms of cortisol secretion in male Eld's deer (*Cervus eldi thamin*). *Journal of Endocrinology* 138, 41–49.

Morris, J.S., Ohman, A. and Dolan, R.J. (1998) Conscious and unconscious emotional learning in the human amygdala. *Nature* 393, 467–470.

Munksgaard, L., Ingvartsen, K.L., Pedersen, L.J. and Nielsen, V.K. (1999) Deprivation of lying down affects behaviour and pituitary-adrenal axis responses in young bulls. *Acta Agriculturae Scandinavica, Section A, Animal Science* 49, 172–178.

Obrenovitch, T.P., Urenjak, J. and Zilkha, E. (1994) Intracerebral microdialysis combined with recording of extracellular field potential: a novel method for investigation of depolarising drugs *in vivo*. *British Journal of Pharmacology* 113, 1295–1302.

Palme, R., Fischer, P., Schildorfer, H. and Ismail, M.N. (1996) Excretion of infused ^{14}C-steroid hormones via faeces and urine in domestic livestock. *Animal Reproduction Science* 43, 43–63.

Pierson, L.L., Gerhardt, S.K., Griffiths, K.J. and Abrams, R.M. (1995) Auditory brainstem response in sheep. Part I: prenatal development. *Developmental Psychobiology* 28, 293–305.

Ramström, O., Ye, L. and Mosbach, K. (1996) Artificial antibodies to corticosteroids prepared by molecular imprinting. *Chemistry and Biology* 3, 471–477.

Reimers, T.J., McCann, J.P. and Cowan, R.G. (1983) Effects of storage times and temperature on T3, T4, LH, prolactin, insulin, cortisol and progesterone concentrations in blood samples from cows. *Journal of Animal Science* 57, 683–691.

Rushen, J. (1990) Use of aversion-learning techniques to measure distress in sheep. *Applied Animal Behaviour Science* 28, 3–14.

Sapolsky, R. (1992) *Stress, the Ageing Brain, and Mechanisms of Neuron Death.* Bradford Books, MIT Press, Cambridge, Massachusetts.

Seal, U.S., Ozoga, J.J., Erickson, A.W. and Verme, L.J. (1972) Effects of immobilization on blood analyses of white-tailed deer. *Journal of Wildlife Management* 36, 1034–1040.

Smith, R.F. and Dobson, H. (1990) Effect of preslaughter experience on behaviour, plasma cortisol and muscle pH in farmed deer. *Veterinary Record* 126, 155–158.

Stephan, E. and Cybik, M. (1989) Evaluation of information obtained from a series of blood samples from stress-free roaming cattle taken with the aid of radio remote controlled small portable blood (or other body fluids) sampling device. *Bovine Practitioner* 24, 35–37.

Van Liere, D.W., Kooijman, J. and Wiepkema, P.R. (1990) Dustbathing behaviour of laying hens as related to quality of dustbathing material. *Applied Animal Behaviour Science* 26, 127–141.

Verkerk, G.A., Phipps, A.M., Carragher, J.F., Matthews, L.R. and Stelwagen, K. (1998) Characterization of milk cortisol concentrations as a measure of short-term stress responses in lactating dairy cows. *Animal Welfare* 7, 77–86.

Waas, J.R., Ingram, J.R. and Matthews, L.R. (1997) Physiological responses of red deer (*Cervus elaphus*) to conditions experienced during road transport. *Physiology and Behavior* 61, 931–938.

Wilson, G.F., Fullenkamp, P. and Davis, I. (1994) Evoked potential, cardiac, blink and respiration measures of pilot workload in air to ground missions. *Aviation, Space and Environmental Medicine* 65, 100–105.

Wise, R.A. (1980) The dopamine synapse and the notion of pleasure centres in the brain. *Trends in Neurosciences* 3, 91–94.

Accumulation and Long-term Effects of Stress in Fish

7

C.B. Schreck

Oregon Cooperative Fish and Wildlife Research Unit, Biological Resources Division, USGS, Oregon State University, Corvallis, Oregon, USA

Introduction

Vertebrates respond to stressful situations through a neuroendocrine cascade that leads to physiological and often behavioural alarm responses. Unless the duration of the stressor is very brief, a physiological resistance phase follows that leads to either compensation or exhaustion (death) (Selye, 1950, 1973). This chapter draws on research conducted on stress in fish to discuss basic principles of accumulation and long-term exposure. We have little understanding of how severity and duration of stress relate to produce response patterns. We are also unable to differentiate the boundaries between stressors to which the animal can compensate and those that lead to exhaustion. The environment can regulate and set boundaries on physiological performance in a variety of ways. Fry (1947) proposed that some environmental variables govern metabolic rate. These 'controlling factors', such as temperature, operate on the internal medium (i.e. affect internal processes) and influence the state of activation of metabolites. Other factors that govern metabolic rate, 'limiting factors', operate within the metabolic chain and set the bounds within which normal physiological operation and function can take place. Gradients of environmental conditions can be 'directive factors', allowing or necessitating a response by the organism along the gradient. Environmental factors, 'accessory factors', can also place a metabolic load on an organism beyond that governed by overall metabolic rate. I believe that these concepts are important for understanding how fish respond to and cope with stressors of differing types, severities, exposure rates and durations that push physiological systems beyond the rather narrow bounds dictated by homeostasis (see Johnson *et al.*, 1992, for an overview of stress and homeostasis).

Situations that extend the organism beyond its homeostatic capacities result in allostasis, which is the ability of the body to achieve an alternative form of stasis through change rather than maintenance of prior operation (Sterling and Eyer, 1988). The long-term consequences of exposure to a stressor result in an allostatic load, which is the price of accommodation to stress (McEwen and Stellar, 1993). From an ecological perspective, the currency of this price may be thought of in terms of energy (Schreck, 1982; Barton and Schreck, 1987a; Davis and Schreck, 1997). In his wonderful review of long-term effects of the physiological response to stress, McEwen (1998) points out that 'the core of the body's response to a challenge . . . is twofold, turning on an allostatic response that initiates a complex adaptive pathway, and then shutting off this response when the threat is past.' He describes the temporal dynamics associated with the stress response and the allostatic load that could be imposed by sequential stressors to which mammals can or cannot adapt.

Unfortunately, there is little information on the physiological response of vertebrates, and particularly cold-blooded taxa such as fish, to a persistent stressor. We also know little about how fish respond when exposed simultaneously to two or more different stressors, or to sequential stressors. Questions related to the importance of the duration of time between discrete stressors and the number of stressful events need to be addressed. How fish respond to single and multiple stressors is obviously important for considering fish health and for environmental risk assessment. Nevertheless, if and how these sorts of stressful situations affect the ability of fish to cope with necessary, routine life functions is also poorly understood. In this chapter, I attempt to address these issues by developing models of the physiological response of teleosts to stressors that are: (i) persistent (continuous stressor); (ii) experienced during the presence of other stressors (multiple stressors); or (iii) experienced sequentially (sequential stressors). I also present models describing the effects of such stressors on performance phenotypes at the organism level; that is, what the response means in terms of fitness. For this purpose I draw on data from many published and/or recent experiments conducted primarily in my laboratory, synthesizing this diverse information into a conceptual framework. Assessment of stress was based on responses of the hypothalamic–hypophyseal–interrenal (HPI) axis and secondary and tertiary stress response factors (see Barton and Iwama, 1991).

Continuous stressor

The physiological response to a stressor is almost immediate once the stressor is perceived (Mazeaud *et al.*, 1977). If the stressor is persistent, fish either die or adapt. Using concentrations of plasma cortisol to judge adaptation, the plasma levels of cortisol or the activation of the HPI axis appears to return to pre-stress conditions within a week or so in salmonids (Schreck, 1981). However, Pacific salmon (*Oncorhynchus* spp.) stressed by low hierarchical standing (Ejike and Schreck, 1980), sublethal concentrations of heavy metals (Schreck and Lorz, 1978), elevated rearing density or reduced water flow (Schreck *et al.*, 1985; Patino *et al.*, 1986) have slightly elevated cortisol secretion rates above prior 'resting' levels for weeks or months. Stress from long-term (months) infection with causative agents of diseases like bacterial kidney disease (*Renibacterium salmoninarum*) lead to extended increased plasma cortisol and lactate concentrations and depressed glucose levels in salmon (Mesa *et al.*, 1999). Surgical implantation of a radiotransmitter into Pacific lamprey (*Lampetra tridentata*) resulted in a rapid surge in plasma glucose levels followed by a decrease, but differences from unperturbed control fish were still noticeable over 3 months later (D. Close, Oregon, 1998, personal communication). It thus appears that the onset of the physiological stress response to continuous stressors is similar to that caused by a discrete stressful event, but if the fish does not die, the stress response indicators eventually return to near pre-stress conditions (Fig. 7.1).

Even though fish may appear to compensate physiologically or recover from the presence of a perturbing situation, their ability to perform necessary functions at the whole organism level can be diminished for extended periods of time. This is evident from several lines of research. Heavy metals that result in a physiological stress response (Schreck and Lorz, 1978) also diminish the ability of salmon to migrate or adapt to sea water (Lorz and McPherson, 1976). Development of salmon (smoltification) is impaired or retarded by high rearing density and low water flow conditions which affect the HPI axis (Schreck *et al.*, 1985; Patino *et al.*, 1986). Acidification of rainbow trout, *Oncorhynchus mykiss,* which caused a transient rise in plasma cortisol levels and depression in sodium concentrations resulted in a glucose stress response that lasted at least 5 days (Barton *et al.*, 1985), as well as impaired reproduction (Weiner *et al.*, 1986). Subordinate salmon that appear mildly stressed do not convert food as efficiently and exhibit reduced growth rates (Ejike and Schreck, 1980). In addition to causing a physiological stress response, bacterial kidney disease also increases the vulnerability of salmon to predation (Mesa *et al.*, 1998) but, interestingly, has no measurable effect on development at the smoltification stage (Mesa *et al.*, 1999).

It thus appears that once continuously exposed to a stressor, fish either eventually become exhausted and die or achieve a level of whole organism performance that is similar to or moderately diminished from that shown before the onset of stress (Fig. 7.2). In other words, the allostatic load can be such that even if fish can adapt to their changed environmental conditions, their ability to do such routine things as develop, grow, reproduce and avoid predators may be impaired to some extent.

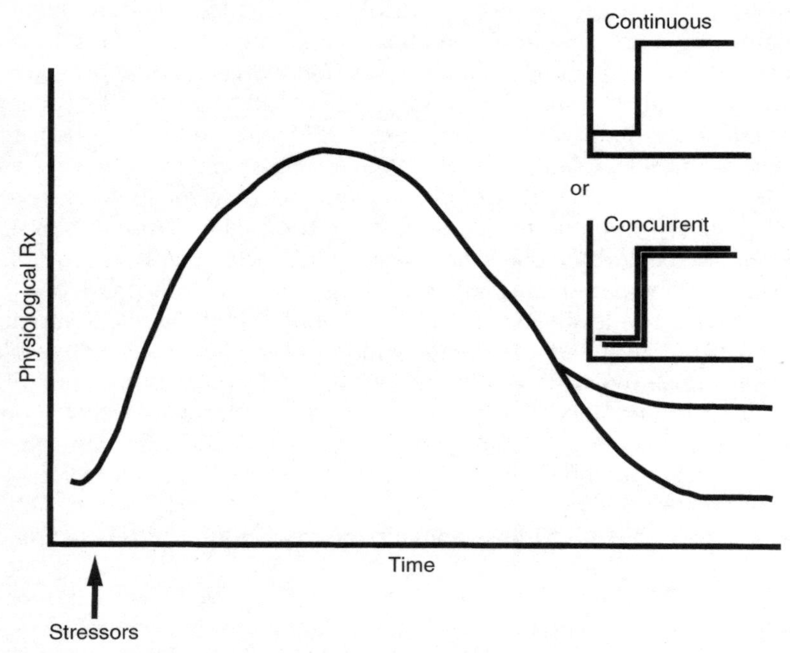

Fig. 7.1. Conceptual physiological response pattern (primary and secondary stress response factors) when fish are exposed continuously to one or concurrent stressors for which they can compensate. The new, final 'steady state' reached is similar to or slightly different from that shown before the stressor. Depending on the physiological parameter measured, the pattern may be reversed, i.e. a factor may initially decrease then increase.

Multiple stressors

It is not uncommon for fish to be exposed to more than one stressor at the same time, although even less is known about such events than is known for continuous exposure to a single stressor. In the face of multiple stressors, it appears that there can be a cumulative effect on the physiological response. For example, rainbow trout exposed to sublethal concentrations of acidification experience an acid dose-dependent increase in their cortisol and glucose and a decrease in their sodium concentrations in response to a secondary handling stressor (Barton *et al.*, 1985). Salmon reared for extended periods of time in sublethal copper concentrations that result in elevated cortisol titres show further elevated cortisol levels in response to a crowding stressor (Schreck and Lorz, 1978). Salmon acclimatized to temperatures ranging from cold to near the upper incipient lethal level and then exposed to handling or crowding stressors in general exhibited response rates and recovery dynamics directly correlated with temperature, as judged

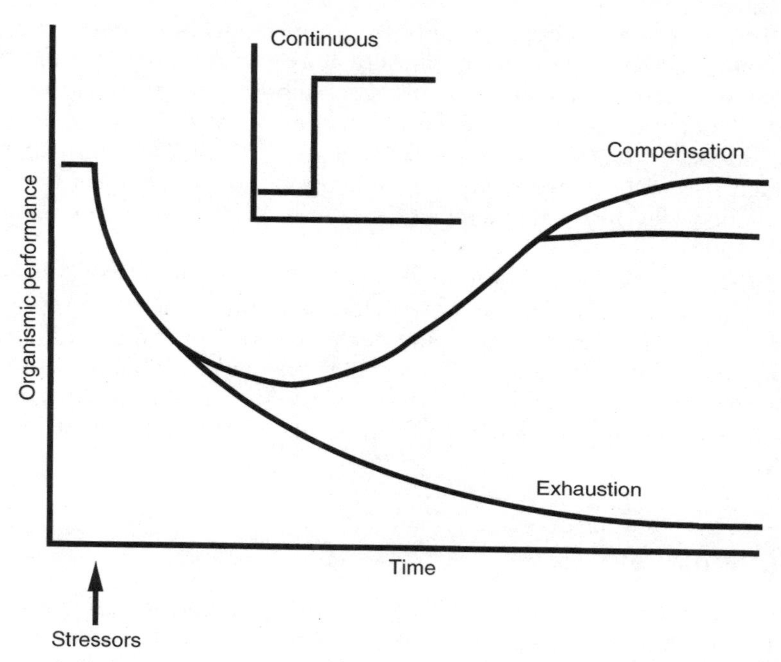

Fig. 7.2. Conceptual organismic response (measures of performance of physiological functions) to a continuous stressor. If the stressor is severe enough, performance drops to a level from which the animal cannot recover (exhaustion). If compensation is achieved, then performance returns to levels similar to or relatively close to that seen prior to the stressor.

from plasma cortisol and glucose concentrations (Barton and Schreck, 1987b). In addition, the magnitude of the cortisol stress response is greater in warm water.

Temperature also modifies how sablefish, *Anoplopoma fimbria*, respond to additional stressors. Sablefish captured in a simulated trawl in cold water but raised to the surface through a warm water surface layer initiate a cortisol, glucose and lactate stress response that is accentuated by the warm water. The capture stress response in above-ambient temperatures is temperature dependent (Olla *et al.*, 1998). Several weeks of fasting influence how salmon respond to handling or confinement stressors. Hyperglycaemia in response to handling is not as great in the fasted fish, but cortisol dynamics are unaffected (Barton *et al.*, 1988). It thus appears that the physiological response to multiple, concurrent stressors follows a general pattern similar to that of fish stressed by a single continuous challenging situation (Fig. 7.1). However, the magnitude of the response may be greater than if only one stressor were present. Recovery may be complete or partial.

The simultaneous occurrence of two or more stressors may jeopardize the ability of fish to perform routine life activities to a greater extent than would exposure to either stressful agent alone. For example, confinement-induced mortality was highest in fish acclimatized near their upper incipient level (Barton and Schreck, 1987b). Similarly, sablefish that are stressed by capture have lower survivorship following release if they also experience elevated water temperature during the capture process (Olla *et al.*, 1998).

Based on this information and on other general literature such as Fry (1947) and Precht (1958), I propose that organismic-level responses to one stressor are magnified for fish living near their tolerance limits of another environmental variable (Fig. 7.3). Overall survival and probably most other performance traits would be compromised by exposure to the presence of one stressor when other environmental conditions are suboptimal for that individual.

Sequential stressors

Fish are frequently exposed to a series of stressful events in succession. We know little about the effects of temporal scaling of sequential stressors. How

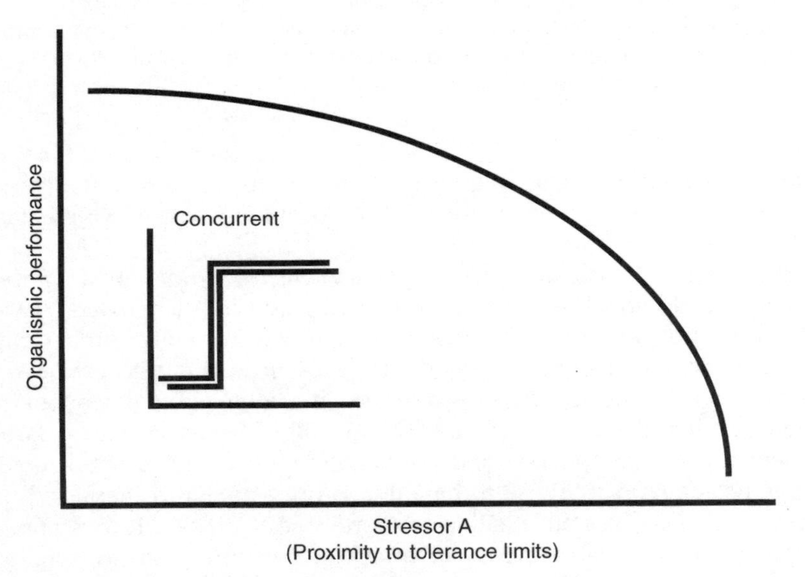

Fig. 7.3. Conceptual organismic response (measures of performance of physiological functions) to a stressor concurrent with another continuous stressor. The ability of a fish to perform necessary functions decreases directly, probably exponentially, the closer the individual is to its tolerance limit of another concurrent stressor.

close together in time do stressors need to be experienced before they are perceived as one single, perhaps more severe, stressor? Or how far apart do stressors need to be before the experience with the first stressor does not influence the effect of a subsequent stressor? If fish have some sort of physiological memory of prior exposures to stressors, does the number of such exposures affect subsequent responses?

Time between stressors is clearly important. Barton *et al.* (1985) demonstrated that salmon have a cumulative stress response, as measured by elevated plasma cortisol and glucose, to very brief handling stressors spaced 3 h apart; the effects of each subsequent stressor were cumulative to those caused by the previous stressor or stressors. Similarly, discrete stressful events encountered by juvenile salmon as they pass through a dam that probably takes minutes to hours for complete passage result in cumulative stress effects (Maule *et al.*, 1988). When chinook salmon are exposed to brief handling, crowding or disturbance stressors twice a day sequentially over a period of several weeks, those fish experiencing the stressors in a positive conditioning situation (i.e. food reward associated with each stressor) or experiencing the stressor independently of positive conditioning have lower cortisol, glucose and lactate responses and more rapid recovery than fish with no prior experience with stress when exposed to subsequent severe handling/crowding for several hours (Schreck *et al.*, 1995). Fish that were positively conditioned perform best and are followed by those having had stressful experiences without reward. Similarly, when salmon are moderately to severely stressed for several minutes on eight separate occasions, with a few days between each stressor, fish with prior experience also appear to recover more rapidly than naive fish with no history of stress (Schreck *et al.*, unpublished data). The length of time between stressors can thus affect the physiological response to a subsequent stressor (Fig. 7.4). For salmon, if the inter-stressor interval is only a few hours then the stress response appears to be cumulative. If the stressor interval is a day or days, however, then compensation or habituation can occur such that a subsequent stressor produces a similar or even diminished response relative to exposure to a single stressor.

The time between discrete stressors is important in affecting an organism-level response. Healthy juvenile chinook salmon are able to cumulate responses to exposure to at least three sequential stressors separated by 3-h intervals, while unhealthy individuals are not (Barton *et al.*, 1985). Survival, sea water adaptation and disease resistance were much enhanced in salmon stressed daily as compared with naive controls when they were challenged with a more severe stressor; positive conditioning provides fish with greater fitness than experience with stressors not associated with reward (Schreck *et al.*, 1995). In mature female rainbow trout, handling and disturbance stressors administered briefly on a daily basis for 45 days or longer immediately prior to ovulation resulted in earlier ovulation

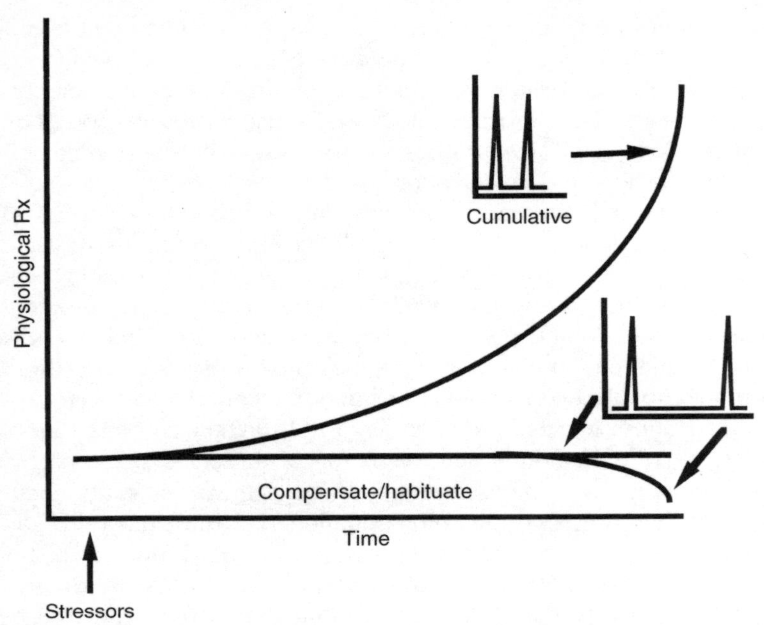

Fig. 7.4. Conceptual physiological response pattern (primary and secondary stress response factors) when fish are exposed to sequential stressors that are not directly lethal. If the stressors occur relatively close together temporally, then there can be cumulative effects. If the temporal spacing is greater, then compensation or habituation may occur.

and smaller embryos than in unstressed individuals (Contreras-Sanchez *et al.*, 1998). Campbell *et al.* (1992, 1994) also found negative effects of repeated acute or chronic stressors on reproductive fitness of adult rainbow and brown trout (*Salmo trutta*) in terms of gamete quality and survival of progeny, but in this case ovulation in the rainbow trout was delayed by the stressors (Cambell *et al.*, 1992). Thus, the nature of stressors appears to affect the timing of reproduction in different ways, albeit always negatively. Separation of eight discrete stressors by several days each, while likely to result in physiological compensation, apparently retards or reverses juvenile development since unstressed control fish voluntarily enter sea water when given a choice (as appropriate for this life history stage) while stressed fish do not (Seals *et al.*, unpublished data).

Thus, it appears that sequential stressors separated by hours can be maladaptive with regard to organism performance, while greater temporal spacing can lead to enhanced performance if compensation or habituation occurs (Fig. 7.5). I speculate that there may be a cost associated with such compensation, however, perhaps resulting in deleterious developmental or reproductive consequences.

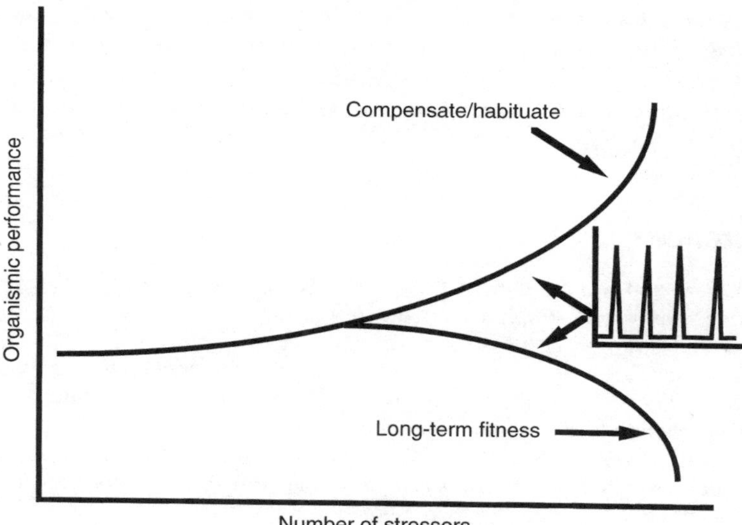

Fig. 7.5. Conceptual organismic response (measures of performance of physiological functions) to sequentially applied stressors. Performance traits that provide immediate or short-term adaptive value (e.g. immune function, behaviour, osmoregulation) may be enhanced but traits that provide long term fitness (e.g. growth, development and reproduction) may be diminished as the number of stressful experiences increases.

Conclusions

Physiological performance and hence fitness are regulated by environmental conditions (see Fry, 1947), including those that are directly or indirectly anthropogenic. Rapid shifts to new environmental conditions or to extremes in environmental conditions can lead to a physiological stress response. Understanding how the duration, severity and frequency of change of such experiences affect physiological processes is critical for animal health considerations. Knowing how fish cope with different stressful situations is also essential for environmental risk assessment.

It appears that there may be trade-offs between the immediate benefits of the stress response and the long-term consequences. Prolonged or repeated exposure to stressors can shift the performance capacities of fish (Schreck, 1981; Schreck and Li, 1991); prolonged or repeated stress responses impose an allostatic load (McEwen, 1998) that tends in the long run to be maladaptive to general organism performance. While fish may compensate and adapt to the perturbing situation(s) and even become conditioned or habituated such that they can perform necessary physiological and behavioural functions, there may be an energetic cost associated with this compensation leading to deleterious effects on growth, development and reproduction. Ultimate effects of stressors on the physiological stress response and organism fitness are dependent on the duration and severity of the stressor, the frequency of stressful situations and the number and temporal duration between exposures, and the number of contemporaneous stressors experienced.

References

Barton, B.A. and Iwama, G.K. (1991) Physiological changes in fish from stress in aquaculture with emphasis on the response and effects of corticosteroids. *Annual Review of Fish Diseases* 1, 3–26.

Barton, B.A. and Schreck, C.B. (1987a) Influence of acclimation temperature on interrenal and carbohydrate stress responses in juvenile chinook salmon (*Oncorhynchus tshawytscha*). *Aquaculture* 62, 299–310.

Barton, B.A. and Schreck, C.B. (1987b) Metabolic cost of acute physical stress in juvenile steelhead trout. *Transactions of the American Fisheries Society* 116, 257–263.

Barton, B.A., Weiner, G.S. and Schreck, C.B. (1985) Effect of prior acid exposure on physiological responses of juvenile rainbow trout (*Salmo gairdneri*) to acute handling stress. *Canadian Journal of Fisheries and Aquatic Sciences* 42, 710–717.

Barton, B.A., Schreck, C.B. and Fowler, L.G. (1988) Fasting and diet content affect stress-induced changes in plasma glucose and cortisol in juvenile chinook salmon. *Progressive Fish Culturist* 50, 16–22.

Campbell, P.M., Pottinger, T.G. and Sumpter, J.P. (1992) Stress reduces the quality of gametes produced by rainbow trout. *Biology of Reproduction* 47, 1140–1150.

Campbell, P.M., Pottinger, T.G. and Sumpter, J.P. (1994) Preliminary evidence that chronic confinement stress reduces the quality of gametes produced by brown and rainbow trout. *Aquaculture* 120, 151–169.

Contreras-Sanchez, W.M., Schreck, C.B., Fitzpatrick, M.S. and Pereira, C.B. (1998) Effects of stress on the reproductive performance of rainbow trout (*Oncorhynchus mykiss*). *Biology of Reproduction* 58, 439–447.

Davis, L.E. and Schreck, C.B. (1997) The energetic response to handling stress in juvenile coho salmon. *Transactions of the American Fisheries Society* 126, 248–258.

Ejike, C. and Schreck, C.B. (1980) Stress and social hierarchy rank in coho salmon. *Transactions of the American Fisheries Society* 109, 423–426.

Fry, F.E.J. (1947) *Effects of the Environment on Animal Activity. University of Toronto Studies Biology Series* No. 55. Publication of the Ontario Fisheries Research Laboratory, No. 68. University Press, Toronto.

Johnson, E.O., Kamilars, T.C., Chrousos, G.P. and Gold, P.W. (1992) Mechanisms of stress: a dynamic overview of hormonal and behavioral homeostasis. *Neuroscience and Biobehavioral Reviews* 16, 115–130.

Lorz, H.W. and McPherson, B.P. (1976) Effects of copper or zinc in fresh water on the adaptation to sea water and ATPase activity, and the effects of copper on migratory disposition of coho salmon (*Oncorhynchus kisutch*). *Journal of the Fisheries Research Board of Canada* 33, 2023–2030.

Maule, A.G., Schreck, C.B., Bradford, C.S. and Barton, B.A. (1988) Physiological effects of collecting and transporting emigrating juvenile chinook salmon past dams on the Columbia River. *Transactions of the American Fisheries Society* 117, 245–261.

Mazeaud, M.M., Mazeaud, F. and Donaldson, E.M. (1977) Primary and secondary effects of stress in fish: some new data with a general review. *Transactions of the American Fisheries Society* 106, 201–212.

McEwen, B.S. (1998) Protective and damaging effects of stress mediators. *New England Journal of Medicine* 338, 171–179.

McEwen, B.S. and Stellar, E. (1993) Stress and the individual: mechanisms leading to disease. *Archives of Internal Medicine* 153, 2093–2101.

Mesa, M.G., Poe, T.P., Maule, A.G. and Schreck, C.B. (1998) Vulnerability to predation and physiological stress responses in juvenile chinook salmon (*Oncorhynchus tshawytscha*) experimentally infected with *Renibacterium salmoninarum*. *Canadian Journal of Fisheries and Aquatic Sciences* 55, 1599–1606.

Mesa, M.G., Maule, A.G., Poe, T.P. and Schreck, C.B. (1999) Influence of bacterial kidney disease on smoltification in salmonids: is it a case of double jeopardy? *Aquaculture* 174, 25–41.

Olla, B.L., Davis, M.W. and Schreck, C.B. (1998) Temperature magnified post-capture mortality in adult sablefish after simulated trawling. *Journal of Fish Biology* 53, 743–751.

Patino, R., Bradford, C.S. and Schreck, C.B. (1986) Adenylate cyclase activators and inhibitors, cyclic nucleotide analogs, and phosphatidylinositol: effects on interrenal function of coho salmon (*Oncorhynchus kisutch*) *in vitro*. *General and Comparative Endocrinology* 62, 230–235.

Precht, H. (1958) Concepts of the temperature adaptation of unchanging reaction systems of cold-blooded animals. In: Prosser, C.L. (ed.) *Physiological Adaptations*. Lord Baltimore Press Inc., Baltimore, Maryland, pp. 50–78.

Schreck, C.B. (1981) Stress and compensation in teleostean fishes: response to social and physical factors. In: Pickering, A.D. (ed.) *Stress and Fish*. Academic Press, London, pp. 295–321.

Schreck, C.B. (1982) Stress and rearing of salmonids. *Aquaculture* 28, 241–249.

Schreck, C.B. and Li, H.W. (1991) Performance capacity of fish: stress and water quality. In: Brune, D.E. and Tomasso, J.R. (eds) *Aquaculture and Water Quality. Advances in World Aquaculture*, vol. 3. World Aquaculture Society, Baton Rouge, Louisiana, pp. 21–29.

Schreck, C.B. and Lorz, H.W. (1978) Stress response of coho salmon (*Oncorhynchus kisutch*) elicited by cadmium and copper and potential use of cortisol as an indicator of stress. *Journal of the Fisheries Research Board of Canada* 35, 1124–1129.

Schreck, C.B., Patino, R., Pring, C.K., Winton, J.R. and Holway, J.E. (1985) Effects of rearing density on indices of smoltification and performance of coho salmon, *Oncorhynchus kisutch*. *Aquaculture* 45, 345–358.

Schreck, C.B., Jonsson, L., Feist, G. and Reno, P. (1995) Conditioning improves performance of juvenile chinook salmon, *Oncorhynchus tshawytscha*, to transportation stress. *Aquaculture* 135, 99–110.

Selye, H. (1950) Stress and the general adaptation syndrome. *British Medical Journal* 1, 1383–1392.

Selye, H. (1973) The evolution of the stress concept. *American Scientist* 61, 692–699.

Sterling, P. and Eyer, J. (1988) Allostasis: a new paradigm to explain arousal pathology. In: Fisher, S. and Reason, J. (eds) *Handbook of Life Stress, Cognition and Health.* John Wiley & Sons, New York, pp. 629–649.

Weiner, G.S., Schreck, C.B. and Li, H.W. (1986) Effects of low pH on reproduction of rainbow trout. *Transactions of the American Fisheries Society* 115, 75–82.

Chronic Intermittent Stress: A Model for the Study of Long-term Stressors

8

J. Ladewig

Department of Animal Science and Animal Health, Division of Ethology and Health, The Royal Veterinary and Agricultural University, Denmark

> Stress is a dangerous and useless word. It may seem useful because it is a unifying concept, but it unifies our ignorance rather than our knowledge
>
> (Zanchetti, 1972)

Introduction

Despite decades of intense research, our understanding of long-term stress is poor. We still know too little about what aspects of, for instance, long-term confinement act as stressors and how they affect the various stress responses of the organism. Consequently, we have great difficulty in identifying animals that are experiencing chronic stress. We do not have specific tests or other methods that can be used to diagnose chronic stress. Neither behavioural tests (such as the open-field test; Rushen, Chapter 2, this volume), nor physiological tests (such as determination of baseline cortisol secretion; Ladewig and Smidt, 1989; Rushen, 1991), nor the adrenal challenge test (Von Borell and Ladewig, 1992) have helped us so far.

One reason for these difficulties is that we tend to treat chronic stress as a constant, unvarying state. Instead, we need to consider long-term stress as a succession of repeated acute stressors, a state that has been termed chronic intermittent stress (Burchfield, 1979; Ladewig, 1994). Another reason is that an organism that is subjected to long-term intermittent stress changes its responses to the stressors over time. Some responses diminish because of adaptation at the cognitive level (e.g. behavioural responses). Other responses may be suppressed or return to normal (e.g. basal cortisol secretion). Yet other responses may increase by a process called sensitization (e.g. cortisol response to a new stressor or to adrenocorticotropic hormone (ACTH) challenge). Furthermore, the type of change depends on how a specific stressor affects an animal. It is thus logical to assume that repeated exposure to a mild stressor of short duration (e.g. restraint, Fig. 8.1) is more likely to result in a gradually diminished stress response than repeated exposure to a severe (e.g. painful) stressor. In addition, since the stressor effect depends on the animal's subjective experience of the situation, we need to realize that the same stress situation can affect individual animals or species of animals differently. For instance, early experience with a slatted floor reduces heifers' response to later housing on this floor type (Pougin, 1981), and tether housing is likely to affect pigs completely differently from the way in which it affects cattle.

To advance our understanding of long-term stress, I suggest that we use a model of repeated exposure to acute stressors. The model may enable us to analyse the behavioural and physiological mechanisms in much greater detail and, possibly, point out ways in which specific procedures can be developed to diagnose chronic stress. Improvement in the welfare of

Fig. 8.1. Example of chronic intermittent stress: daily 30 min restraint of pigs in a hammock over a 14 day period. Blood samples are collected continuously (Ladewig and Stribrny, 1988) on day 0, 1, 3, 7 and 14, and behaviour observed daily from video recordings (C. Eberbach, unpublished Masters thesis, Freie Universität Berlin, 1990).

animals kept under intensive conditions will only be possible if proper diagnostic methods are available.

Unifying concepts

All too often, aversive treatments are lumped together into one category under the term stress. By doing so, the term becomes a unifying concept. However, as indicated in the quotation at the beginning of this chapter, unifying concepts can be dangerous. They can lull us into the belief that, once we have put a label on a phenomenon, we understand what is going on and can clarify and explain matters by using the term. Moreover, when we lump different phenomena together into one category we tend to overemphasize the similarities and ignore the differences. Because of these dangers, it has been suggested that the term stress should be abandoned. This, however, is hardly possible. Instead, we should use the term appropriately. It is all right to use the term at some level (such as giving a scientific meeting the title 'Animal Stress'), as long as we realize that the information given by the term is limited.

To illustrate how a unifying concept can be used appropriately, consider for a moment another unifying concept with which we have much less of a problem, namely the term health (or its opposite: disease). We have no problem using this term because we know that the information given is limited. We know that we can use it safely in some situations but that it is insufficient in others. If we call our employer one morning to say that we cannot come to work because we are sick, it may be sufficient explanation for not showing up for work. However, if we visit our physician because we feel ill, and he after having done a check-up says: 'You are sick!', we will not be very satisfied. We will want to know more specifically what kind of disease we are suffering from, i.e. we want an exact diagnosis. Only if we know more specifically what ails us will it be possible to think about how to correct the problem, i.e. what kind of therapy can cure us.

The same is true of the unifying concept of stress. In everyday language, the term conveys some degree of information. However, as soon as we want to study 'stress' in some detail, we must abandon the term and be more specific. Only if we know more specifically what kind of stress we are dealing with, i.e. only if we can give an exact diagnosis, will we be able to investigate the cause and think about what can be done to improve the situation. Although the specification 'long-term stress' is a step in the right direction away from the unifying concept 'stress', the degree of specification is far from enough. The term 'long-term stress' is also a unifying concept that unifies many different situations. If the information conveyed by the term 'stress' is comparable with the information given by the term 'disease', the term 'long-term confinement' is no more specific than a term such as 'bacterial infection'. We still do not know what kind of bacteria we are dealing with, in what state the infected body is, and with what kind of antibiotic and at what dosage the infection shall be treated.

A comparable example is tether housing of cattle and pigs. Although the confinement in this housing system is the same for the two species, its effect on them may be vastly different. Cattle spend a large part of their time ruminating, an activity that they can also perform when they are tethered. Pigs spend a large part of their active time moving around rooting, a behaviour that they cannot perform satisfactorily when tethered. On the other hand, we know that cattle have problems lying down on a hard surface, something that does not seem to disturb sows to the same degree (Ladewig and Smidt, 1989; Müller *et al.*, 1989). Thus, even though the housing system is very similar for the two species, it probably affects the animals in completely different ways. Cattle are disturbed by tethering every time they want to lie down, not so much because of the confinement but because of the hard floor. Sows, on the other hand, are disturbed by tethering when they are active and want to move around to root. Application of the adrenocorticotropic hormone (ACTH) challenge test, in which the corticosteroid response of the adrenal gland is measured after a standard application of synthetic ACTH, has shown that tethered cattle react with a reduced cortisol response compared with loose housed cattle (Ladewig and Smidt, 1989), and that tethered pigs react with an increased response compared with loose housed pigs (Von Borell and Ladewig, 1989). It is possible that the difference in response pattern (whether reduced or increased) is not a species difference but is due to the fact that the stressor is completely different for the two species of animals.

Chronic intermittent stress

As indicated, long-term confinement such as tethering does not exert its effect on the animal 'chronically', i.e. with a constant intensity over a long time, but rather in bouts depending upon the behaviour of the animal. At least we must assume a variation in the stressor intensity to which the animal is exposed. There is no reason to think that a tethered animal that is lying sleeping or standing eating considers the confinement aversive at that particular time. Not until the animal wants to move about to root or wants to find a better place to lie down to ruminate does it find tethering aversive. For this reason it makes more sense to think of chronic stress in terms of chronic intermittent stress (Burchfield, 1979; Ladewig, 1994), i.e. a succession of repeated acute stressors. Applied to the example above, this suggests that tethered cattle are subjected to a relatively frequent stressor of relatively high intensity (i.e. discomfort every time they lie down, Fig. 8.2E), whereas tethered pigs are subjected to a relatively infrequent stressor of longer duration (i.e. frustration every time they want to root after feeding, Fig. 8.2D). Along this line of thinking, it is possible to study the behavioural

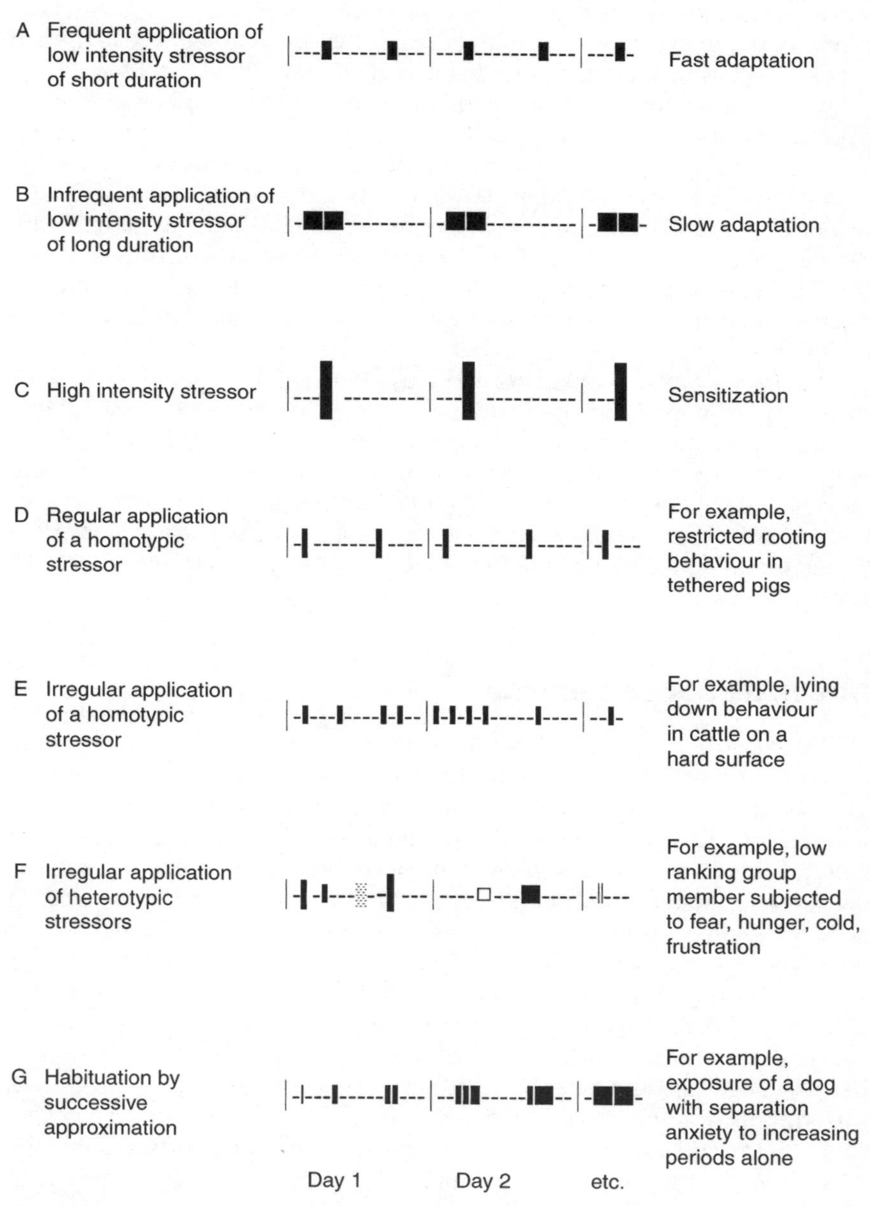

Fig. 8.2. Schematic examples of variations of stressor intensity, stressor duration, stressor frequency and interstressor interval, and possible connection to 'real life' situations.

and physiological responses to tether housing by subjecting experimental animals to acute-type stressors of different intensity and duration. In addition, it becomes possible to subject animals repeatedly to one type of stressor and then analyse how they respond to a different type of stressor. More importantly, it is possible to vary the interval between stressor exposure, the so-called interstressor interval (De Boer *et al.*, 1990).

I believe (although I have no proof of it) that the time between stress exposures is a very important factor when considering the damaging effects of stress. During the intermissions between stress exposures the organism recuperates. Cholesterol stores build up in the adrenals, enzymes and receptors are synthesized, and the systems are replenished and prepared for the next onslaught. If so, we should expect that the repeated stressor exposure becomes less damaging to the organism the longer the interstressor interval or, stated differently, any stressor may exert a negative effect on the organism if it is applied often enough.

In the past it has often been discussed whether stressors should be divided into 'good' and 'bad' stressors (or, as Selye (1979) called them, 'eustress' and 'distress'). A common argument has been that a male animal may show a similar corticosteroid response to mating as it shows to restraint. However, if the interstressor interval is of importance, making such a distinction between stressors becomes unnecessary in that any stressor (including frequent mating) can have a detrimental effect on the organism if it occurs often enough.

Stress response systems

Subjecting experimental animals to repeated acute stressors enables us to analyse changes in several different stress response systems simultaneously as well as to analyse the interaction between systems. Such systems may include the 'traditional' ones (the hypothalamic–pituitary–adrenal system and the autonomic nervous system), but also less 'traditional' ones, such as the immune system, the opioid system and various neurotransmitters. In light of the possible importance of the interstressor interval, the interaction between stress hormones and the immune system is of major interest. In this connection, it is important to point out a further complication that we need to consider in this type of research, namely that different stress response systems may differ in their manner of response alteration. One system may become sensitized concurrently with another stress system becoming desensitized. This discrepancy might depend on the type of stressor. In a recent study on the effect of repeated social isolation in pigs, it was found that the ACTH and cortisol responses gradually diminished with repeated stressor exposure, whereas the adrenaline, noradrenaline and heart rate responses remained unchanged (Schrader and Ladewig, 1999). Furthermore, the interaction between different response systems may change with repeated stressor exposure. For instance, in pigs submitted to 15 min nose snare restraint daily over 9 days it was found that naloxone enhanced the ACTH, cortisol and prolactin increase during restraint on the first day but had less effect during restraint on the 9th day, indicating different opioid modulation at the beginning and end of the experiment (Rushen *et al.*, 1993).

Sensitization versus desensitization

Repeated exposure to a stressor can result in alteration of the response over time. The response can increase in intensity, a process often called sensitization (Fig. 8.2C), or it can decrease in intensity, a process called desensitization (or habituation, see later) (Fig. 8.2A and B). The third possibility is, of course, that no alteration of the response occurs. Whether an increase or a decrease in response occurs depends on various factors. A highly intense stressor is more likely to result in sensitization than a stressor of low intensity (Konarska *et al.*, 1990). Imagine an animal subjected repeatedly to a painful stimulus compared with an animal subjected repeatedly to a novel environment. Behaviourally, the former animal will become increasingly anxious, whereas the latter will gradually respond less intensely. Furthermore, it has been shown that desensitization to one type of repeated stressor does not extend to a different type of stressor (a so-called heterotypic stressor), i.e. desensitization is stressor specific (Kant *et al.*, 1985). In addition, there are indications that exposure to a stressor of low intensity will cause desensitization to that particular stressor, but sensitization to another type of stressor. One indication of this phenomenon is the fact that pigs subjected to tethering show a different cortisol reaction to exogenous ACTH from that of loose housed pigs (e.g. Von Borell and Ladewig, 1989).

Interestingly, the process of desensitization is an important aspect of behaviour modification therapy in both human and veterinary behaviour therapy. Desensitization of, for instance, a fear response to a certain stimulus (or stressor) can be achieved either by repeated exposure to the stimulus at full intensity (the so-called flooding approach) or by repeated exposure to the stimulus with gradually increasing intensity (also called habituation by successive approximation, Hart and Hart, 1985) (Fig. 8.2G).

Desensitization versus normalization

A further complication when considering desensitization is that it is possible for a stress system to stop responding, not because of actual adaptation (i.e. at the cognitive level) but because of some mechanism that suppresses the response. We have shown that increased cortisol secretion in bulls due to tethering disappears within 1 month and that the basal episodic secretory pattern returns to normal, despite the fact that the bulls do not show any behavioural adaptation and that they respond differently to an ACTH challenge than do control bulls (Ladewig and Smidt, 1989). Considering the dramatic effects corticosteroids have on many body functions, it is not surprising that powerful mechanisms have evolved that dampen excess corticosteroid release.

The changes in stress responses after repeated stressor exposure may thus involve different mechanisms, such as central biochemical changes (learning, changes in neurotransmitters or their receptors, e.g. whether the stressor is predictable or unpredictable; De Boer *et al.*, 1989) or peripheral biochemical changes (e.g. altered biosynthesis and storage of hormones or neurotransmitters, receptor modulation in target organs; De Boer *et al.*, 1988). The diversity of these changes leads to yet another problem that needs to be approached, namely that of terminology. Should we call a reduced response to a repeated acute type stressor a desensitization, a normalization or an adaptation? Obviously, we need further detailed information about the processes involved before this question can be answered. Considering the importance of these phenomena in the welfare debate, however, the answers are urgently needed. It would be a grave fault to claim that an animal had adapted to a specific housing system if its lack of a stress response was, in fact, due to receptor down-regulation in the adrenal gland and not due to changes (adaptation) at the cognitive level.

Sequence of sensitization and desensitization

When an animal is subjected to a repeated stressor, the first few times it usually shows an increasing response. Not until later does the response decrease with each stressor exposure. In other words, sensitization is followed by desensitization. Although chronic intermittent stress has not yet been studied over a sufficiently long period, we have reason to believe that these shifts in sensitization and desensitization may continue. If so, it means that after the first sensitization followed by desensitization, a further period of sensitization may follow later on, and so forth (Schwarze *et al.*, 1992).

Another aspect of chronic intermittent stress that must be considered is the possible impact of frequent stress responses on the diurnal cycle of hormone secretion. Thus, unpredictable, inescapable electroshocks administered intermittently to fattening pigs over a 31 day period cause a normalization of baseline ACTH secretion but a shift in the time course of diurnal ACTH secretion (Jensen *et al.*, 1996).

Towards behavioural and physiological tests as a diagnostic tool

The effect of repeated exposure to one type of stressor and its effect on the response pattern to another type of stressor is an area that has been studied in some detail. One example of such studies is the stimulation of the adrenal glands of experimental and control animals with a standard dosage of synthetic ACTH (the ACTH challenge test) and comparison of the corticosteroid response of the two groups of animals. Another example is studies in which differently treated animals are subjected to a 'standard stressor' and the resulting response analysed (e.g. Jensen *et al.*, 1995). Applying the model of repeated acute-type stressor exposure and control of (or controlled variation of) stressor intensity, stressor duration and interstressor interval will enable a much more detailed analysis of the biochemical changes involved in the response alteration. It is not unrealistic that, at some point in the future, these changes can be used as a diagnostic tool, and that the welfare status of an animal can be evaluated by, among other things, specific measurements of such biochemical changes.

Conclusion

In order to advance our understanding of long-term stress, a useful way to proceed is to use the model of repeated stress. The model enables us to analyse systematically the effects of the intensity and duration of each stressor and of the interstressor interval on the sensitization–desensitization processes over time. It further enables us to analyse the responses of the various systems (the hypothalamic–pituitary–adrenal axis, the autonomic nervous system, the immune system, the opioid system, serotonin and other neurotransmitters) as well as behaviour, and enables us to compare such responses in different species. By studying stress in such detail, we may be able to develop specific procedures by which it will be possible to diagnose 'stress'. Only if we have a specific diagnosis will we be able to institute a therapy and thus improve the welfare of animals kept under intensive conditions.

References

Burchfield, S.R. (1979) The stress response: a new perspective. *Psychosomatic Medicine* 41, 661–672.

De Boer, S.F., Slangen, J.L. and van der Gugten, J. (1988) Adaptation of catecholamine and corticosterone responses to short-term repeated noise stress in rats. *Physiology and Behavior* 44, 273–280.

De Boer, S.F., van der Gugten, J. and Slangen, J.L. (1989) Plasma catecholamine and corticosterone responses to predictable and unpredictable noise stress in rats. *Physiology and Behavior* 45, 789–795.

De Boer, S.F., Koopmans, S.J., Slangen, S.L. and van der Gugten, J. (1990) Plasma catecholamine, corticosterone and glucose responses to repeated stress in rats: effect of interstressor interval length. *Physiology and Behavior* 47, 1117–1124.

Hart, B.L. and Hart, L.A. (1985) *Canine and Feline Behavioral Therapy.* Lea and Febiger, Philadelphia.

Jensen, K.H., Pedersen, L.J., Giersing Hagelsø, A.M., Heller, K.E., Jørgensen, E. and Ladewig, J. (1995) Intermittent stress in pigs: behavioural and pituitary-adrenocortical reactivity. *Acta Agricultura Scandinavia, Section A, Animal Science* 45, 276–285.

Jensen, K.H., Pedersen, L.J., Nielsen, E.K., Heller, K.E., Ladewig, J. and Jørgensen, E. (1996) Intermittent stress in pigs: effects on behavior, pituitary-adrenocortical axis, growth, and gastric ulceration. *Physiology and Behavior* 59, 741–748.

Kant, G.J., Eggleston, T., Landman-Roberts, L., Kenion, C.C., Driver, G.C. and Meyerhoff, J.L. (1985) Habituation to repeated stress is stressor specific. *Pharmacology, Biochemistry and Behavior* 22, 631–634.

Konarska, M., Stewart, R.E. and McCarty, R. (1990) Habituation and sensitization of plasma catecholamine responses to chronic intermittent stress: effects of stressor intensity. *Physiology and Behavior* 47, 647–652.

Ladewig, J. (1994) Stress. In: Döcke, F. (ed.) *Veterinärmedizinische Endokrinologie,* 3rd edn. Verlag Gustav Fischer, Jena, pp. 379–398.

Ladewig, J. and Smidt, D. (1989) Behavior, episodic secretion of cortisol, and adrenocortical reactivity in bulls subjected to tethering. *Hormones and Behavior* 23, 344–360.

Ladewig, J. and Stribrny, K. (1988) A simplified method for the stress free continuous blood collection in large animals. *Laboratory Animal Science* 38, 333–334.

Müller, C., Ladewig, J., Thielscher, H.H. and Smidt, D. (1989) Behavior and heart rate of heifers housed in tether stanchions without straw. *Physiology and Behavior* 46, 751–754.

Pougin, M. (1981) Zur Anpassung von Jungrindern an die Spaltenbodenhaltung aus etologischer Sicht. *Kuratorium für Technik und Bauwesen in der Landwirtschaft* 281, 32–45.

Rushen, J. (1991) Problems associated with the interpretation of physiological data in the assessment of animal welfare. *Applied Animal Behaviour Science* 28, 381–386.

Rushen, J., Schwarze, N., Ladewig, J. and Foxcroft, G. (1993) Opioid modulation of the effect of repeated stress on ACTH, cortisol, prolactin, and growth hormone in pigs. *Physiology and Behavior* 53, 923–928.

Schrader, L. and Ladewig, J. (1999) Temporal differences in the responses of the pituitary adrenocortical axis, the sympathoadrenomedullar axis, heart rate, and behavior to a chronic intermittent stressor in domestic pigs. *Physiology and Behavior* 66, 775–783.

Schwarze, N., Ladewig, J. and Smidt, D. (1992) Chronisch intermittierender Stress – Bedeutung für Verhalten und Haltung von Schweinen. *Kuratorium für Technik und Bauwesen in der Landwirtschaft eV (KTBL)* 351, 149–157.

Selye, H. (1979) The stress concept and some of its implications. In: Hamilton, V. and Warburton, D.M. (eds) *Human Stress and Cognition: An Information Processing Approach.* Wiley, New York.

Von Borell, E. and Ladewig, J. (1989) Altered adrenocortical response to acute stressors or ACTH(1–24) in intensively housed pigs. *Domestic Animal Endocrinology* 6, 299–309.

Von Borell, E. and Ladewig, J. (1992) The relationship between behaviour and adrenocortical response pattern in domestic pigs. *Applied Animal Behaviour Science* 34, 195–206.

Zanchetti, A. (1972) Expectancy and the pituitary–adrenal system. In: *Physiology, Emotion and Psychosomatic Illness.* Ciba Foundation Symposium 8. Elsevier, Amsterdam.

Quantifying Some Responses to Pain as a Stressor

9

D.J. Mellor,[1] C.J. Cook[3] and K.J. Stafford[2]

[1]Animal Welfare Science and Bioethics Center, Institute of Food, Nutrition and Human Health and [2]Animal Welfare Science and Bioethics Centre, Institute of Veterinary Animal and Biomedical Science, Massey University, Palmerston North, New Zealand; [3]Technology Development Group, Horticultural and Food Research Institute of New Zealand, Hamilton, New Zealand

Introduction

The International Society for the Study of Pain has provided a systematic description of pain (Table 9.1). Pain warns animals (including people) that tissue damage might occur, is occurring or has occurred, thereby eliciting or allowing immediate escape, withdrawal or other behaviour. Pain also

Table 9.1. Major attributes of pain (modified from Merskey, 1979).

Attribute	Description
Purpose	Pain is understood to have evolutionary survival value
Detection	Pain sensations depend on activation of a discrete set of receptors (nociceptors) by noxious stimuli
Perception	Further processing via nerve pathways enables the noxious stimuli to be perceived as pain
Character	Pain perception varies according to site, duration and intensity of stimulus and can be modified by previous experience, emotional state and perhaps innate individual differences
Definition	Pain is defined as an unpleasant sensory and emotional experience associated with actual or potential tissue damage, or is describable in terms of such damage
Variation	The pain detection threshold is apparently uniform across species, whereas the pain tolerance threshold may be more species specific and subject to modification

(eds G.P. Moberg and J.A. Mench)

creates the opportunity for the individual to learn to avoid, if possible, similar pain-causing circumstances in future. Specific pain receptors detect harmful or potentially harmful (noxious) stimuli that can cause tissue damage. Impulse processing via nerve pathways in the spinal cord and the lower and higher centres of the brain converts or transduces nociceptor input into perceived pain. The character of perceived pain varies according to the features of the nociceptor input (site, duration, intensity) and according to other factors that can affect the way in which the central nervous system processes that input (experience, emotional state, individual variation). Pain is an unpleasant, subjective experience usually linked to tissue damage. Across species the pain detection apparatus is apparently equally sensitive, but pain tolerance may vary.

Tissue damage causing pain may occur through trauma or disease which, as undesired states, are avoided if possible. However, husbandry practices causing tissue damage are used on farms because they are considered to be an essential part of managing an efficient and humane livestock enterprise (Stafford and Mellor, 1993). The routine use of these practices provides two opportunities: first, to assess the pain caused by these practices and to devise strategies for alleviating it (Stafford and Mellor, 1993; Molony and Kent, 1997); and second, to improve our knowledge of pain and its control by using such livestock to model physiological and behavioural responses without the need deliberately to inflict pain on other animals (Mellor and Murray, 1989a; Molony and Kent, 1997). While these practices continue to be used routinely, therefore, they offer the opportunity to study acute and chronic responses to injuries caused by cautery, cryocautery, cutting, crushing, constriction and corrosion of ears, skin, bone, horn, scrotum, testes and/or tail (Mellor and Stafford, 1999). Likewise, the clinical setting offers the opportunity to study pain and its control in companion animals (e.g. Sawyer and Rech, 1987; Taylor, 1989, 1990; Flecknell, 1994; Fox *et al.*, 1994, 1998; Lascelles *et al.*, 1994, 1995; Hansen *et al.*, 1997).

The main purpose of this chapter is to outline a range of principles and caveats to guide the quantitative and qualitative evaluation of physiological and behavioural responses to painful stimuli. However, before proceeding it is necessary to distinguish between the terms 'stress', 'distress' and 'pain-induced distress'. We assign distinct meanings to the words 'stress' and 'distress'. 'Stress' responses refer to the full range of *physiological reactions*, from the small deviations just beyond everyday homeostatic adjustments seen under benign circumstances to the maximum physiological changes of which body systems are capable under extreme challenge. The magnitudes of these physiological stress responses, indicated by changes in measured variables, can often be described using qualitative terms such as 'minor', 'moderate', 'marked' and 'very marked'. In contrast, we use the word 'distress' to acknowledge the *emotional content* of noxious experiences that elicit physiological stress responses in animals, whether that noxiousness is predominantly emotional (e.g. fear), predominantly physical (e.g. vigorous exercise) or a combination of both (e.g. pain). The levels of 'distress' are assessed by reference to the same variables as are used to assess physiological stress, and are therefore also describable as 'minor', 'moderate', 'marked' and 'very marked'. However, it is noteworthy that any conclusions reached about the subjective content of noxious experiences remain educated judgements and cannot be regarded as statements of fact. The term 'pain-induced distress', whether referring to the consequences of injury caused by husbandry practices, or in clinical settings, is used to indicate that the physiological responses reflect the interacting emotional and physical facets of the noxious experience.

How useful are neurohumoral variables for monitoring pain-induced distress?

Recent advances in methodology have made it possible to monitor central nervous system (CNS) neurohumoral function in conscious, freely behaving animals (Cook 1997a, b). While this has provided a greater understanding of pain, whether it will also be useful for monitoring pain-induced distress remains to be seen. As with changes in plasma cortisol concentration (see below), neurohumoral variables do not measure pain as such but can assist understanding and interpretation of a continuum of negative states ranging in intensity from minor to very marked distress.

Neurohumoral approaches have provided the foundation for our current understanding of pain and of rationales for its pharmacological alleviation (Melzack and Wall, 1965; Kuhar *et al.*, 1973; Hughes *et al.*, 1975; Johnson, 1989). Moreover, the known stereochemical and dose–response relationships between analgesics and analgesia offer promise for monitoring purposes. At the level of the spinal cord a range of effects on pain sensation or perception have been attributed to the amino acid neurotransmitters glutamate and gamma amino-4-butyric acid (GABA), purines (adenosine, adenosine triphosphates) and the opioid peptides (Keil and De Lander, 1996; Dickenson *et al.*, 1997). Noradrenaline, dopamine, serotonin, tachykinins (substance P and neurokinin A), prostaglandins, bradykinins and histamine also show profiles of release associated with pain, as do more 'exotic' peptides such as pituitary adenylate cyclase-activating peptide (Yang *et al.*, 1996; Zhang *et al.*, 1996, 1997; Clauw and Chrousos, 1997; Henderson and McKnight, 1997; Xu *et al.*, 1997).

These neurohumoral factors clearly have broad roles in discrimination of sensory input in addition to pain (Keil and De Lander, 1996) and in the setting of spinal cord responsiveness and wide dynamic range neuronal behaviour (Filaretov *et al.*, 1996; Urban and Nagy, 1997). By measuring the changes in a number of the above-mentioned neurohumoral substances we may be able to discriminate between different types of pain-induced distress. For example, glutamate acting at *N*-methyl-D-aspartate (NMDA) receptors appears to be strongly implicated in generation and maintenance of spinal states of hypersensitivity, and this may underlie 'wind-up' – a phenomenon in which nociceptor impulse transmission is potentiated such that physiological and behavioural responses to pain are exaggerated in relation to the stimulus. In inflammatory states 'wind-up' appears to be inhibited, but with neuropathy it is not (Dickenson *et al.*, 1997). Assessment at this level may offer opportunities to quantify distress, irrespective of the stimulus.

At a higher level of the CNS, to talk of a 'pure' pain CNS response may be meaningless. Both acute and chronic stress or pain can induce affective states of anxiety or depression, and the contribution of the CNS to these states may overshadow any discriminatory specificity to pain or stress (Filaretov *et al.*, 1996; Urban and Nagy, 1997). Again the use of a mixture of CNS effects may provide discriminatory power under suitable experimental conditions. This approach could include magnetic resonance or positron emission imaging techniques to assess activated brain regions (Eldeman, 1990; Gyulai *et al.*, 1997), monitoring of evoked potentials in the CNS (Crawford *et al.*, 1998) and monitoring changes in corticotropin-releasing factor (CRF), urocortin, orphanin and nociceptin, oxytocin, glutamate, GABA, serotonin, catecholamines and opioid-like peptides (Matsumoto *et al.*, 1996; Betancur *et al.*, 1997; Henderson and McKnight, 1997; Jensen, 1997; Hao *et al.*, 1998; Helmstetter *et al.*, 1998). The assessment of specific changes within areas of the CNS linked to analgesic control may also be useful (Helmstetter *et al.*, 1998).

An understanding of longer term changes in areas such as the periaqueductal gray, in receptor function and genome expression (e.g. c-*fos* and c-*jun*), may help to assess chronic pain. However, our technical capacity to monitor these changes currently outstrips understanding of the phenomena revealed.

How useful are hormonal variables for monitoring pain-induced distress?

More familiar variables have been used to assess the levels of distress apparently experienced by animals exposed to noxious situations (Table 9.2). These variables are usually direct or indirect measures of the activities of the sympathetic adrenomedullary system, which is primarily concerned with fast-acting 'fight–fright–flight' responses, and of the hypothalamic–pituitary–adrenocortical (HPA) system, which initiates more protracted metabolic and anti-inflammatory responses that promote healing. Reports on catecholamine responses to a range of stressors including isolation, weaning, cold exposure, herding, handling, transport, branding and slaughter are available (e.g. Graham *et al.*, 1981; Mitchell *et al.*, 1988; Hattingh *et al.*, 1989; Lay *et al.*, 1992; Parrott *et al.*, 1994; Lefcourt and Elsasser, 1995), but the most commonly used indices are those representing changes

Table 9.2. Some physiological and behavioural indices of distress responses to noxious stimuli in ruminants and other species.

Physiological indices	Behavioural indices
Blood hormone concentrations	Vocalization
adrenaline	whimpers, howls, growls,
noradrenaline	screams, grunts, moans, squeaks,
corticotropin-releasing factor	squeals, chirps, silent
adrenocorticotropic hormone	Posture
glucocorticoids (e.g. cortisol)	cowers, crouches, huddled, hiding,
prolactin	lying (legs extended, all or some legs
Blood metabolite concentrations	tucked in), standing (on all or not all
glucose	legs, rigid, head against wall, drooping)
lactic acid	Locomotion
free fatty acids	reluctant to move, awkward, shuffles,
β-hydroxybutyrate	staggers, falls, stands up /lies down
Other variables	repeatedly, circles, escape/avoidance
heart rate	movements, pacing, restless, writhing
breathing (rate and depth)	Temperament
packed cell volume	withdrawn, depressed, quiet, docile,
sweat production	miserable, agitated, anxious,
muscle tremor	frightened, terrified, aggressive
body temperature	
plasma α-acid glycoprotein levels	
blood leukocyte levels	
cellular immune responses	
humoral immune responses	

Sources: Stephens (1980), Dantzer *et al.* (1983), Duncan and Dawkins (1983), Laden *et al.* (1985), Morton and Griffiths (1985), Sanford *et al.* (1986), Sawyer (1988), Griffin (1989), Lay *et al.* (1992), Flecknell (1994), Parrott *et al.* (1994), Carragher *et al.* (1997a, b), Waas *et al.* (1997).

in the activity of the HPA system, i.e. concentrations of CRF, adrenocorticotropic hormone (ACTH), and especially cortisol, the use of which is emphasized here. The four questions posed below to aid the present consideration of cortisol as an index of pain-induced distress are equally applicable to assessments of the value of other physiological indices.

What do changes in plasma cortisol concentration represent?

Changes in plasma cortisol concentration appear to be particularly useful as an index of acute distress, whether it is pain-induced or not, because the activity of the HPA system increases, often in a graded way, in response to both emotionally and physically noxious experiences (Table 9.3). This

Table 9.3. Noxious, unpleasant or challenging experiences known to stimulate the hypothalamic–pituitary–adrenocortical system.

Physical injuries	Emotional challenges
Branding	Anger/rage
cautery (hot iron)	Anxiety/fear
cryocautery (freezing)	Anticipating/remembering challenge
Burns	Electroimmobilization
Castration	Strange environments/isolation
cutting (knife)	Unusual handling/restraint
constriction (rings)	Shearing
clamp (e.g. Burdizzo)	Mustering/yarding/barking dogs
chemical	Transport, loading and unloading
Disbudding, cautery (hot iron)	Predator–prey interactions
Dehorning, cutting (amputation)	Social dominance expression
Mulesing, cutting	
Tailing	**Physiological challenges**
cutting (knife)	Extreme cold or heat
constriction (rings)	Hypotension
cautery (docking iron)	Hypoxaemia
Tooth grinding	Vigorous exercise
Surgical injuries, postanaesthetic	Metabolic disease
Other physical injuries	pregnancy
Some disease states	toxaemia

Sources: Alvarez and Johnson (1973), Johnston and Buckland (1976), Pearson and Mellor (1975, 1976), Graham *et al.* (1981), Fulkerson and Jamieson (1982), Moberg *et al.* (1980), Eales and Small (1986), Fell *et al.* (1986), Jephcott *et al.* (1986), Kent and Ewbank (1986), Macaulay and Friend (1987), Shutt *et al.* (1987), Engler *et al.* (1988), Parrott *et al.* (1988, 1994), Boandl *et al.* (1989), Hattingh *et al.* (1989), Herd (1989), Mellor and Murray (1989a, b), Taylor (1989, 1990), Cohen *et al.* (1990), Lester *et al.* (1991a, b, 1996), Mellor *et al.* (1991), Kent *et al.* (1993, 1995, 1998), Fox *et al.* (1994, 1998), Robertson *et al.* (1994), Mellor and Molony (1995), Molony *et al.* (1995, 1997), Morisse *et al.* (1995), Taschke and Folsch (1995), Petrie *et al.*, (1996a, b), Carragher *et al.* (1997a, b), Dinniss *et al.* (1997a, b), Hansen *et al.* (1997), McMeekan *et al.* (1997, 1998a, b), Molony and Kent (1997), Waas *et al.* (1997), Sylvester *et al.* (1998a, b).

non-specificity of the HPA system – its responsiveness in such a wide range of noxious or challenging situations – suggests a common foundation, which adds credibility to the use of cortisol to assess distress.

The pain-induced distress caused by a range of husbandry and clinical practices has been assessed extensively using cortisol (Table 9.3), but it is important to note two points. First, changes in plasma cortisol concentrations do not measure pain as such, but they do provide an indication of the overall noxiousness of the experience which, in the case of pain-induced distress, includes both physical and emotional components. Second, the relatively slow response time of the HPA axis may make it insensitive as a means of discriminating different levels of distress elicited within the first few minutes of a noxious stimulus (Mellor and Stafford, 1997). The physiological changes elicited by the sympathetic-adrenomedullary system (heart rate, plasma concentrations of adrenaline and noradrenaline) may therefore be more useful in assessing the early stages of distress responses (e.g. Graham *et al.*, 1981; Hattingh *et al.*, 1989; Parrott *et al.*, 1994; Lefcourt and Elsasser, 1995).

How can the cortisol distress response best be characterized quantitatively?

Cortisol concentration–time curves derived by repeated blood sampling as the response manifests and recedes remain a major tool for quantifying the response (Fig. 9.1). They allow the magnitude and speed of change, and the duration and pattern of the whole response or each part of it, to be determined. Differences between groups in initial or later concentration changes, peak concentration and time to reach it, and time of return to pretreatment values are informative. Three examples will illustrate this. First, a faster rise to peak cortisol values after surgical or clamp castration of lambs than after ring castration is attributed to a marked nociceptor barrage due to cutting or crushing of tissues compared with a slower onset of intense nociceptor input during the progression from hypoxia to anoxia in the tissues distal to the ring (Lester *et al.*, 1991a; Kent *et al.*, 1993; Cottrell and Molony, 1995; Dinniss *et al.*, 1997a). Second, the virtual absence of differences between the cortisol responses of control and ring-castrated lambs that received local anaesthetic 15–20 min before treatment indicates successful blockade of noxious sensory input from the scrotum and testes (Dinniss *et al.*, 1997a). Third, compared with the cortisol response that follows ring only castration, a lower or earlier cortisol peak and a faster return to pretreatment concentrations seen in 1-week-old lambs when ring castration is followed immediately by application of a castrating clamp across the full width of the scrotum indicates that the clamp disables sensory nerves from the scrotum and testes, thereby minimizing nociceptor input due to tissue hypoxia/anoxia distal to the ring (Kent *et al.*, 1995, 1998).

Cortisol responses vary in complexity. They may be simple (e.g. rising to a peak and then returning to pretreatment values, as is usual with castration and/or tailing of lambs; Fig. 9.1A), or they may be more complex (e.g. first rising to a peak, then declining to a plateau and finally returning to pretreament values, as usually occurs with amputation dehorning of calves; Fig. 9.1B). Further complexity arises when responses include two (or more) peaks. For example, when local anaesthetic is given before amputation dehorning there is an initial small cortisol peak due to the onset of handling,

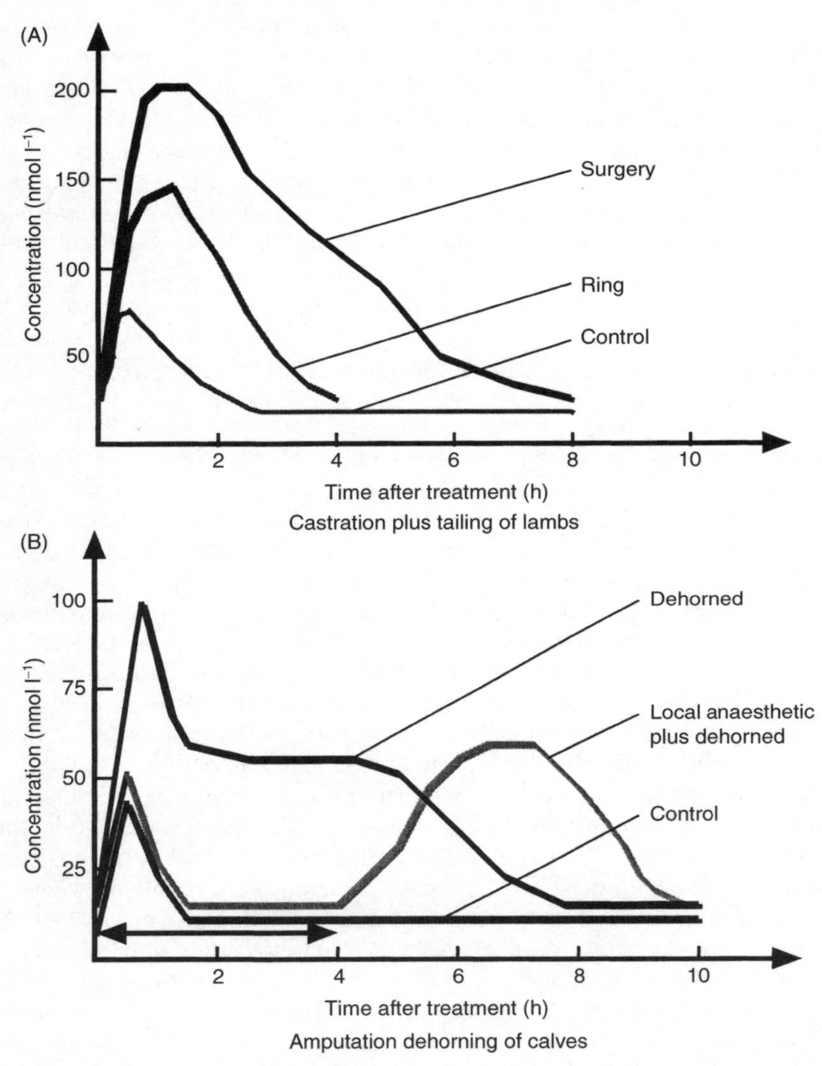

Fig. 9.1. Changes in plasma cortisol concentrations after treatment: (A) surgical or rubber ring castration plus tailing, and control handling (derived from Mellor and Murray, 1989a; Lester *et al.*, 1991a, b); and (B) amputation dehorning with or without local anaesthesia lasting 4 h (double-headed arrow), and control handling (derived from Petrie *et al.*, 1996b; McMeekan *et al.*, 1998a, b; Sylvester *et al.*, 1998a, b).

a return to pretreatment values until the anaesthesia wears off and then a large and protracted rise in cortisol concentration (Fig. 9.1B).

Quantitative tools for characterizing cortisol distress responses include numerical representation of individual facets of the response (e.g. peak height, response duration) and statistical evaluation of concentration–time curves to detect within-group deviations from pretreatment values and between-group differences after treatment (e.g. Kent *et al.*, 1995, 1998; Dinniss *et al.*, 1997a; McMeekan *et al.*, 1998a). There is no single numerical factor that adequately defines distress responses, even simple ones, and it is obvious that the more complex the response, the less likely it is that a single number could represent it effectively. For simple responses, the peak height and the time after treatment to reach the peak, and peak height and response duration are often poorly correlated, so that the use of any one of these parameters may be uninformative or even misleading. The integrated cortisol response (i.e. the area under the cortisol curve that lies above the pretreatment concentration), which includes the response magnitude and duration, can be useful (e.g. Lester *et al.*, 1991b; Dinniss *et al.*, 1997a; McMeekan *et al.*, 1997), but if used alone may mislead as a short-lived large response and a long-duration low response can integrate to a similar number (e.g. Petrie *et al.*, 1996a). Also, it may be better to represent different phases of complex responses by deriving integrated areas for particular parts of the cortisol curve (e.g. Petrie *et al.*, 1996b; McMeekan *et al.*, 1998a). Quantitative definition of cortisol responses may best be achieved, therefore, by using several of these approaches, with the chosen combination depending on the characteristics of the particular response.

How well correlated is the cortisol distress response with the anticipated noxiousness of different procedures?

Cortisol has been used extensively to assess distress because its response magnitude, as indicated by peak height, response duration and/or integrated response, usually accords with the predicted noxiousness of different procedures. For example: (i) tame sheep when handled have lower cortisol responses than do wilder sheep (Pearson and Mellor, 1976); (ii) control and local anaesthetic control treatments cause similar cortisol responses in lambs (Wood *et al.*, 1991; Dinniss *et al.*, 1997a); (iii) control handling, ring tailing and ring castration plus tailing of young lambs cause progressively greater cortisol responses (Mellor and Murray, 1989a), as do control handling, short scrotum creation (ring placed on scrotum distal to testes) and ring castration of young lambs (Molony and Kent, 1997) and of older lambs (Dinniss *et al.*, 1997a); and (iv) in calves, control handling, cautery disbudding (causing localized epidermal and dermal burns around each horn bud) and amputation dehorning (leaving gouge wounds in skull bone and overlying skin) also cause progressively greater cortisol responses (Petrie *et al.*, 1996b).

At each end of the response range, however, interpretation may be less straightforward. At the lower end, for instance, control handling, tailing with a ring and tailing with a docking iron (tail transected by cautery) cause similar cortisol responses in older lambs (Lester *et al.*, 1991a, 1996). This does not mean that tailing by these methods is pain-free. Rather, it suggests that the distress of handling in lambs unfamiliar with it is similar but of a different character from that caused by handling plus tailing, and that under these circumstances the tailing distress dominates the experience, substituting for, rather than adding to, the handling distress (Lester *et al.*, 1991a). Use of the hands-off approaches to determining cortisol responses outlined by Cook *et al.* (Chapter 6, this volume) could help to clarify this issue.

At the upper end of the range, despite the expectation that cortisol responses would progressively increase in proportion to the severity of different treatments, there are several examples where this apparently does not occur. For instance, in lambs, surgical tailing, castration and castration plus tailing all cause near maximum cortisol responses (Lester *et al.*, 1991a), as do ring castration and castration plus tailing (Lester *et al.*, 1991a), and unilateral ring castration, ring castration (both testes) and ring castration plus tailing (Molony and Kent, 1997), contrary to expectations. This may arise because the most painful factor dominates, so that the cortisol responses validly indicate that the overall noxiousness of these linked experiences is similar (Mellor *et al.*, 1991). There is some physiological support for this notion (Sapolsky and Meaney, 1986; Dallman *et al.*, 1987). It may also arise because there is a 'ceiling effect' on cortisol responses (Molony and Kent, 1997) such that maximum cortisol secretion occurs at submaximal levels of noxiousness, leading to underestimates of the negative effects of the more invasive treatments. Although a 'ceiling effect' is of significance theoretically, in welfare terms it is less consequential because most cortisol responses near the maximum give cause for concern, and practices causing them would be targets for review.

It is noteworthy that total plasma cortisol concentrations have been measured in most studies to date, there being an implicit assumption that free and total cortisol levels change in parallel. Although this is likely to be true over the full range of concentrations, deviations from parallelism within small segments of the full range, if they occur, could contribute to 'ceiling effects' and also raise the possibility that animals experiencing low-level distress might exhibit near baseline total cortisol concentrations although their free cortisol levels remain elevated.

What are the implications of within-group variability in cortisol distress responses?

Wide variation in cortisol responses to treatment is common (e.g. Kent *et al.*, 1993; Petrie *et al.*, 1996a; Dinniss *et al.*, 1997a). As this variability rises our capacity to detect small, between-group differences diminishes and greater numbers of animals are required in each group. Variation in pretreatment plasma cortisol concentrations within the non-stressed range is also usual. Such individual differences are often allowed for by subtracting the pretreatment value from all subsequent concentrations and presenting results as changes in concentration. A small proportion of animals have high pretreatment cortisol concentrations, indicating significant prior distress, and are usually excluded from subsequent data analysis because their responses to control or noxious treatments are distorted by steeply declining baseline values as the effects of the prior distress wane. Another small proportion of animals have pretreatment values in the unstressed range but exhibit exaggerated cortisol responses. Their responses are sometimes excluded to avoid distortion of mean results, but this is hard to justify when data from low responders (i.e. no greater than control responses) are included. All such exclusions need to be reported. Finally, while it may be acceptable within groups to distinguish between animals with responses that are no greater than the greatest control response and those with larger responses (Petrie *et al.*, 1996a), a division of greater than control responses into low and high types (Kent *et al.*, 1993), which will almost invariably produce significant differences, is harder to justify.

Quantitative age-related differences have been found in cortisol responses of lambs to castration and/or tailing by particular methods (Mellor and Murray, 1989b; Kent *et al.*, 1993). However, as with between-species differences in cortisol distress responses (see below), these age effects are difficult to interpret because of the potential for differences in the operational dynamics of inputs to and outputs from the HPA axis (Mellor and Murray, 1989b).

How useful is behaviour for monitoring pain-induced distress?

A major advantage of behaviour, whether manifesting a rapid or slow onset, is that it is immediately seen, thereby allowing speedy assessment, unlike plasma parameters which, even if they reflect rapidly responding physiological processes, are typically not measured until some time after an event.

Behavioural responses to pain have been noted as achieving four purposes that enhance an animal's chances of survival (Molony and Kent, 1997): (i) those often automatic responses that protect the whole animal or parts of it (e.g. withdrawal reflexes); (ii) those that minimize pain and assist healing (e.g. lying and standing still); (iii) those that are designed to elicit help or stop other animals (including people) from inflicting more pain (e.g. communicating vocally, by posture or by other means including smell); and (iv) those that induce learning and, by modifying an animal's behaviour, enable it to avoid recurrence of the noxious experience. The linkages between physiology and behaviour that underlie these responses allow behaviour to be used to assess the significance of noxious experiences for animals. Thus, as in the clinical setting (Morton and Griffiths 1985; Sanford *et al.*, 1986), in the context of farm animal welfare behaviour has been used extensively to assess the responses of calves, deer and lambs to painful or potentially painful husbandry procedures. This experience now allows the strengths and weaknesses of behavioural indices of distress to be reviewed by considering six questions.

Is the behaviour linked explicitly to noxious sensory input?

A behaviour is likely to be a useful index of noxious sensory input leading to pain and distress if it is not seen in control animals but is seen in treated animals while the pain-induced distress manifests and recedes (Mellor and Murray, 1989a; Mellor *et al.*, 1991; Molony *et al.*, 1993; Lester *et al.*, 1996). Further validation is obtained when effective local anaesthesia in treated animals maintains behaviour similar to that seen in controls (Wood *et al.*, 1991; Petrie *et al.*, 1995; Dinniss *et al.*, 1999) or when partly effective local anaesthesia significantly reduces the occurrence of pain-related behaviour (Graham *et al.*, 1997; Molony *et al.*, 1997; Kent *et al.*, 1998). The effectiveness of local anaesthesia can be gauged by 'needle prick' tests and by anaesthetic-induced elimination or reduction of physiological responses (e.g. cortisol) to injury (Wood *et al.*, 1991; Petrie *et al.*, 1996a, b; Dinniss *et al.*, 1997a; Graham *et al.*, 1997; Kent *et al.*, 1998; McMeeken *et al.*, 1998a, b; Sylvester *et al.*, 1998a).

How injury-specific is the behaviour?

Different noxious treatments can elicit unique behavioural responses (Molony *et al.*, 1993, 1995; Lester *et al.*, 1996; Dinniss *et al.*, 1999) because the sensations experienced by the animals are probably not the same when different tissues are damaged or when the same tissues are damaged in different ways (Wood and Molony, 1992; Lester *et al.*, 1996). Thus, with castration and tailing of lambs the dominant behavioural feature following the surgical method is reduced activity or immobility, whereas the constricting rubber ring method initially leads to markedly increased physical activity (Molony *et al.*, 1993; Lester *et al.*, 1996; Dinniss *et al.*, 1999). The surgical removal of tissues and ischaemic death of those same tissues probably cause markedly different noxious sensory inputs.

The existence of unique behavioural responses to different treatments prevents behaviour alone from being used to compare the relative levels of pain-induced distress in the different groups, and questions conclusions drawn on that basis (Mellor and Holmes, 1988). For such behavioural comparisons of distress intensity to be credible, there needs to be a validating continuity between the situations being compared, with the behavioural responses to all of them being part of one continuum (Lester *et al.*, 1996), i.e. the different noxious treatments need to elicit similar behaviours but not necessarily with the same frequency of occurrence. Thus, when behavioural responses to different treatments are injury-specific, it is necessary to monitor physiological distress responses (e.g. cortisol) in order to assess the relative effects of such treatments (Lester *et al.*, 1991a, 1996).

Can behaviour be used to identify different features of the pain-induced distress response?

The features of interest here are the duration and intensity of the response and its different phases. Clearly, the presence of a pain-related behaviour throughout a cortisol distress response can, although a shorter-lived behaviour cannot, be used alone to indicate the *response duration*. However, short-lived behaviours may be used to indicate different phases of a response, and under such circumstances particular sequences of pain-related behaviours can be used to indicate the response duration. Some behaviours (like 'statue standing' and other forms of abnormal standing/walking by lambs after surgical castration and tailing) do occur throughout cortisol distress responses (Lester *et al.*, 1991a, b, 1996). In contrast, other behaviours (like restlessness, then immobile lateral lying while awake, then normal lying while asleep, all elicited by ring castration and tailing of young lambs) are limited to early, middle or later phases of cortisol distress responses (Mellor and Murray, 1989a; Mellor *et al.*, 1991; Molony *et al.*, 1993).

Assessment of between-treatment differences in *response magnitude* using behaviour, even when all noxiously stimulated groups exhibit common behaviours, is difficult for two reasons. First, the recording of behaviour is often intermittent, with each behaviour noted simply as present or absent, so that quantitative representations of behaviour may be imprecise or even distorted. Second, individual quantitative features of a cortisol distress response (peak height, duration of elevated concentration, area under the cortisol response curve) often do not represent the whole response accurately (see above). It is not surprising, therefore, that correlations between the recorded incidences of particular behaviours and different individual quantitative features of the cortisol distress response have rarely been significant (e.g. Lester *et al.*, 1996), but a statistical approach that scores behavioural clusters, as opposed to individual behaviours, shows promise (Fox, 1995; Molony and Kent, 1997). Such scoring is acceptable when done statistically, but there is a danger of exaggerating effects if the incidences of dependent behaviours are summed, for example to give a total activity score (Molony *et al.*, 1997).

How sensitive is the behaviour as an index of pain-induced distress?

A behaviour that occurs only when animals experience pain would presumably be a more sensitive index of pain-induced distress than would a pain-related behaviour that also occurs at lower frequencies in pain-free control animals. For instance, abnormal standing/walking in lambs occurs with a frequency of 40–80% of all behaviours recorded at each time after castration and tailing but also occurs in 10–20% of all recorded behaviours in control lambs (Lester *et al.*, 1996). Such control values represent false-positive results and can arise either because the behaviour is difficult to discriminate or because observers are inexperienced, or both.

There is also the question of whether behaviour is more or less sensitive as an index of pain-induced distress than a physiological parameter like cortisol. This question arises when there are between-group differences in particular behaviours that are common to all groups, but no difference in plasma cortisol concentrations (e.g. Lester *et al.*, 1996; Molony and Kent, 1997). This question is not resolvable because the interpretation depends on an arbitrary choice between emphasizing behaviour or cortisol; when behaviour is emphasized, cortisol is regarded as less sensitive, and vice versa.

How clearly defined is the behaviour?

It is necessary to consider whether or not the features of complex behaviour patterns can be defined or described accurately. Typically, each behaviour is identified by distinct, easily recognized characteristics which are either present or absent or occur at different frequencies that can be quantified. However, care must be taken when the same descriptor is applied to behaviours that are broadly similar but that are nevertheless distinctly different. This can occur when different noxious stimuli elicit the behaviour and when those stimuli change during the development of a distress response. Three examples will illustrate this. First, behaviour identified as *abnormal standing/walking* follows both surgical and ring castration plus tailing of lambs. However, after surgery, the lambs stand still for long periods in characteristic pain-associated postures and lie down rarely, whereas after ring application lambs initially stand up and lie down repeatedly, and when walking regularly do so unsteadily, in circles and with hindleg kicks and jumps (e.g. Molony *et al.*, 1993; Lester *et al.*, 1996). Thus, although standing/walking can be identified as abnormal in both cases, there are evidently at least two types. Second, a *restlessness score* (the number of times a lamb stands up and lies down during a particular period) has been devised and used successfully to quantify the characteristically heightened physical activity of lambs after ring castration plus tailing (e.g. Molony *et al.*, 1993; Lester *et al.*,

1996). However, when the same measurement is used to quantify the characteristic absence of such behaviour after surgical castration plus tailing (e.g. Molony *et al.*, 1993; Lester *et al.*, 1996) it should perhaps be called an *inactivity score* to emphasize that the distinct behavioural responses to these different stimuli (surgery and ring) do not have a common foundation (Lester *et al.*, 1996). Third, *abnormal lying* is commonly observed during the first 1–2 h after ring castration plus tailing of lambs aged less than 1 week (Mellor and Murray, 1989a). However, during the first 45 min, many short periods of abnormal lying and standing alternate during the restlessness which dominates that period, whereas that phase is followed by one of continuous abnormal lying in what may be characterized as pain-induced immobile lying. Simply reporting these occurrences as abnormal lying does not distinguish between these distinct forms of it. It is therefore necessary to describe as fully as practicable the development and waning of the often complex behaviour patterns elicited by noxious stimuli and not simply quantify the incidences of a few broadly defined individual behaviours. In addition, identifying subclasses of particular behaviours, like lateral lying with head up or head down or with rolling and kicking, within the class of abnormal lying (Molony *et al.*, 1993; Robertson *et al.*, 1994) can be more informative.

Finally, it is worth noting that some behavioural effects may be too subtle to detect easily or consistently. This may be more likely when the predominant effect of a noxious stimulus is to decrease an animal's general physical activity or cause postures or activities that occur infrequently to disappear.

Can specific behaviours be assigned to a scale corresponding to different levels of pain?

Such a scaling has been attempted by Molony *et al.* (1993) who advanced six hypotheses regarding possible relationships between the behaviour and pain that follow castration and tailing of lambs. Four of these hypotheses appear to be sound, but the first and fifth cannot be accepted without qualification. Note that these hypotheses are specific to lambs undergoing castration and/or tailing and therefore are not intended as generalizations, even to other pain-producing stimuli in lambs. Nevertheless, they are helpful. Comparable lists would need to be developed for each species and situation.

- 'An increase in restlessness generally indicates an increase in pain.' This is true for individuals when increased restlessness occurs during the onset of tissue hypoxia/anoxia following rubber ring application to the scrotum, and/or tail. However, it is less clear that higher levels of restlessness in one group than in another during similar phases of the response to ring application indicate greater pain in that group. Note also that the subsequent decrease in restlessness does not necessarily indicate a decrease in distress, because with castration or castration plus tailing of young lambs the initial restlessness evolves into a phase of pain-induced immobile lateral lying (Mellor and Murray, 1989a; Mellor *et al.*, 1991).

- 'Lateral recumbency generally indicates more pain than ventral recumbency.'
- 'Extension, rather than flexion, of the hind limb indicates more pain.'
- 'More abnormality of standing and/or walking including ataxia, swaying and falling indicates more pain.'
- 'Standing still or lying still may reduce pain and at any particular time a lamb is considered to be suffering less pain when standing still than when moving abnormally.' This may be true, especially when lambs are standing, but it is also possible that the character and intensity of the pain could immobilize the lambs and that the 'choice' implied in the proposition does not arise. Note also that this proposition will apply to individual lambs treated in particular ways, but not to comparisons of lambs exhibiting different injury-specific behaviours.
- 'Behaviour rarely observed in the control group can be referred to as abnormal.'

What else can behaviour be used to monitor?

Behavioural responses during aversion tests allow the noxiousness of various repeatable husbandry practices (e.g. handling, restraint, shearing, electroimmobilization, electroejaculation) to be compared (Rushen, 1986; Stafford *et al.*, 1996). Noxious husbandry practices can also be assessed using behavioural tests to monitor elasticity of demand (Matthews and Ladewig, 1994) but care needs to be taken in interpreting the results (Mason *et al.*, 1998). Moreover, as these tests rely upon repeated exposure to husbandry procedures, they are not suitable for practices such as dehorning or castration that damage tissues and are carried out once on each animal. Similar opportunities and constraints attend the use of behavioural methods to monitor the welfare of laboratory animals (Mench, 1998).

How species-specific are corticosteroid and behavioural responses to painful procedures?

Emotionally and/or physically noxious experiences stimulate the HPA axis across a wide range of species leading to acute cortisol responses in those where cortisol is the dominant corticosteroid (e.g. most mammalian species) and to acute corticosterone responses in those where corticosterone predominates (e.g. birds and rodents). This suggests that corticosteroid responses could be used to evaluate stress and/or distress within a wide range of individual species (noting the limits indicated above). However, potential species-specific differences in the operational dynamics of the HPA system, including neural and neurohumoral inputs, hormonal outputs and metabolic clearance rates indicate that interpreting quantitative differences between species in the features of their corticosteriod responses is problematical. In addition, in the absence of direct observations, predicting that qualitative features of responses (e.g. their patterns) to a particular noxious procedure will be similar in different but related species could also be misleading. For instance, cortisol responses of lambs, kids and calves to ring castration are not as quantitatively or qualitatively uniform as might otherwise be anticipated (Mellor *et al.*, 1991; Robertson *et al.*, 1994).

Behavioural responses to pain show marked differences between divergent species, reflecting the unique behavioural repertoire of the group to which each species belongs (e.g. carnivores, herbivores), whereas responses within a group may be more uniform (Sanford *et al.*, 1986). Nevertheless, responses of related species to similar stimuli can also differ. For instance, lambs, kids and calves castrated with rubber rings exhibit different behaviours (Mellor *et al.*, 1991). During the first 30 min, lambs are restless, standing up and lying down frequently; they lie down in lateral recumbency and sometimes roll on their backs. In contrast, calves and kids stand for the first 30 min and if they lie down they do not do so laterally. The differences in behavioural response to ring castration suggest that it is problematical to extrapolate from one species to another regarding expected behavioural responses or the significance of these behaviours. The age of an animal and its experience may also influence behavioural responses to particular procedures. Lambs or calves castrated before weaning would be expected to behave differently from weaned animals, as would hand- versus naturally reared animals. Other factors may also influence behaviour. For instance, the presence of a predator (e.g. dog) can reduce the incidence of lateral recumbency in lambs after ring castration (K.J. Stafford, New Zealand, 1998, personal communication).

Can sex steroids affect cognition and pain-induced distress responses?

We turn now from the quantification of distress responses to the impact on those responses of the hormonal state of the animal, with particular regard to sex steroids.

Perception of, and relief from, pain differs markedly between individual animals (Bhargava, 1994). Both gender and hormone-related differences have been well documented (Dawson Basoa and Gintzler, 1997; Nakamura *et al.*, 1997), as have the effects of late pregnancy and parturition (Cook, 1997c). There is evidence of hormonal effects whereby oestradiol and testosterone act in opposite directions to influence the presence and function of either kappa or mu opioid receptors, respectively (Cook, 1997c, 1998). The CNS location of these receptors and their relative numbers apparently can influence the overall pattern of response to a nociceptive stimulus (Cook, unpublished data).

During oestrus, late pregnancy, and immediately after birth, nociceptive threshold shows an apparent increase, and in anoestrous-barren females and in castrated males this can be mimicked by injecting oestradiol. In entire males, or in anoestrous-barren females and castrates given testosterone concurrently with oestradiol, this increase is not seen (Cook, 1998). Males and anoestrous, non-pregnant females and non-parturient pregnant females have similar nociceptive thresholds (Wood and Shors, 1998), but analgesic efficacy of different opioids may still differ on the basis of sex (Cook, 1998).

Individual sensitivity to different forms of pain stimuli also varies during the oestrous cycle (Fillingim *et al.*, 1997), which further highlights the complexity of individual differences. Neurotransmitter changes apparently occur during the oestrous cycle and in response to oestradiol administration (Fink *et al.*, 1996; Sumner and Fink, 1997), and changes in stress and anxiety perception and response have also been documented (Mora *et al.*, 1996; Cook, 1997d; Neumann *et al.*, 1998). Moreover, cognitive ability and perceived mental state vary with hormonal changes in people (Fink and Sumner, 1996), as does learning ability in rats (Wood and Shors, 1998). These observations raise the question of whether changes in pain perception are due to antinociceptive activity or are a reflection of an overall change in mental state. Studies addressing this question should advance understanding of the 'affective' or cognitive components of animal pain.

There is further evidence that sex steroids can influence CNS activity. Behavioural arousal and responsiveness to stimuli (benign or noxious) in fetal or newborn lambs are increased by oestrogen injections (Mellor *et al.*, 1972) and by lowering plasma progesterone concentrations (Nicol *et al.*, 1997), and are inhibited by progesterone injections (Crenshaw *et al.*, 1966) and by increasing progesterone concentrations (Nicol *et al.*, 1997). Moreover, judged behaviourally and physiologically (e.g. Pearson and Mellor, 1976), an adult ewe's perception of threat, which presumably reflects some components of her cognitive state, can also apparently be affected by sex steroids: (i) mid-pregnant ewes that have high circulating progesterone concentrations and low oestrogen levels (Mellor *et al.*, 1987) become tame more rapidly than do concurrently handled barren ewes with very low progesterone and oestrogen levels (Pearson and Mellor, 1976; Mellor *et al.*, 1993; unpublished observations); and (ii) barren ewes injected with oestradiol and progesterone nuzzle and lick each other and facilitate handling of the udder by postural changes, whereas ewes treated with progesterone plus cortisol are aggressive to others and vigorously resist attempts to touch their udders (Mellor *et al.*, 1993). Finally, motivation of parturient ewes is also apparently influenced by sex steroids as high circulating oestradiol and progesterone concentrations combined with cervico-vaginal stimulation are major physiological triggers to the onset of mothering behaviour and successful ewe–lamb bonding at birth (Poindron *et al.*, 1988; Kendrick and Keverne, 1991; Mellor *et al.*, 1993).

Differences in pain (and stress) perception and response not related to the individual's sex are also well documented (Sapolsky, 1992; Bellgowan and Helmstetter, 1996; Cook *et al.*, 1996; King *et al.*, 1997). Position within a social hierarchy, previous experiences, ability to learn adaptative behaviours and concurrent stressors all appear to impart variation on to the response seen to a standard stimulus. Such individual differences may hinder interpretation of inter-animal variation in distress response as indicated by markers such as cortisol (Cook, 1996; see above). On the other hand, clarifying the bases of these individual variations may also enable the development of better cognitive and welfare-based models of pain-induced distress.

Summary

Pain as a stressor elicits a range of physiological and behavioural responses. These responses are commonly used to assess the impact of pain-causing stimuli on animals to determine whether or not significant pain is caused and to devise strategies for alleviating pain. We have considered here a range of principles and caveats to guide the quantitative and qualitative evaluation of physiological and behavioural responses to painful stimuli. Although most examples are drawn from studies of farm animals responding to tissue removal or damage, the guidance provided is more generally applicable. The potential for neurohumoral variables to be used to improve understanding and to monitor responses to painful stimuli was addressed first. Then the use of cortisol to monitor pain-induced distress was considered with regard to what cortisol responses represent, how those responses can be characterized quantitatively, correlations of responses with the presumed noxiousness of stimuli and the implications of variability in responses. In evaluating behavioural monitoring of pain-induced distress, issues raised were whether observed behaviour is linked to noxious sensory input, is injury-specific, can be used to identify different features of distress responses, is a sensitive index of pain-induced distress, is clearly defined, and can be scaled according to different levels of pain. Finally, the species specificity of corticosteroid and behavioural responses to pain was outlined, as was evidence for the impact of sex steroids on cognition and pain-induced distress responses.

References

Alvarez, M.B. and Johnson, H.D. (1973) Effects of environmental heat exposure on cattle plasma catecholamine and glucocorticoids. *Journal of Dairy Science* 56, 189–194.

Bellgowan, P.S. and Helmstetter, F.J. (1996) Neural systems for the expression of hypoalgesia during nonassociative fear. *Behavioural Neuroscience* 110, 727–736.

Betancur, C., Azzi, M. and Rostene, W. (1997) Nonpeptide antagonists of neuropeptide receptors; tools for research and therapy. *Trends in Pharmacological Sciences* 18, 372–383.

Bhargava, H.N. (1994) Diversity of agents that modify opioid tolerance, physical dependence, abstinence syndrome and self administrative behavior. *Pharmacological Reviews* 46, 293–324.

Boandl, K.E., Wohlt, J.E. and Carsia, R.V. (1989) Effects of handling, administration of a local anesthetic and electrical dehorning on plasma cortisol in Holstein calves. *Journal of Dairy Science* 72, 2193–2197.

Carragher, J.F., Ingram, J.R. and Matthews, L.R. (1997a) Effects of yarding and handling procedures on stress responses of red deer stags (*Cervus elaphus*). *Applied Animal Behaviour Science* 51, 143–158.

Carragher, J.F., Knight, T.W., Death, A.F., Fisher, A.D. and Matthews, L.R. (1997b) Measures of stress and growth suppression in surgically castrated bulls. *Proceedings of the New Zealand Society of Animal Production* 57, 100–104.

Clauw, D.J. and Chrousos, G.P. (1997) Chronic pain and fatigue syndromes; overlapping clinical and neuroendocrine features and potential pathogenic mechanisms. *Neuroimmunomodulation* 4, 134–153.

Cohen, R.D.H., King, B.D., Thomas, L.R. and Janzen, E.D. (1990) Efficacy and stress of chemical versus surgical castration of cattle. *Canadian Journal of Animal Science* 70, 1063–1072.

Cook, C.J. (1996) Basal and stress response cortisol levels and stress avoidance learning in sheep. *New Zealand Veterinary Journal* 44, 162–163.

Cook, C.J. (1997a) Real-time measurement of corticosteroids in conscious animals using an antibody-based electrode. *Nature Biotechnology* 15, 467–472.

Cook, C.J. (1997b) Real-time extracellular measurement of neurotransmitters in conscious sheep. *Journal of Neuroscience Methods* 72, 161–166.

Cook, C.J. (1997c) Sex-related differences in analgesia in sheep. *New Zealand Veterinary Journal* 45, 169–170.

Cook, C.J. (1997d) Oxytocin and prolactin suppress cortisol responses to acute stress in both lactating and non-lactating sheep. *Journal of Dairy Research* 64, 327–339.

Cook, C.J. (1998) Steroidal hormones determine sex-related differences in opioid-induced elevation of nociceptive threshold in sheep. *New Zealand Veterinary Journal* 46, 68–71.

Cook, C.J., Maasland, S.A. and Devine, C.E. (1996) Social behaviour in sheep relates to behaviour and neurotransmitter responses to nociceptive stimuli. *Physiology and Behavior* 60, 741–751.

Cottrell, D.F. and Molony, V. (1995) Afferent activity in the superior spermatic nerve of lambs – the effects of application of rubber castration rings. *Veterinary Research Communications* 19, 503–515.

Crawford, H.J., Knebel, T., Kaplan, L., Vendemia, J.M., Xie, M., Jamison, S. and Pribram, K.H. (1998) Hypnotic analgesia: 1. Somatosensory event-related potential changes to noxious stimuli and 2. Transfer learning to reduce chronic low back pain. *International Journal of Clinical and Experimental Hypnosis* 46, 92–132.

Crenshaw, M.C., Meschie, G. and Barron, D.H. (1966) Role of progesterone in inhibition of muscle tone and respiratory rhythm in foetal lambs. *Nature* 212, 842.

Dallman, M., Akana, S., Cascio, C., Darlington, D., Jacobson, L. and Levin, N. (1987) Regulation of ACTH secretion: variations on a theme. *Recent Progress in Hormone Research* 43, 113–127.

Dantzer, R., Mormede, P. and Henry, J.P. (1983) Significance of physiological criteria in assessing animal welfare. In: Smidt, D. (ed.) *Indicators Relevant to Farm Animal Welfare.* Commission of the European Community, Martinus Nijhoff Publishers, Brussels, pp. 29–37.

Dawson Basoa, M. and Gintzler, A.R. (1997) Involvement of spinal cord delta opiate receptors in the antinociception of gestation and its hormonal stimulation. *Brain Research* 757, 37–42.

Dickenson, A.H., Chapman, V. and Green, G.M. (1997) The pharmacology of excitatory and inhibitory amino acid-mediated events in the transmission and modulation of pain in the spinal cord. *General Pharmacology* 28, 633–638.

Dinniss, A.S., Mellor, D.J., Stafford, K.J., Bruce, R.A. and Ward, R.N. (1997a) Acute cortisol responses of lambs to castration using a rubber ring and/or a castrating clamp with or without local anaesthetic. *New Zealand Veterinary Journal* 45, 114–121.

Dinniss, A.S., Stafford, K.J., Mellor, D.J., Bruce, R.A. and Ward, R.N. (1997b) Acute cortisol responses of lambs castrated and docked using rubber rings with or without a castrating clamp. *Australian Veterinary Journal 75*, 494–497.

Dinniss, A.S., Stafford, K.J., Mellor, D.J., Bruce, R.A. and Ward, R.N. (1999) The behaviour of lambs after castration using a rubber ring and/or castrating clamp with or without local anaesthetic. *New Zealand Veterinary Journal* 47, 198–203.

Duncan, I.J.H. and Dawkins, M.S. (1983) The problem of assessing 'well-being' and 'suffering' in animals. In: Smidt, D. (ed.) *Indicators Relevant to Farm Animal Welfare.* Commission of the European Community, Martinus Nijhoff Publishers, Brussels, pp. 13–25.

Eales, F.A. and Small, J. (1986) Plasma hormone concentrations in newborn Scottish Blackface lambs during basal and summit metabolism. *Research in Veterinary Science* 40, 339–343.

Eldeman, R.R. (1990) Magnetic resonance imaging of the nervous system. *Discussions in Neuroscience* 7, 11–63.

Engler, D., Pham, T., Fullerton, M.J., Funder, J.W. and Clarke, I.S. (1988) Studies of the regulation of the hypothalamic-pituitary-adrenal axis in sheep with hypothalamic–pituitary disconnection. I. Effect of an audio-visual stimulus and insulin-induced hypoglycaemia. *Neuroendocrinology* 48, 551–560.

Fell, L.R., Wells, R. and Shutt, D.A. (1986) Stress in calves castrated surgically or by application of rubber rings. *Australian Veterinary Journal* 63, 16–18.

Filaretov, A.A., Bagdanov, A.I. and Yarushkina, N.I. (1996) Stress-induced analgesia. The role of hormones produced by the hypophyseal–adrenocortical system. *Neuroscience and Behavioural Physiology* 26, 572–578.

Fillingim, R.B., Maixner, W., Firdler, S.S., Light, K.C., Harris, M.B., Sheps, D.S. and Mason, G.A. (1997) Ischemic but not thermal pain sensitivity varies across the menstrual cycle. *Psychosomatic Medicine* 59, 512–520.

Fink, G. and Sumner, B.E. (1996) Oestrogen and mental state. *Nature* 383, 306.

Fink, G., Sumner, B.E., Rosie, R., Grace, O. and Quinn, J.P. (1996) Estrogen control of central neurotransmission: effect on mood, mental state, and memory. *Cellular and Molecular Neurobiology* 16, 325–344.

Flecknell, P.A. (1994) Advances in the assessment and alleviation of pain in laboratory and domestic animals. *Journal of Veterinary Anaesthesia* 21, 98–105.

Fox, S.M. (1995) Pain-induced distress and its alleviation using butorphanol after ovariohysterectomy of bitches. PhD thesis, Massey University, New Zealand.

Fox, S.M., Mellor, D.J., Firth, E.C., Hodge, H. and Lawoko, C.R.O. (1994) Changes in plasma cortisol concentrations before, during and after analgesia, anaesthesia and anaesthesia plus ovariohysterectomy in bitches. *Research in Veterinary Science* 57, 110–118.

Fox, S.M., Mellor, D.J., Lawoko, C.R.O., Hodge, H. and Firth, E.C. (1998) Changes in plasma cortisol concentrations in bitches in response to different combinations of halothane and butorphanol, with or without ovariohysterectomy. *Research in Veterinary Science* 65, 125–133.

Fulkerson, W.J. and Jamieson, P.A. (1982) Pattern of cortisol release in sheep following administration of synthetic ACTH or imposition of various stressor agents. *Australian Journal of Biological Science* 35, 215–222.

Graham, A.D., Christopherson, R.J. and Thompson, J.R. (1981) Endocrine and metabolic changes in sheep associated with acclimation to constant or intermittent cold exposure. *Canadian Journal of Animal Science* 61, 81–90.

Graham, M.J., Kent, J.E. and Molony, V. (1997) Effects of four analgesic treatments on the behavioural and cortisol responses of 3-week-old lambs to tail docking. *The Veterinary Journal* 153, 87–97.

Griffin, J.F.T. (1989) Stress and immunity: a unifying concept. *Veterinary Immunology and Immunopathology* 20, 263–312.

Gyulai, F.E., Firestone, L.L., Mintun, M.A. and Winter, P.M. (1997) *In vivo* imaging of nitrous oxide-induced changes in cerebral activation during noxious heat stimuli. *Anesthesiology* 86, 538–548.

Hansen, B.D., Hardie, E.M. and Carroll, G.S. (1997) Physiological measurements after ovariohysterectomy in dogs: what's normal? *Applied Animal Behaviour Science* 51, 101–109.

Hao, J.X., Yu, W. and Xu, X.J. (1998) Evidence that spinal endogenous opioidergic systems control the expression of chronic pain-related behaviours in spinally injured rats. *Experimental Brain Research* 118, 259–268.

Hattingh, J., Ganhao, M. and Kay, G. (1989) Blood constituent responses of cattle to herding. *Journal of the South African Veterinary Association* 60, 219–220.

Helmstetter, F.J., Tershner, S.A., Poore, L.H. and Bellgowan, P.S. (1998) Antinociception following opioid stimulation of the basolateral amygdala is expressed through the periaqueductal gray and rostral ventromedial medulla. *Brain Research* 779, 104–118.

Henderson, G. and McKnight, A.T. (1997) The orphan opioid receptor and its endogenous ligand-nociceptin/orphanin FQ. *Trends in Pharmacological Sciences* 18, 293–300.

Herd, R.M. (1989) Serum cortisol and 'stress' in cattle, *Australian Veterinary Journal* 10, 341–342.

Hughes, J., Smith, T., Kosterlitz, H., Fothergill, L., Morgan, B. and Morris, H. (1975) Identification of two related pentapeptides from the brain with potent opiate agonist activity. *Nature* 258, 577–579.

Jensen, T.S. (1997) Opioids in the brain: supraspinal mechanisms in pain control. *Acta Anaesthesiologica Scandinavica* 41, 123–132.

Jephcott, E.H., McMillen, I.C., Rushen, J., Hargreaves, A. and Thorburn, G.D. (1986) Effect of electroimmobilisation on ovine plasma concentrations of β-endorphin/β-lipotropin, cortisol and prolactin. *Research in Veterinary Science* 41, 371–377.

Johnson, S.M. (1989) Opioid tolerance and dependence. *Pharmacological Reviews* 41, 435–488.

Johnston, J.D. and Buckland, R.B. (1976) Responses of male Holstein calves from seven sires to four management stresses as measured by plasma cortisol levels. *Canadian Journal of Animal Science* 56, 727–732.

Keil, G.J. and De Lander, G.E. (1996) Altered sensory behaviours in mice following manipulation of endogenous spinal adenosine neurotransmission. *European Journal of Pharmacology* 312, 7–14.

Kendrick, K.M. and Keverne, E.B. (1991) Importance of progesterone and estrogen priming for the induction of maternal behavior by vaginocervical stimulation in sheep: effects of maternal experience. *Physiology and Behaviour* 49, 745–750.

Kent, J.E. and Ewbank, R. (1986) The effect of road transportation on the blood constituents and behaviour of cattle. III. Three months old. *British Veterinary Journal* 142, 326–335.

Kent, J.E., Molony, V. and Robertson, I.S. (1993) Changes in plasma cortisol concentration in lambs of three ages after three methods of castration and tail docking. *Research in Veterinary Science* 55, 246–251.

Kent, J.E., Molony, V. and Robertson, I.S. (1995) Comparison of the Burdizzo and rubber ring methods for castrating and tail docking lambs. *Veterinary Record* 136, 192–196.

Kent, J.E., Molony, V. and Graham, M.J. (1998) Comparison of methods for the reduction of acute pain produced by rubber ring castration or tail docking of week-old lambs. *The Veterinary Journal* 155, 39–51.

King, T.E., Hoynes, R.L. and Grau, J.W. (1997) Tail-flick test: II. The role of supraspinal systems and avoidance learning. *Behavioural Neuroscience* 111, 754–767.

Kuhar, M.J., Pert, C.B. and Snyder, S.H. (1973) Regional distribution of opiate receptor binding in monkey and human brain. *Nature* 245, 447–450.

Laden, S.A., Wohlt, J.E., Zajac, P.K. and Carsia, R.V. (1985) Effects of stress from electrical dehorning on feed intake, growth and blood constituents of Holstein heifer calves. *Journal of Dairy Science* 68, 3062–3066.

Lascelles, B.D.X., Butterworth, S.J. and Waterman, A.E. (1994) Postoperative analgesic and sedative effects of carprofen and pethidine in dogs. *Veterinary Record* 134, 187–191.

Lascelles, B.D.X., Cripps, P., Mirchandani, S. and Waterman, A.E. (1995) Carprofen as an analgesic for postoperative pain in cats: dose titration and assessment of efficacy in comparison to pethidine hydrochloride. *Journal of Small Animal Practice* 36, 535–541.

Lay, D.C., Friend, T.H., Bower, C.L., Grissom, K.K. and Jenkins, O.C. (1992) A comparative physiological and behavioural study of freeze and hot-iron branding using dairy cows. *Journal of Animal Science* 70, 1121–1125.

Lefcourt, A.M. and Elsasser, T.H. (1995) Adrenal responses of Angus × Hereford cattle to the stress of weaning. *Journal of Animal Science* 73, 2669–2676.

Lester, S.J., Mellor, D.J. and Ward, R.N. (1991a) Effects of repeated handling on the cortisol responses of young lambs castrated and tailed surgically. *New Zealand Veterinary Journal* 39, 147–149.

Lester, S.J., Mellor, D.J., Ward, R.N. and Holmes, R.J. (1991b) Cortisol responses of young lambs to castration and tailing using different methods. *New Zealand Veterinary Journal* 39, 134–138.

Lester, S.J., Mellor, D.J., Holmes, R.J., Ward, R.N. and Stafford, K.J. (1996) Behavioural and cortisol responses of lambs to castration and tailing using different methods. *New Zealand Veterinary Journal* 44, 45–54.

Macaulay, A.S. and Friend, T.H. (1987) Use of hormonal responses, open-field tests, and blood cell counts of beef calves to determine relative stressfulness of different methods of castration. *Journal of Animal Science* 65 (Suppl. 1), 436.

Mason, G., McFarland, D. and Garner, J. (1998) A demanding task: using economic techniques to assess animal priorities. *Animal Behaviour* 55, 1071–1075.

Matsumoto, K., Mizowaki, M., Suchitra, T., Takayama, H., Sakai, S., Aimi, N. and Watanabe, H. (1996) Antinociceptive action of mitragynine in mice: evidence for the involvement of supraspinal opioid receptors. *Life Sciences* 59, 1149–1155.

Matthews, L.R. and Ladewig, J. (1994) Environmental requirements of pigs measured by demand functions. *Animal Behaviour* 47, 713–719.

McMeekan, C.M., Mellor, D.J., Stafford, K.J., Bruce, R.A., Ward, R.N. and Gregory, N.G. (1997) Effects of shallow scoop and deep scoop dehorning on plasma cortisol concentrations in calves. *New Zealand Veterinary Journal* 45, 72–74.

McMeekan, C.M., Mellor, D.J., Stafford, K.J., Bruce, R.A., Ward, R.N. and Gregory, N.G. (1998a) Effects of local anaesthesia of 4 or 8 hours duration on the acute cortisol response to scoop dehorning in calves. *Australian Veterinary Journal* 76, 281–285.

McMeekan, C.M., Stafford, K.J., Mellor, D.J., Bruce, R.A., Ward, R.N. and Gregory, N.G. (1998b) Effects of regional analgesia and/or a non-steroidal anti-inflammatory analgesic on the acute cortisol response to dehorning in calves. *Research in Veterinary Science* 64, 147–150.

Mellor, D.J. and Holmes, R.J. (1988) Rings *versus* knife for docking and castration of lambs. *Australian Veterinary Journal* 65, 403–404.

Mellor, D.J. and Molony, V. (1995) Castration and/or tail docking of lambs. *Veterinary Record* 137, 227.

Mellor, D.J. and Murray, L. (1989a) Effects of tail docking and castration on behaviour and plasma cortisol concentrations in young lambs. *Research in Veterinary Science* 46, 387–391.

Mellor, D.J. and Murray, L. (1989b) Changes in the cortisol responses of lambs to tail docking, castration and ACTH injection during the first seven days after birth. *Research in Veterinary Science* 46, 392–395.

Mellor, D.J. and Stafford, K.J. (1997) Interpretation of cortisol responses in calf disbudding studies. *New Zealand Veterinary Journal* 45, 126–127.

Mellor, D.J. and Stafford, K.J. (1999) Assessing and minimising the distress caused by painful husbandry procedures in ruminants. *In Practice* 21, 436–446.

Mellor, D.J., Mackay, J.M.K. and Williams, J.T. (1972) Effects of oestrogen on activity and survival of lambs delivered by hysterectomy. *Research in Veterinary Science* 13, 399–401.

Mellor, D.J., Flint, D.J., Vernon, R.G. and Forsyth, I.A. (1987) Relationships between plasma hormone concentrations, udder development and the production of early mammary secretions in twin-bearing ewes on different planes of nutrition. *Quarterly Journal of Experimental Physiology* 72, 345–356.

Mellor, D.J., Molony, V. and Robertson, I.S. (1991) Effects of castration on behaviour and plasma cortisol concentrations in young lambs, kids and calves. *Research in Veterinary Science* 51, 149–154.

Mellor, D.J., Dingwall, R.A. and King, T. (1993) Successful fostering of orphan lambs by nonpregnant ewes induced to lactation using exogenous hormone treatment. *New Zealand Veterinary Journal* 41, 200–204.

Melzack, R. and Wall, P.D. (1965) Pain mechanisms: a new theory. *Science* 150, 971–979.

Mench, J. (1998) Why it is important to understand animal behavior. *Institute of Laboratory Animal Resources Journal* 39, 20–26.

Merskey, H. (1979) Pain terms: a list with definitions and notes on usage. *Pain* 6, 249–252.

Mitchell, G., Hattingh, J. and Ganhao, M. (1988) Stress in cattle assessed after handling, after transport and after slaughter. *Veterinary Record* 123, 201–205.

Moberg, G.P., Anderson, C.O. and Underwood, T.R. (1980) Ontogeny of the adrenal and behavioural responses of lambs to emotional stress. *Journal of Animal Science* 51, 138–142.

Molony, V. and Kent, J.E. (1997) Assessment of acute pain in farm animals using behavioural and physiological measurements. *Journal of Animal Science* 75, 266–272.

Molony, V., Kent, J.E. and Robertson, I.S. (1993) Behavioural responses of lambs of three ages in the first three hours after three methods of castration and tail docking. *Research in Veterinary Science* 55, 236–245.

Molony, V., Kent, J.E. and Robertson, I.S. (1995) Assessment of acute and chronic pain after different methods of castration of calves. *Applied Animal Behaviour Science* 46, 33–48.

Molony, V., Kent, J.E., Hosie, B.D. and Graham, M.J. (1997) Reduction in pain suffered by lambs at castration. *The Veterinary Journal* 153, 205–213.

Mora, S., Dussaubat, N. and Diaz-Veliz, G. (1996) Effects of the estrous cycle and ovarian hormones on behavioural indices of anxiety in female rats. *Psychoneuroendocrinology* 21, 609–620.

Morisse, J.P., Cotte, J.P. and Huonnic, D. (1995) Effect of dehorning on behaviour and plasma cortisol responses in young calves. *Applied Animal Behaviour Science* 43, 239–247.

Morton, D.B. and Griffiths, P.H.M. (1985) Guidelines on the recognition of pain, distress and discomfort in experimental animals and an hypothesis for assessment. *Veterinary Record* 116, 431–436.

Nakamura, I., Ohta, Y. and Kemmotsu, O. (1997) Characterisation of adenosine receptors mediating spinal sensory transmission related to nociceptive information in the rat. *Anesthesiology* 87, 577–584.

Neumann, I.D., Johnstone, H.A., Hatzinger, M., Liebsch, G., Shipston, M., Russell, J.A., Landgraf, R. and Douglas, A.J. (1998) Attenuated neuroendocrine responses to emotional and physical stressors in pregnant rats involve adenohypophysical changes. *Journal of Physiology* 508, 289–300.

Nicol, M.B., Hirst, J.J., Walker, D.W. and Thorburn, G.D. (1997) Effect of alteration of maternal plasma progesterone concentrations on fetal behavioural state during late gestation. *Journal of Endocrinology* 152, 379–386.

Parrott, R.F., Thornton, S.N. and Robinson, J.E. (1988) Endocrine responses to acute stress in castrated rams: no increase in oxytocin but evidence for an inverse relationship between cortisol and vasopressin. *Acta Endocrinologica* 117, 381–386.

Parrott, R.F., Misson, B.H. and De La Riva, C.F. (1994) Differential stressor effects on the concentration of cortisol, prolactin and catecholamines in the blood of sheep. *Research in Veterinary Science* 56, 234–239.

Pearson, R.A. and Mellor, D.J. (1975) Some physiological changes in pregnant sheep and goats before, during and after surgical insertion of uterine catheters. *Research in Veterinary Science* 19, 102–104.

Pearson, R.A. and Mellor, D.J. (1976) Some behavioural and physiological changes in pregnant goats and sheep during adaptation to laboratory conditions. *Research in Veterinary Science* 20, 215–217.

Petrie, N., Stafford, K.J., Mellor, D.J., Bruce, R.A. and Ward, R.N. (1995) The behaviour of calves tail docked with a rubber ring used with or without local anaesthetic. *Proceedings of the New Zealand Society of Animal Production* 55, 58–60.

Petrie, N., Mellor, D.J., Stafford, K.J., Bruce, R.A. and Ward, R.N. (1996a) Cortisol responses of calves to two methods of tail docking used with or without local anaesthetic. *New Zealand Veterinary Journal* 44, 4–8.

Petrie, N., Mellor, D.J., Stafford, K.J., Bruce, R.A. and Ward, R.N. (1996b) Cortisol responses of calves to two methods of disbudding used with or without local anaesthetic. *New Zealand Veterinary Journal* 44, 9–14.

Poindron, P., Levy, F. and Krehbiel, D. (1988) Genital, olfactory and endocrine interactions in the development of maternal behaviour in the parturient ewe. *Psychoneuroendocrinology* 13, 99–125.

Robertson, I.S., Kent, J.E. and Molony, V. (1994) Effect of different methods of castration on behaviour and plasma cortisol in calves of three ages. *Research in Veterinary Science* 56, 8–17.

Rushen, J. (1986) The validity of behavioral measures of aversion: a review. *Applied Animal Behaviour Science* 46, 309–324.

Sanford, J., Ewbank, R., Molony, V., Tavenor, W.D. and Uvarov, O. (1986) Guidelines for the recognition and assessment of pain in animals. *Veterinary Record* 118, 334–338.

Sapolsky, R.M. (1992) *Stress, the Aging Brain and the Mechanisms of Neuron Death.* Bradford Books, MIT Press, Massachusetts, pp. 261–299.

Sapolsky, R. and Meaney, M. (1986) Maturation of the adrenocortical stress response: neuroendocrine control mechanisms and the stress hyporesponsive period. *Brain Research Review* 11, 65–82.

Sawyer, D.C. and Rech, R.H. (1987) Analgesia and behavioral effects of butorphanol, nalbuphine, and pentazocine in the cat. *Journal of the American Animal Hospital Association* 23, 438–446.

Shutt, D.A., Fell, L.R., Connell, R., Bell, A.K., Wallace, C.A. and Smith, A.I. (1987) Stress-induced changes in plasma concentrations of immunoreactive β-endorphin and cortisol in response to routine surgical procedures in lambs. *Australian Journal of Biological Science* 40, 97–103.

Stafford, K.J. and Mellor, D.J. (1993) Castration, tail docking and dehorning – what are the constraints? *Proceedings of the New Zealand Society of Animal Production* 53, 189–195.

Stafford, K.J., Spoorenberg, J., West, D.M., Vermunt, J.J., Petrie, N. and Lawoko, C.R.O. (1996) The effects of electroejaculation on aversive behaviour and plasma cortisol in rams. *New Zealand Veterinary Journal* 44, 95–98.

Stephens, D.B. (1980) Stress and its measurement in domestic animals: a review of behavioural and physiological studies under field and laboratory situations. *Advances in Veterinary Science and Comparative Medicine* 24, 179–210.

Sumner, B.E. and Fink, G. (1997) The density of 5-hydroxytryptamine 2A receptors in forebrain is increased at pro-oestrus in intact female rats. *Neuroscience Letters* 234, 7–10.

Sylvester, S.P., Mellor, D.J., Stafford, K.J., Bruce, R.A. and Ward, R.N. (1998a) Acute cortisol responses of calves to scoop dehorning using local anaesthesia and/or cautery of the wound. *Australian Veterinary Journal* 76, 118–122.

Sylvester, S.P., Stafford, K.J., Mellor, D.J., Bruce, R.A. and Ward, R.N. (1998b) Acute cortisol responses of calves to four methods of dehorning by amputation. *Australian Veterinary Journal* 76, 123–126.

Taschke, A.C. and Folsch, D.W. (1995) Effects of electrical dehorning on behaviour and on salivary cortisol in calves. In: Nichelmann, M., Nierenga, H.K. and Braun, S. (eds) *Proceedings of International Congress of Applied Ethology, 3rd Joint Meeting,* Humbolt University Berlin, p. 99.

Taylor, P.M. (1989) Equine stress responses to anaesthesia. *British Journal of Anaesthesia* 63, 702–709.

Taylor, P.M. (1990) The stress response to anaesthesia in ponies: barbiturate anaesthesia. *Equine Veterinary Journal* 22, 307–312.

Urban, L. and Nagy, I. (1997) Is there a nociceptive carousel? *Trends in Pharmacological Sciences* 18, 223–224; Erratum. *Trends in Pharmacological Sciences* 18, 345.

Waas, J.R., Ingram, J.R. and Matthews, L.R. (1997) Physiological responses of red deer (*Cervus elaphus*) to conditions experienced during road transport. *Physiology and Behavior* 61, 931–938.

Wood, G.E. and Shors, T.J. (1998) Stress facilitates classical conditioning in males, but impairs classical conditioning in females through activational effects of ovarian hormones. *Proceedings of the National Academy of Sciences of the United States of America* 95, 4066–4071.

Wood, G.N. and Molony, V. (1992) Welfare aspects of castration and tail docking of lambs. *In Practice* 14, 2–7.

Wood, G.N., Molony, V., Fleetwood-Walker, S.M., Hodgson, J.C. and Mellor, D.J. (1991) Effects of local anaesthesia and intravenous naloxone on the changes in behaviour and plasma concentrations of cortisol produced by castration and tail docking using tight rubber rings in young lambs. *Research in Veterinary Science* 15, 193–199.

Xu, X.J., Elfvin, A., Hao, J.X., Fournie-Zaluski, M.C., Roques, B.P. and Wiesenfeld-Hallin, Z. (1997) CI 988, an antagonist of the cholecystokinin B receptor, potentiates endogenous opioid-mediated antinociception at spinal level. *Neuropeptides* 31, 287–291.

Yang, S.W., Zhang, C., Zhang, Z.H., Qiao, J.T. and Dafny, N. (1996) Sequential mediation of norepinephrine and dopamine induced antinociception at the spinal level: involvement of different local neuroactive substances. *Brain Research Bulletin* 41, 105–109.

Zhang, Y., Malmberg, A.B., Sjolund, B. and Yaksh, T.L. (1996) The effect of pituitary adenylate cyclase activating peptide (PACAP) on the nociceptive formalin test. *Neuroscience Letters* 207, 187–190.

Zhang, Y.Q., Tang, J.S., Yuan, B. and Jia, H. (1997) Inhibitory effects of electrically evoked activation of ventrolateral orbital cortex on the tail-flick reflex are mediated by periaqueductal gray in rats. *Pain* 72, 127–135.

Multiple Factors Controlling Behaviour: Implications for Stress and Welfare

10

F. Toates

Department of Biology, The Open University, Milton Keynes, UK

Introduction

In general terms, stress might be defined as a chronic disturbance of the processes that underlie adaptive behaviour, i.e. a long-term failure of regulation by systems that normally exhibit effective negative feedback (Toates, 1995; Jensen and Toates, 1997). A broad consensus suggests that stress can be associated with a failure of the normally adaptive processes of either active goal-directed strategies or passive strategies (Toates, 1995). Logically then, to understand stress we need to build upon insight into the controls of 'normal', non-stressed behaviour. With the help of a recently developed model of behaviour (Toates, 1998a, b, c), this chapter looks at some processes underlying behavioural control and suggests ways in which they act under conditions of stress.

To explain both normal and disturbed animal behaviour, applied ethology has attached itself to the 'cognitive revolution', placing weight upon cognitive (Wiepkema, 1987; Duncan and Petherick, 1991), purposive and rational goal-seeking (Dawkins, 1990) processes. In doing so, it coincides with a perspective within human psychology: that behaviour can be understood in terms of its *consequences*, even in the case of self-injurious behaviour (Carr, 1977). A similar purposive logic involving goal-seeking is involved in physiological explanations of stress, the principle of homeostasis commonly being used (Jones *et al.*, 1989; Le Moal and Simon, 1991). Thus, animals are said to find various ways of lowering stress as indexed by, for example, activity of the pituitary adrenocortical system (Antelman and

Caggiula, 1980; Jones *et al.*, 1989). Similarly, it is said that animals try to find an optimal level of arousal (Jones *et al.*, 1989).

Although the existence of a cognitive, goal-seeking aspect of behaviour seems beyond dispute, the contention of this chapter is that behaviour and its malfunction arise from interactions between several processes of control, only one of which is the cognitive process (cf. Glassman, 1976; Wemelsfelder, 1993). The chapter will describe several different processes involved in behavioural control. A challenge comes in trying to define the rules of interaction between these processes in the determination of behaviour. The chapter will first present the basics of a broad theoretical approach that has been developed to explain behaviour and then its possible implications for stress and welfare will be explored. The ideas should be seen as a stimulus to theory building.

Processes controlling behaviour

Background

Behaviour is determined by many interacting factors (e.g. learning, motivational, developmental, etc.) but the chapter will focus upon just four sets of factors:

1. *Cognitive processes* involving representations of the environment (Gallistel, 1980) and associated things such as goals, expectations and affect (Epstein, 1982; Panksepp, 1998a). This type of process allows flexibility in behaviour and is similar to the 'locale process' proposed by O'Keefe and Nadel (1978). By 'cognitive', I mean information that relates to the world but is not linked directly to behaviour. Cognitive processes allow an animal to extrapolate beyond current sensory input and are, by definition, 'offline' (Hirsh, 1974). In hierarchical terms (Glassman, 1976; Ridley, 1994; Toates, 1998a), they are 'high-level controls'. The ability to form (action) → (outcome) cognitions (Ursin, 1988), e.g. lever-pressing terminates shock, relates process 1 to process 3, below.
2. *External stimuli* that have a strong tendency to trigger particular behaviour. One such process is a stimulus–response (S–R) association (Hirsh, 1974; cf. Ridley, 1994), which specifies a relationship between currently present stimuli and behaviour. In Ridley's (1994) terms, some processes of this kind constitute 'preset packages of behaviour' that are relatively inflexible. They are similar to O'Keefe and Nadel's 'taxon processes' and in hierarchical terms constitute relatively low-level controls.
3. *The performance of behaviour* itself can be important to the animal, in addition to any end-point achieved. There are intrinsic tendencies to perform certain types of behaviour when in a particular motivational state and context (Breland and Breland, 1961; Bolles, 1970). As a result of

intrinsic organization, the strength of the tendency to perform a particular behaviour can increase as a function of time since last performed, e.g. dust bathing in hens (Vestergaard, 1980). More controversially, in some cases intrinsic processes appear to be able to produce behaviour even in the absence of external triggers (Lorenz, 1981).

4. *Transition-biasing processes* whose role, given the occurrence of one behaviour, is to bias towards sequences of functionally related behaviours, e.g. grooming sequences (Aldridge *et al.*, 1993).

Cognitive processes enable the animal to extrapolate beyond physically present stimuli and proceed to goals (Morris, 1983), something for which an intact cortex (Whishaw and Kolb, 1984) and hippocampus (Eichenbaum, 1994) are required. It is suggested that there is hierarchical control and that goal-directed behaviour is often implemented by cognitive processes exerting top-down control over a bank of S–R processes. At a given moment, the tendency to behave in a particular way, as triggered by S–R processes, can be compatible (cf. Whishaw *et al.*, 1987) or incompatible with the tendency suggested by cognitive processes. Sometimes a behavioural tendency caused by an S–R association needs to be inhibited in the interests of future goal achievement (Cohen and Servan-Schreiber, 1992).

It is proposed that the dynamics of the interaction between these four processes are relevant for understanding some of the abnormalities of behaviour studied by stress researchers (cf. Fentress, 1976). Changes in weighting between processes 1–4 can profoundly alter behaviour. Although all four processes can be involved simultaneously in the control of behaviour, a difference in weighting between them can occur as a function of development, learning or pathology (Toates, 1998a, b, c).

Initially during development (e.g. fetal stage), behaviour is generated by intrinsic spinal processes (Corner, 1990). Behaviour then tends to come under the control of S–R processes (reviewed by Toates, 1998a). With further maturation, cognitive controls assume a greater weighting. This provides an increased capacity to inhibit behaviour that would otherwise tend to be caused by the S–R process. Conditions of impoverished rearing disrupt the emergence of the top-down controls that would normally be able to restrain lower-level controls (Ridley and Baker, 1983). With ageing, cognitive control can diminish in efficacy compared with S–R control (Gallagher, 1997).

As an animal becomes more accomplished at a task by utilizing a particular behaviour, the weighting attached to the S–R process involved tends to increase (O'Keefe and Nadel, 1978). That is, actions become habits (Adams, 1982; Dickinson, 1985; Serruya and Eilam, 1996) and the capacity of stimuli to take control increases, something relevant to the study of stress (e.g. Wiepkema, 1987) and to welfare (Walker, 1990). For example, under some conditions, even what 'should' constitute the punishment of a response (e.g. shock delivered contingent upon the response) tends not to diminish the tendency to exhibit this response (Brown *et al.*, 1964). The weighting of control can also shift following disruption to the brain. For example, rats with damage to the hippocampus exhibit behaviour that largely conforms to principles of S–R control (Hirsh, 1974). In humans, schizophrenia is associated with a loss of contextual control (Cohen and Servan-Schreiber, 1992).

Although the adaptive significance of different levels of control is to work in coordination in pursuit of goals, the theme of this chapter is that, under certain conditions, variation in weighting between processes can give rise to behavioural anomalies. Thus, behaviour cannot always be understood simply in rational goal-seeking terms. In defining processes of behavioural control, the notions of reward and reinforcement are of central importance and are considered next.

Reward and reinforcement

Within applied ethology one sees expressions of the kind 'performing behaviour X might be rewarding' or 'stereotypies could be reinforcing' (Dantzer, 1986; Segal, 1990; Turkkan, 1990; Mason, 1993). Arguments about welfare sometimes depend upon such considerations. However, as White (1989) notes, even within psychology there is little or no consensus on what 'reward' and 'reinforcement' mean. The distinctions between the four processes described as underlying behavioural control could prove relevant here (cf. White, 1989).

In the terms developed here, rewards are affectively positive and act at a high level of control to modify the animal's cognitions (e.g. tasty food is at site A). Subsequently, such representations do not tell an animal *exactly* what to do (e.g. neither specifically to run nor to swim to a goal-box). Rather, rewards (as part of process 1) help to determine very broad classes of different behaviour (e.g. to approach the goal-box), the exact form depending upon current circumstances (e.g. whether the maze is dry or flooded). The term reinforcement can be used in a theoretically neutral way, as in, for example, food placed in the goal-box to the left reinforces the animal's left turn (Bindra, 1978). This can be operationally defined in terms of a subsequently increased frequency of turning left. However, 'reinforcement' can also be used at a theoretical level to refer to one particular process, in White's (1989) terms, to the strengthening of particular S–R associations by repeated triggering of R by S (cf. O'Keefe and Nadel, 1978), i.e. process 2.

In many cases, processes of both reward and reinforcement (as defined in the theoretical sense) would act in parallel and would be entirely compatible. For example, an animal repeatedly turning left in a T-maze to find food might be described as seeking the reward located there (process 1) and, in making repeated left body-turns, an S–R connection gets reinforced (process 2). In some cases, however, these processes act in opposition. For example, a rat performing a foraging task in an 8-arm radial maze for small morsels of food (Olton *et al.*, 1979) needs to abandon an arm (win–shift) after having been reinforced for entering it and to search elsewhere. One can imagine a high-level cognitive control that suggests a shift of maze arm (win–shift) acting in opposition to a lower-level (S–R) reinforcement mechanism (win–stay). Indeed, different brain lesions favour either win–shift or win–stay strategies (Packard *et al.*, 1989).

In applied ethology, a formal distinction between reward and reinforcement could prove useful. If we extrapolate to suppose that animals have subjective feelings, then it would seem that reward and aversion are the appropriate currency, e.g. the theoretical position of Dawkins (1990). She compares the attraction of two different goals and considers the price that an animal will pay to gain or avoid an object. Reinforcement, a strengthening of particular motor acts, might be relevant to understanding some instances of bizarre behaviour (Turkkan, 1990) that appear to defy rational analysis in terms of reward. It would appear to be an explanatory concept that we require for understanding stereotypies, discussed below.

The intrinsic importance of behaviour

Is behaviour simply an arbitrary means to obtain a desirable outcome or is the ability to perform behaviour *per se* important to the animal? The inclusion of process 3 as a component in behavioural control reflects a belief in the latter. On some occasions an animal has learned successfully a goal-directed sequence of behaviour, such as lever-pressing or depositing a token in a slot in order to obtain reward. Then, after apparently becoming accomplished at this, species-typical behaviour such as burying (in this case, the token) intrudes and disrupts the goal-directed sequence (Breland and Breland, 1961).

The performance of specific patterns of behaviour *per se* is important as a factor in the achievement of stress reduction in an aversive context as indexed by reduced gastric ulceration (Weiss, 1971; Weiss *et al.*, 1976) and is also relevant to achieving metabolic homeostasis (de Passillé *et al.*, 1991, 1993). The relevance of behaviour *per se* in achieving consequences is also evident in the phenomenon of electrical stimulation of the brain. Apparently rewarding stimulation, contingent on the animal's own behaviour, can become aversive if it is later applied by the experimenter in a situation where the animal does not have control (Steiner *et al.*, 1969). This suggests that a simple hedonic model of such phenomena is inadequate and emphasizes the importance of the animal being able to perform behaviour (process 3) that exerts control over the situation (process 1). Similarly, although the capacity to show locomotor activity is not necessary for an animal to learn an association between a location and an amphetamine injection made there, it is necessary in order for the animal to exhibit a preference for the location (Swerdlow and Koob, 1984). This again underlines the importance of behaviour *per se* in achieving an end-point.

Transition-biasing processes and modal action patterns

Transition-biasing processes (process 4) help to orchestrate sequences of behaviour. Brainstem nuclei play a role in the organization of individual stereotyped responses, e.g. biting, chewing and swallowing (Klemm and Vertes, 1990). However, the timing and coordination of such responses are under the modulation of higher brain regions (Berntson and Micco, 1976; Berridge, 1989a), i.e. transition-biasing (cf. Fentress, 1976; Rushen *et al.*, 1993).

In rats, sequences of grooming are disrupted by lesions of the striatum, even though the component actions remain intact (Berridge and Fentress, 1987a). Dopamine (DA) projections from the substantia nigra to the striatum are involved in generating such sequences (Berridge, 1989b), which are intrinsically determined in that they do not depend upon somatosensory feedback (Berridge and Fentress, 1987b). The cortex plays a role in these sequences and the importance of its role appears to increase in primates as compared with rodents (Berridge and Whishaw, 1992). In generating a species-typical sequence of actions, A_1, A_2, A_3 . . ., the role of the striatum appears to consist of the control of A_1 biasing towards A_2, and so on, even in the face of competitive tendencies (Berridge and Whishaw, 1992). Aldridge *et al.* (1993, p. 4) suggest that: 'hierarchical neostriatal circuits temporarily switch motor control toward central pattern generators and away from sensory-guided systems at the beginning of a syntactical grooming chain'.

Although sequences of grooming are innately organized, it appears that learning can lead to a very similar solution involving the striatum (Aldridge *et al.*, 1993). For example, even when any one of a number of flexible and varied sequences of learned behaviour could serve to achieve the same end-point, there is a tendency for certain fixed sequences to develop (Schwartz, 1985). Such a process might, at least in part, be the embodiment of what underlies the move to more automatic control with repetition, discussed above.

Classical ethologists placed weight upon fixed action patterns (Tinbergen, 1969; Lorenz, 1981), while later ethologists loosened the concept slightly in the designation 'modal-action pattern' (MAP) (Hoyle, 1984). Either way, the hallmark is that a sequence of actions once triggered tends to carry on to completion. This involves intrinsic organization such that sequences are produced with syntactical structure.

I next consider some more details of the suggested processes in terms of the nervous, endocrine and immune systems.

Mechanisms underlying different levels of control

A double-dissociation

What is the neural embodiment of reward and reinforcement mechanisms? In the 8-arm radial maze, rats were rewarded either for win–shift (change arm after reward) or win–stay (repeat arm choice after reward) strategies. By selective lesions, Packard *et al.* (1989) performed a double-dissociation. For the win–stay task, rats with lesions to the caudate nucleus were disrupted. This basal ganglia structure appears to exert an influence near to the motor output side and to form a basis of S–R learning. Animals with lesions to the fimbria fornix, which disrupts utilization of the hippocampus, on the other hand, improved relative to controls. On a win–shift task, caudate nucleus lesions had no effect but fimbria fornix lesions disrupted behaviour.

The role of dopamine

Introduction

In any discussion of reward, motivation and reinforcement, DA often features large (Wise, 1982; Berridge and Robinson, 1998), although of course behaviour is influenced by many interacting neurochemicals. A more complete account would need to consider the role of noradrenaline, serotonin and opioids.

DA appears to serve a role in homeostasis in that its activation tends to lead to active behaviour, e.g. escape or avoidance (Le Moal and Simon, 1991), which then lowers DA levels. More consideration might usefully be given to DA in the context of theories of stress. For example, the role of DA in stereotypies is established (Schiff, 1982; Kennes *et al.*, 1988). Levels of DA correlate negatively with the tendency to develop gastric ulcers (Glavin, 1993). DA activation appears to counter tendencies to passivity, e.g. in an avoidance task (Le Moal and Simon, 1991). Whereas controllable shock is associated with DA release in the nucleus accumbens, uncontrollable shock is not (Cabib and Puglisi-Allegra, 1994). There are interactions between DA and corticosteroids (Jones *et al.*, 1989).

Sites of dopamine action

To reiterate an earlier point, it is assumed that behaviour is under the joint control of physically present stimuli and cognitive representations. DA appears to act at a number of sites in the brain, possibly corresponding to different aspects of such multi-process control. There are tentative hints (discussed shortly) that one common feature of DA's postsynaptic role at different brain regions is to bias in favour of the candidate for behavioural control that is represented at the region, e.g. prefrontal cortex or basal ganglia (Cohen and Servan-Schreiber, 1992; cf. Salamone *et al.*, 1997). The postsynaptic effect of DA at one brain site might even act in opposition to that at another. It is suggested that, in the interests of goal-directed behaviour, one form of biasing involves strengthening the candidacy of certain (e.g. effort demanding or intrinsically weak) stimuli and memories (e.g. mediated via the prefrontal cortex), while inhibiting that of intrinsically strong stimuli (e.g. acting via subcortical sites).

DA activity in the prefrontal cortex can inhibit DA activity in the nucleus accumbens (Deutch *et al.*, 1990). The expression of cognitive processes in behaviour in the face of incompatible tendencies for behaviour triggered by lower-level processes requires DA activation within the cortex (Cohen and Servan-Schreiber, 1992). DA lesions in the prefrontal cortex disrupt delayed alternation (Brozoski *et al.*, 1979; Diamond *et al.*, 1994), a task that involves behavioural choice based on a memory of the past choice. Cohen and Servan-Schreiber suggest that DA-blocking drugs particularly target subcortical DA systems, biasing towards cognitive control. However, the relationship between cortical DA levels and cognitive control seems not to be a simple one. Thus, noise stress, which elevates prefrontal cortical (PFC) DA release, can impair cognitive functioning (Arnsten and Goldman-Rakic, 1998). The authors suggest (p. 366): 'Selective dysfunction of the PFC during stress may have survival value, favouring well-rehearsed or instinctual behaviours regulated by subcortical structures and posterior cortex rather than slower, more complicated PFC regulation'. Possibly only a narrow band of DA activation at different brain regions is compatible with an adaptive balance between cognitive and S–R controls.

The events that cognitive processes represent are either (i) not physically present in space or time when behaviour is instigated, e.g. foraging; or (ii) their significance needs to be triggered by cues. It is possible that nucleus accumbens DA plays a role in converting cognitions into candidate goals. Thus, loss of DA disrupts foraging, hoarding and active avoidance behaviour (Blackburn *et al.*, 1992) and the behaviour of going to a food niche prompted by a predictive cue. Foraging in an Olton maze is disrupted by DA antagonists injected into the nucleus accumbens (Floresco *et al.*, 1996).

However, acting at other sites (e.g. basal ganglia), DA appears to strengthen the capacity of lower-level (S–R) controls. Thus, in connection with the DA agonist amphetamine, Ridley (1994, p. 226) suggests: 'that descending frontal projections exert an inhibitory effect on striatal activity and counteract the stimulant effects of the rising dopaminergic projections'. Amphetamine appears to act at the nucleus accumbens in promoting activity and exploration but at the ventrolateral striatum in increasing the tendency to stereotypy (Kelley *et al.*, 1988).

It appears that dopaminergic neurons in the substantia nigra are implicated in sequencing, since their stimulation elicits features of modal action patterns (MAPs) described above (Piazza *et al.*, 1989a, b). This is relevant to the suggestion that stereotypies represent an exaggerated response of the processes that normally underlie adaptive MAPs, since dopamine has been a focus of attention in the study of stereotypies. Behaviour shows some autonomy from DA processes when either (i) it is based initially upon the properties of physically present stimuli that have strong established behavioural associations (Keefe *et al.*, 1989) or (ii) it has become habitual (Posluns, 1962; Singh, 1964; Ray and Bivens, 1966; Fibiger *et al.*, 1974, 1975; Sahakian and Robbins, 1977; Robinson and Berridge, 1993).

Dopamine and stress

Theories on stress need to heed the observation that, although the mesotelencephalic DA system is intimately connected with reward (e.g. drug-based), it is also excited by stressors (Cabib, 1993; Robinson and Berridge, 1993). Understanding this might be helped by considering that behaviour has more than one level of determinant. The mesolimbic DA system is activated both by acute and chronic stressors (Hall, 1998) and by drugs such as amphetamines, cocaine and opioids (Bardo, 1998).

Stressors are effective in triggering activity in DA projections to the nucleus accumbens and the medial frontal cortex, two sites of increased DA activity accompanying drug-seeking. Conditional stimuli paired with stressors have a similar effect (Young *et al.*, 1993). Robinson and Berridge (1993) suggest that the mesotelencephalic DA system could mediate attention to all salient stimuli. The valence of the stimulus, i.e. positive or negative, might be differentiated elsewhere in the central nervous system (CNS). Restraint stress triggers a rise in DA release in the nucleus accumbens followed within about 50 min by a fall to below baseline (Cabib, 1993). The end of restraint triggers a rise to above baseline. It is as if DA release corresponds to when active strategies are indicated.

Piazza *et al.* (1990) refer to: 'repeated amphetamine injections (amphetamine sensitization) employed as a pharmacological model of stress'. Herein lies the paradox that an animal will learn an operant task for such injections and, by this definition, learn to stress itself. Similarly, in rats, corticosteroids are positively reinforcing when administered intravenously or orally (Deroche *et al.*, 1993) in a concentration comparable with that seen under conditions described as stressful and that elevates nucleus accumbens DA activity. This is paradoxical given that elevated corticosteroid levels are often used as an index of stress. DA neurons contain corticosteroid receptors (Le Moal and Simon, 1991).

Stress, the hippocampus and changes in weighting of control

Evidence suggests that: (i) the hippocampus has a role in mediating cognitive control; and (ii) this mediation can be disrupted by stress. O'Keefe and Nadel (1978) proposed two distinct processes: a taxon system, similar to an S–R system, and a cognitive map-based system, termed a locale system. In these terms, Jacobs and Nadel (1985) suggest that (p. 518): 'stress disrupts the function of the hippocampally based locale system and its context-specific learning capacities while potentiating taxon systems and their context-free associations'.

According to Altman *et al.* (1973, p. 567), hippocampectomized rats, when aroused: 'be that due to novelty, change in reinforcement schedule or reward magnitude, reversal, extinction, or the like, . . . will tend to act out, or emit overt responses while, in contrast, normal adults will tend to pause in responding'. Thus, hippocampectomized rats are similar to young rats with an immature hippocampus, i.e. lacking restraint. The hippocampus has a rich density of corticosteroid receptors. What is their functional significance?

The role of corticosteroids

Corticosteroid receptors in the CNS appear to influence sensory and learning processes (McEwen *et al.*, 1986). Their loss in the hippocampus is associated with a reduction in what is termed here cognitive control (Oitzl *et al.*, 1997). In group-living birds, a high level of corticosteroids appears to increase the probability that an individual will disperse from the group (Silverin, 1997; Belthoff and Dufty, 1998). This suggests that corticosteroids bias CNS processes away from the production of behaviour determined by physically present stimuli and cause distant locations (e.g. remote migration sites represented cognitively) to be pursued.

Chronically elevated cortisol levels can be damaging to neural tissue, especially in the hippocampus (Uno *et al.*, 1989; Lupien *et al.*, 1998), and might contribute to a decline in the capacity to exert high-level control. Under stress, this could trigger a positive feedback loop with increased weighting towards lower level controls.

The immune system

By means of communication to the CNS, the immune product interleukin-1 (IL-1) triggers a pattern of behaviour characterized by sleep, reduction in exploration, fever and loss of motivation to engage with positive incentives (Hart, 1988; Maier and Watkins, 1998). It also reduces cognitive control in that mice injected with IL-1 are disrupted in finding a submerged platform in a Morris water maze, although their response to physically present stimuli is unimpaired (Gibertini *et al.*, 1995a, b). IL-1 changes the activity of neurons within the hippocampus, which might mediate a bias away from cognitive control. This might make the animal more stimulus bound and less likely to wander prompted by distant goals. Gilbertini and associates suggested that there could be an integration of cytokine and corticosteroid information at the hippocampus.

Having outlined some of the basic ideas of multiple factors in behavioural control, the next section considers some implications of a shift of weighting of control away from cognition (process 1).

Loss of top-down control

It is suggested that a number of features of behaviour might be understood in terms of a reduction in the strength of the cognitive, top-down process relative to lower levels of organization. This section looks at some examples of this.

Anomalous behaviour

Some anomalies of behaviour caution against simple rational interpretations and suggest control by more than one process. These processes can sometimes be in conflict (cf. Turkkan, 1990). For example, animals sometimes perform an operant task that is reinforced by the *presentation* of electric shocks (Kelleher and Morse, 1968). The addition of contingent shocks sometimes appears to strengthen appetitive goal-directed behaviour (Brown *et al.*, 1964). Such behaviour might reveal an S–R link that acts in spite of reward/aversion principles. This could be relevant to the phenomenon of self-injurious behaviour (Harlow and Harlow, 1962; Berkson, 1967), though exactly what is responsible for strengthening a lower-level control under these conditions remains to be explained.

Monkeys can be persuaded to press a lever for intravenous drug infusion and simultaneously learn to press another lever, which, within the same session, terminates this schedule (Spealman, 1979), suggesting elements of automatic 'compulsive' control.

Adjunctive behaviour and displacement activities

In rats, mild stress arising from, say, electrical shock or tail-pinch, has the effect of triggering a variety of different behaviours, e.g. eating, aggression or sexual behaviour (Robbins, 1978; Antelman and Caggiula, 1980). If no suitable objects are available, nail-biting is a typical reaction. These stimuli might have the common focus of triggering DA, which gets channelled into sensitizing whatever outlet is suggested by the stimuli impinging upon the animal at the time. Amphetamine-injected marmosets tend to engage in extensive focused self-grooming with sometimes damaging consequences (Ridley and Baker, 1983), which might model features of self-mutilation under non-injected conditions.

Electrical stimulation of the lateral hypothalamus results in any of a variety of different coordinated acts, depending upon the environmental stimuli present (Valenstein, 1969). With experience in the same situation, the variety gives way to one particular behaviour which becomes increasingly stereotyped (Wayner *et al.*, 1981). Activities that involve orogastric sensory feedback, e.g. chewing and licking water, are very prone to develop such electrically induced stereotypy. Gradually the independent component behaviours become more coordinated.

Displacement activities appear to correspond to behaviour organized at a low level in the hierarchy which is triggered at times of stress, ambivalence, etc., when there could be a failure of top-down control (Le Moal and Simon, 1991; Toates, 1995, 1998a). These activities tend to be species-typical (reflecting process 3) and with a high probability of occurrence in the normal life of the animal (Cabib, 1993). For example, grooming as a displacement activity can show a whole sequence of its component acts, pointing to control in part by transition-biasing. Schedule-induced polydipsia is sometimes seen as a variety of displacement activity and is dependent upon DA activity in the nucleus accumbens (Robbins and Koob, 1980).

Stereotypies

Background

In some cases, stereotypies emerge from a situation in which there is an absence of variation in the environment (Wolff, 1968; Terlouw, 1993). This could mean that there is an inadequate (cognitive) signal top-down to resist the drift to S–R control (cf. Le Moal and Simon, 1991). Davenport and Menzel (1963) noted that chimpanzees raised under impoverished conditions exhibit stereotypies, but that these are interrupted when novel objects are introduced into the cage, at which time exploration occurs. They conclude that stereotypies arise because of 'the absence or insufficient amount of stimulation that the mother ordinarily provides her infant' and is triggered when the subject 'is not actively engaged in 'externally-directed' behaviour'.

It is suggested that, in terms of process 1 (cognition and goal direction), stereotypies can emerge out of what is initially cognitive control and goal-directed behaviour, in some cases involving imitation (Ellinwood and Kilbey, 1975; Cooper and Nicol, 1994). When such behaviour repeatedly fails, the normally negative feedback system shifts into 'open-loop' and behaviour becomes perseverative (Hediger, 1964; Meyer-Holzapfel, 1968; Robbins, 1976; Dantzer, 1986; Kennes *et al.*, 1988). Frustration is a potent trigger to stereotypies (Dantzer, 1986; Wiedenmayer, 1997), suggesting a common feature with displacement activities. For example, stereotypies can emerge from unsuccessful attempts to escape confinement (Dantzer, 1986;

Würbel and Stauffacher, 1997). Amphetamine-induced stereotypies appear to emerge sometimes from a combination of processes 1 and 3: species-typical behaviours used in goal-directed behaviour become strengthened (Ellinwood and Kilbey, 1975). By the repetition of a particular behaviour in a context, stimuli can become attached to it (process 2).

Stereotypies appear to correspond to a shift of weighting from cognitive to S–R processes, with strengthening of transition-biasing between components of the stereotypy (cf. Levy, 1944; Wemelsfelder, 1993; Toates, 1998a). Würbel *et al.* (1998) suggested that stereotypies arise from habit formation and sensitization. According to Ridley (1994, p. 221), they arise as a result of a failure of 'higher order representation' and: 'Pathological repetition may consist of over-excitation of behavioural programmes or may be the 'default' effect of the loss of mechanisms which normally direct more variable self-initiated behaviours'.

Similarly, Dantzer (1986, p. 1785) argues that stereotypy: 'reflects a cut-off of higher nervous functions coupled with a disinhibition of hypothalamic and brain-stem structures where the basic organization of most motor acts is hard-wired' and: 'All the conditions would therefore be met for a positive feedback mechanism in which sensory factors that normally guide behaviour trigger a behavioural sequence that becomes self-organized independently of further environmental guidance'. In these terms, we might expect the stereotypies that emerge to correspond to behaviours that have a high probability of appearing in a given situation (see description of process 3), e.g. oral and rooting stereotypies in pigs and pacing in carnivores (Rubovits. and Klawans, 1972; cf. Mason and Mendl, 1997).

Hediger (1964, p. 75) suggested that: 'Under the influence of certain conditions of captivity (space-confinement, hypertrophy of valances, lack of amusement and occupation), peculiar partial hypertrophies of the space–time pattern may occasionally occur, fixed stereotyped movements for instance'. Stereotypies occasionally appear to lock on to the presence of external stimuli, for example, being performed by zoo animals specifically in the presence of the public (Hediger, 1964). In a restricted environment, the same stimuli are repeatedly encountered, which increases their chances of 'locking-on to' associated responses (Thelen, 1979, 1996; cf. Ridley and Baker, 1982; Serruya and Eilam, 1996).

Ridley and Baker (1982) note that many stereotypies shown by isolated animals represent *habit residuals*. That is to say, they are normal infant forms of behaviour that have not been overridden during the course of development. With experience, stereotypies become difficult to disrupt by environmental modification (Cooper *et al.*, 1996). Thus, they sometimes involve responding on the basis of what was present in the environment but no longer is. For example, an animal that repeatedly jumps over an obstacle in performing a routine might continue to jump at the point where the obstacle used to be (Fentress, 1976). This suggests a role of transition-biasing.

Biochemical bases

Stevens *et al.* (1977, p. 809) suggest that psychomotor stimulant (e.g. amphetamine) injections trigger: 'stereotyped behaviours that in all vertebrates appear to represent an exaggeration and perseveration of fragments of species specific exploratory behaviours'. This is in distinction to, say, simply exaggerated motor automatism, divorced from sensory input: sensory input appears to engage species-typical motor patterns. In response to amphetamine injections, when cats were blindfolded there was a sharp drop in the amount of side-to-side movement of the head, which led Stevens *et al.* (1977) to suggest that amphetamine-induced stereotypy reveals that DA activity (p. 811): 'enhances the perceptions or significance of sensory stimuli, a phenomenon strikingly similar to the heightened sensory awareness and significance of banal stimuli reported by many patients with schizophrenia'.

Kennes *et al.* (1988) found that opioid blocking disrupts the early development of stereotypies but not their occurrence once established. Stereotypies did not develop a similar autonomy from DA blocking with haloperidol. However, injection of haloperidol is a blunt instrument unless targeted into specific brain nuclei by microinjection, since DA blocking can have quite different effects according to the nucleus involved (Bakshi and Kelley, 1991), as discussed in the next section.

Dopamine is implicated in mediating differences between control and loss of control as a factor in the development of stereotypies. Uncontrollable, but not controllable, shocks enhance amphetamine-induced stereotypy (MacLennan and Maier, 1983).

Brain regions

Evidence points to an involvement of the nucleus accumbens in psychomotor activation, e.g. flexible searching, and involvement of the caudate nucleus nearer the motor output side with repetitive behaviour (Rubovits and Klawans, 1972). Thus, Jones *et al.* (1989) report that amphetamine-induced stereotypies are blocked by DA depletion in the caudate nucleus, whereas DA depletion in the nucleus accumbens blocks the locomotor-activating role of amphetamine.

In rats, hippocampal lesions both cause a hyperresponsivity to stimuli (O'Keefe and Nadel, 1978) and an increase in the tendency to stereotypy (Devenport and Holloway, 1980). Amphetamine-induced stereotypies are enhanced by cortical lesions (see Le Moal and Simon, 1991). This suggests the removal of layers of control that could otherwise override the lower level organization. Devenport and Holloway (1980, p. 701) suggest that: 'The possession of hippocampi helps to prevent the evidently compelling tendency to repeat with increasing frequency an accidentally reinforced response or response sequence', and that: 'intact animals adjust their behaviour according to a rule distinct from, or in addition to, contiguity'.

They note the greater development of the hippocampus in most mammals as compared with birds and the greater tendency of birds to exhibit stereotypy. In rats, a combination of (i) presentation of reward; (ii) food deprivation; and (iii) loss of the hippocampus is a very potent one for inducing heightened activation and stereotypy, much like that induced by amphetamine (Devenport *et al.*, 1988). This might capture some of the conditions of loss of functional top-down control suggested here to underlie the development of stereotypy in neurally intact animals.

In a somewhat similar vein, Eichenbaum (1994, p. 179) places a focus upon the:

> inflexibility of *memory representations* supported outside the hippocampal system. In our view the lack of behavioural variability and the abnormally rapid acquisition of superstitious behaviours are consequences of hippocampal-independent conditioning processes, whose influence is more readily exerted in the absence of the sometimes competing objectives of the hippocampal-dependent processing.

Reinforcement and reward

To return to the earlier discussion: are stereotypies reinforcing? In an operational sense they cannot be, since, unlike food or heat, they are not something presented by the experimenter as a result of performing behaviour. If, however, performing them tends to strengthen the future tendency to show them, then, by this criterion, they might be termed 'reinforcing'. Thus, Katz (1982) suggested that 'certain repetitive acts, by virtue of their kinesthetic consequences are intrinsically reinforcing', a similar point being argued by Azrin *et al.* (1973).

Is performing stereotypies rewarding, in the sense that an animal will seek out environmental supports that permit their performance? Some evidence suggests this (cf. Thelen, 1979, 1996; Cooper and Nicol, 1991), although it is not predicted in any obvious way from the present perspective. Ellinwood and Escalante (1970) found that amphetamine-injected cats were attracted to a specific location at which they performed stereotypies, though whether the performance of the stereotypies as such constituted part of the labelling of the site as attractive is unclear. Both human and non-human subjects can show a preference for a particular location at which they exhibit stereotypies (Berkson, 1967). This might have similarities with the conditioned place preference test (Stewart *et al.*, 1984) in which animals prefer a location associated with a drug injection. After apomorphine injections rats perform an operant task rewarded with a block which they then chew (Robinson *et al.*, 1967). This suggests a reward process associated with the capacity to perform chewing under these conditions (i.e. process 3).

In terms of prior models, the present one is perhaps closest to the idea that stereotypies are the outcome of a disinhibition process (McFarland, 1966). It does not suggest a model in terms of a spillover of energy (cf.

Lorenz, 1981). Consideration of the ideas expressed here would leave open the question of whether, placed in a certain situation, performing stereotypies improves the animal's welfare. However, stereotypies point to a failure of cognitive control and their appearance would support the widely held view that they emerge as a result of environmental inadequacy.

Are there genetic and developmental factors that play a role in differences in the relative strengths of the contributory processes 1–4? The next section briefly considers this issue.

Factors affecting weightings between control processes

Comparing different animals, are there differences in the weighting of the processes involved in behavioural control? Are such differences revealed in a given context, e.g. an environment that tends to trigger stereotypies?

Individual differences

Both in 'pure' behavioural science (Mittleman and Valenstein, 1985; Cools, 1988) and in applied ethology (Wiepkema and Schouten, 1992) there is interest in individual differences in behaviour. Future research could usefully consider how individual differences might map on to differences in the relative strengths of the four processes introduced earlier.

For a given species and test situation, some individuals tend to show an active coping strategy and others a passive strategy (Benus, 1988). There are also individual differences in the tendency to show stereotypy. Feed-restricted pigs tend to develop polydipsia, particularly so in more dominant animals (Terlouw *et al.*, 1991). In rats, there are also large individual differences in schedule-induced polydipsia, electrical stimulation-induced behaviour and avoidance (Mittleman and Valenstein, 1985). Confronted with a threat, rats and mice tend either to freeze or to show the active strategy of fighting/fleeing. Animals can be classified according to which of these tendencies is dominant (Benus *et al.*, 1991). In tree-shrews, different types of stress are associated with protracted and ineffective exercise of one or the other strategy (von Holst, 1986).

There are strain differences in DA activity. In mice, whether immobilization stress enhances frontal cortex DA activity depends upon the mouse strain (Cabib *et al.*, 1988). Following stress, whether apomorphine induces an increase or a decrease in activity also depends upon strain (Cabib *et al.*, 1985). Differences in the extent to which behaviour is stimulus bound or flexible are associated with differences in DA activity in the nucleus accumbens (Cools, 1988). Stressors of various kinds elevate DA activity in the prefrontal cortex and nucleus accumbens in high avoidance, but not low avoidance, rats (Scatton *et al.*, 1988), suggesting that DA activation is associated with the initiation of an active behavioural strategy.

In rats, a positive correlation exists between the tendency to flee and the amount of gnawing induced by the DA agonist apomorphine (Cools, 1988). Benus *et al.* (1991) observed the relationship between attack latency and the tendency to stereotypy induced by apomorphine. Animals with a short attack latency (SAL mice) had a much higher stereotypy response than did long attack latency mice (LAL mice). Benus *et al.* (1991) suggest that their results can be understood in terms of Cools' theorizing. Thus, a high sensitivity to apomorphine reveals intrinsically low DA activity in the neostriatum (with up-regulation of receptors), whereas a low sensitivity to apomorphine reveals intrinsically high DA activity (with down-regulation). Benus *et al.* present evidence that SAL mice are more routinized in their behaviour than are LAL mice, which exhibit more flexibility. It remains a challenge to explain this result.

There are important individual differences in the level of central DA activity which appear to correlate positively with the tendency to active behavioural strategies and negatively with the development of gastric ulcers (Cools, 1988; Glavin *et al.*, 1991; Taché, 1991; Glavin, 1993).

Early rearing

Early rearing experience might bias the weighting of cognitive and S–R controls and the subsequent development of stereotypies (Sahakian *et al.*, 1975; Ridley and Baker, 1983). Isolation affects the role of catecholamines in cortical development (Kraemer *et al.*, 1983). Social interactions during development (e.g. play) appear to influence the function of the prefrontal cortex and thereby its later potential to exert a top-down inhibitory role (Panksepp, 1998b). It seems that an intrinsic tendency to programme bouts of play (process 3) can serve the development of cognitive control (process 1). Although rearing rats in social isolation increases the later tendency to stereotypy, in guinea pigs it decreases this tendency (Sahakian and Robbins, 1975; Kehoe *et al.*, 1998). Harlow and Harlow (1962) observed that infant monkeys raised in isolation show exaggerated stereotypies and self-injurious behaviour, suggesting a weak top-down influence.

In the context of the developmental background of stereotypies, Ridley (1994, p. 223) suggests that: 'where behaviour has little effect on the environment, the environment has only a limited and crude influence on behaviour'. One might interpret 'crude' in terms of increased weighting of control by lower-level processes (e.g. innate programmes).

Differences in experience are associated with differences in DA systems, which might create a bias towards or against acquiring particular behaviour patterns when the animal is an adult. Lewis *et al.* (1996) suggested that 'early social deprivation resulted in loss of dopamine innervation to striatal areas with a subsequent dopamine receptor supersensitivity'. Socially deprived monkeys show an elevated stereotypy response to the DA agonist apomorphine (Lewis *et al.*, 1996). Socially reared monkeys exhibit the so-called 'Kamin blocking effect', i.e. they ignore redundant information (Lewis *et al.*, 1996), whereas socially deprived monkeys fail to exhibit this. In the case of the latter, this is suggestive of an increased weighting attached to stimuli *per se* and a failure of top-down modulation based upon the stimulus being placed in context. Rats reared in social isolation show a reduced tendency to habituate to stimuli (Einon *et al.*, 1975) and (by some indices) increased exploration (Sahakian *et al.*, 1977).

Welfare implications

The ideas developed here are as yet somewhat speculative. They invite further research and theorizing. However, they already suggest a number of implications for animal welfare:

1. One of the processes proposed to underlie behavioural control is cognitive, active, purposive and goal-orientated, involving affective states. This reinforces the need for housing conditions to be such that the animal can maintain affective states within acceptable boundaries. Stress can arise not just from physically damaging stimuli but from frustration and thwarting, etc. The suggestion of processes additional to the cognitive goal-directed process (process 1) does not detract from the central role of process 1 in the control of behaviour.
2. If lower-level controls can 'take over' under some conditions, an animal might not always 'know' or reveal what is in its own best interests. Other indices might be needed to assess welfare. It might be misleading to see behaviour as always being goal-directed, e.g. to lower stress levels, although that is not to deny that goal direction is a feature of behaviour.
3. The use of corticosteroids as a gold-standard for stress needs some qualification. Chronic elevated levels are indicative of disturbed welfare but any simple equality between corticosteroids and stress needs to accommodate the observation that animals sometimes work to increase their corticosteroid levels.
4. The ability to perform species-typical behaviour is of some moment in achieving end-points (e.g. to perform normal feeding patterns in the gaining of nutrients) and welfare needs to take this into account by allowing a facility for their performance.
5. It is suggested that stereotypies appear when there is inadequate restraint on lower levels of behavioural organization. This could provide a rationale for environmental enrichment. The ideas developed here appear to be neutral on the issue of whether, in a given context, the performance of stereotypies (as opposed to their non-performance) has any implications for welfare.

6. Meeting welfare needs means that individual differences in levels of control need consideration.

Conclusion

A multi-level control model involving both cognitive and S–R aspects might help to explain a number of features of behaviour and could prove applicable to some of the 'anomalies' that are of the greatest concern to students of stress and welfare. Evidence from both stereotypies (Rubovits and Klawans, 1972) and maze-learning tasks (Packard *et al.*, 1989) suggests that the caudate nucleus of the basal ganglia is an important site for the organization of S–R processes, whereas the hippocampus is involved in mediating cognitive control. This chapter provides a tentative organizing framework rather than a definitive statement and it is hoped that the ideas expressed here can serve as a stimulus to further research.

Acknowledgements

I am most grateful to Kent Berridge of the University of Michigan, Joy Mench and two anonymous referees for their very useful comments on an earlier version of the chapter.

References

Adams, C.D. (1982) Variations in the sensitivity of instrumental responding to reinforcer devaluation. *Quarterly Journal of Experimental Psychology* 34B, 77–98.

Aldridge, J.W., Berridge, K.C., Herman, M. and Zimmer, L. (1993) Neuronal coding of serial order: syntax of grooming in the neostriatum. *Psychological Science* 4, 391–395.

Altman, J., Brunner, R.L. and Bayer, S.A. (1973) The hippocampus and behavioural maturation. *Behavioural Biology* 8, 557–596.

Antelman, S.M. and Caggiula, A.R. (1980) Stress-induced behaviour: chemotherapy without drugs. In: Davidson, J.M. and Davidson, R.J. (eds) *The Psychobiology of Consciousness*. Plenum Press, New York, pp. 65–104.

Arnsten, A.F.T. and Goldman-Rakic, P.S. (1998) Noise stress impairs prefrontal cortical cognitive function in monkeys. *Archives of General Psychiatry* 55, 362–368.

Azrin, N.H., Kaplan, S.J. and Foxx, R.M. (1973) Autism reversal: eliminating stereotyped self-stimulation of retarded individuals. *American Journal of Mental Deficiency* 78, 241–248.

Bakshi, V.P. and Kelley, A.E. (1991) Dopaminergic regulation of feeding behaviour: I. Differential effects of haloperidol microinfusion into three striatal subregions. *Psychobiology* 19, 223–232.

Bardo, M.T. (1998) Neuropharmacological mechanisms of drug reward: beyond dopamine in the nucleus accumbens. *Critical Reviews in Neurobiology* 12, 37–67.

Belthoff, J.R. and Dufty, A.M. (1998) Corticosterone, body condition and locomotor activity: a model for dispersal in screetch-owls. *Animal Behaviour* 55, 405–415.

Benus, I. (1988) Aggression and coping. PhD dissertation, Rijksuniversiteit, Groningen, The Netherlands.

Benus, R.F., Bohus, B., Koolhaas, J.M. and van Oortmerssen, G.A. (1991) Behavioural differences between artificially selected aggressive and non-aggressive mice: response to apomorphine. *Behavioural Brain Research* 43, 203–208.

Berkson, G. (1967) Abnormal stereotyped motor acts. In: Zubin, J. and Hunt, H.F. (eds) *Comparative Psychopathology – Animal and Human*. Grune and Stratton, New York, pp. 76–94.

Berntson, G.G. and Micco, D.J. (1976) Organization of brainstem behavioural systems. *Brain Research Bulletin* 1, 471–483.

Berridge, K.C. (1989a) Progressive degradation of serial grooming chains by descending decerebration. *Behavioural Brain Research* 33, 241–253.

Berridge, K.C. (1989b) Substantia nigra 6-OHDA lesions mimic striatopallidal disruption of syntactic grooming chains: a neural systems analysis of sequence control. *Psychobiology* 17, 377–385.

Berridge, K.C. and Fentress, J.C. (1987a) Disruption of natural grooming chains after striatalpallidal lesions. *Psychobiology* 15, 336–342.

Berridge, K.C. and Fentress, J.C. (1987b) Deafferentation does not disrupt natural rules of action syntax. *Behavioural Brain Research* 23, 69–76.

Berridge, K.C. and Robinson, T.E. (1998) What is the role of dopamine in reward: hedonic impact, reward learning or incentive salience? *Brain Research Reviews* 28, 309–369.

Berridge, K.C. and Whishaw, I.Q. (1992) Cortex, striatum and cerebellum: control of serial order in a grooming sequence. *Experimental Brain Research* 90, 275–290.

Bindra, D. (1978) How adaptive behaviour is produced: a perceptual–motivational alternative to response-reinforcement. *Behavioral and Brain Sciences* 1, 41–91.

Blackburn, J.R., Pfaus, J.G. and Phillips, A.G. (1992) Dopamine functions in appetitive and defensive behaviours. *Progress in Neurobiology* 39, 247–279.

Bolles, R.C. (1970) Species-specific defense reactions and avoidance learning. *Psychological Review* 77, 32–48.

Breland, K. and Breland, M. (1961) The misbehaviour of organisms. *American Psychologist* 16, 681–684.

Brown, J.S., Martin, R.C. and Morrow, M.W. (1964) Self-punitive behaviour in the rat: facilitative effects of punishment on resistance to extinction. *Journal of Comparative and Physiological Psychology* 57, 127–133.

Brozoski, T.J., Brown, R.M., Rosvold, E. and Goldman, P.S. (1979) Cognitive deficit caused by regional depletion of dopamine in prefrontal cortex of rhesus monkey. *Science* 205, 929–931.

Cabib, S. (1993) Neurobiological basis of stereotypies. In: Lawrence, A.B. and Rushen, J. (eds) *Stereotypic Animal Behaviour – Fundamentals and Applications to Welfare*. CAB International, Wallingford, UK, pp. 119–145.

Cabib, S. and Puglisi-Allegra, S. (1994) Opposite responses of mesolimbic dopamine system to controllable and uncontrollable aversive experiences. *Journal of Neuroscience* 14, 3333–3340.

Cabib, S., Puglisi-Allegra, S. and Oliverio, A. (1985) A genetic analysis of stereotypy in the mouse: dopaminergic plasticity following chronic stress. *Behavioral and Neural Biology* 44, 239–248.

Cabib, S., Kempf, E., Schleef, C., Oliverio, A. and Puglisi-Allegra, S. (1988) Effects of immobilization stress on dopamine and its metabolites in different brain areas of the mouse: role of genotype and stress duration. *Brain Research* 441, 153–160.

Carr, E.G. (1977) The motivation of self-injurious behaviour: a review of some hypotheses. *Psychological Bulletin* 84, 800–816.

Cohen, J.D. and Servan-Schreiber, D. (1992) Context, cortex, and dopamine: a connectionist approach to behaviour and biology in schizophrenia. *Psychological Review* 99, 45–75.

Cools, A.R. (1988) Transformation of emotion into motion: role of mesolimbic noradrenaline and neostriatal dopamine. In: Hellhammer, D., Florin, I. and Weiner, H. (eds) *Neurobiological Approaches to Human Disease.* Hans Huber, Toronto, pp. 15–28.

Cooper, J.J. and Nicol, C.J. (1991) Stereotypic behaviour affects environmental preference in bank voles, *Clethrionomys glareolus. Animal Behaviour* 41, 971–977.

Cooper, J.J. and Nicol, C.J. (1994) Neighbour effects on the development of locomotor stereotypies in bank voles, *Clethrionomys glareolus. Animal Behaviour* 47, 214–216.

Cooper, J.J., Ödberg, F. and Nicol, C.J. (1996) Limitations on the effectiveness of environmental improvement in reducing stereotypic behaviour in bank voles (*Clethrionomys glareolus*). *Applied Animal Behaviour Science* 48, 237–248.

Corner, M.A. (1990) Brainstem control of behaviour: ontogenetic aspects. In: Klemm, W.R. and Vertes, R.P. (eds) *Brainstem Mechanisms of Behaviour.* Wiley, Chichester, UK, pp. 239–268.

Dantzer, R. (1986) Behavioural, physiological and functional aspects of stereotyped behaviour: a review and a re-interpretation. *Journal of Animal Science* 62, 1776–1786.

Davenport, R.K. and Menzel, E.W. (1963) Stereotyped behaviour of the infant chimpanzee. *Archives of General Psychiatry* 8, 99–104.

Dawkins, M.S. (1990) From an animal's point of view: motivation, fitness, and animal welfare. *Behavioral and Brain Sciences* 13, 1–61.

De Passillé, A.M.B., Christopherson, R.J. and Rushen, J. (1991) Sucking behaviour affects the post-prandial secretion of digestive hormones in the calf. In: Appleby, M.C., Horrell, R.I., Petherick, J.C. and Rutter, S.M. (eds) *Applied Animal Behaviour: Past, Present and Future* Universities Federation for Animal Welfare, Potters Bar, UK, pp. 130–131.

De Passillé, A.M.B., Christopherson, R. and Rushen, J. (1993) Nonnutritive sucking by the calf and postprandial secretion of insulin, CCK, and gastrin. *Physiology and Behavior* 54, 1069–1073.

Deroche, V., Piazza, P.V., Deminière, J.-M., Le Moal, M. and Simon, H. (1993) Rats orally self-administer corticosterone. *Brain Research* 622, 315–320.

Deutch, A.Y., Clark, W.A. and Roth, R.H. (1990) Prefrontal cortical dopamine depletion enhances the responsiveness of mesolimbic neurons to stress. *Brain Research* 521, 311–315.

Devenport, L.D. and Holloway, F.A. (1980) The rat's resistance to superstition: role of the hippocampus. *Journal of Comparative and Physiological Psychology* 94, 691–705.

Devenport, L.D., Hale, R.L. and Stidham, J.A. (1988) Sampling behaviour in the radial maze and operant chamber: role of the hippocampus and prefrontal area. *Behavioural Neuroscience* 102, 489–498.

Diamond, A., Ciaramitaro, V., Donner, E., Djali, S. and Robinson, M.B. (1994) An animal model of early-treated PKU. *Journal of Neuroscience* 14, 3072–3082.

Dickinson, A. (1985) Actions and habits: the development of behavioural autonomy. *Philosophical Transactions of the Royal Society of London, Series B* 308, 67–78.

Duncan, I.J.H. and Petherick, J.C. (1991) The implications of cognitive processes for animal welfare. *Journal of Animal Science* 69, 5017–5022.

Eichenbaum, H. (1994) The hippocampal system and declarative memory in humans and animals: experimental analysis and historical origins. In: Schacter, D.L. and Tulving, E. (eds) *Memory Systems 1994*. The MIT Press, Cambridge, Massachusetts, pp. 147–201.

Einon, D., Morgan, M.J. and Sahakian, B.J. (1975) The development of intersession habituation and emergence in socially reared and isolated rats. *Developmental Psychobiology* 8, 553–559.

Ellinwood, E.H. and Escalante, O. (1970) Chronic amphetamine effect on the olfactory forebrain. *Biological Psychiatry* 2, 189–203.

Ellinwood, E.H. and Kilbey, M.M. (1975) Amphetamine stereotypy: the influence of environmental factors and prepotent behavioural patterns on its topography and development. *Biological Psychiatry* 10, 3–16.

Epstein, A.N. (1982) Instinct and motivation as explanations for complex behaviour. In: Pfaff, D.W. (ed.) *The Physiological Mechanisms of Motivation*. Springer, New York, pp. 25–58.

Fentress, J.C. (1976) Dynamic boundaries of patterned behaviour: interaction and self-organization. In: Bateson, P.P.G. and Hinde, R.A. (eds) *Growing Points in Ethology*. Cambridge University Press, Cambridge, pp. 135–169.

Fibiger, H.C., Phillips, A.G. and Zis, A.P. (1974) Deficits in instrumental responding after 6-hydroxydopamine lesions of the nigro-neostriatal dopaminergic projection. *Pharmacology, Biochemistry and Behaviour*, 2, 87–96.

Fibiger, H.C., Zis, A.P. and Phillips, A.G. (1975) Haloperidol-induced disruption of conditioned avoidance responding: attenuation by prior training or by anticholinergic drugs. *European Journal of Pharmacology*, 30, 309–314.

Floresco, S.B., Seamans, J.K. and Phillips, A.G. (1996) A selective role for dopamine in the nucleus accumbens of the rat in random foraging but not delayed spatial win-shift based foraging. *Behavioural Brain Research*, 80, 161–168.

Gallagher, M. (1997) Animal models of memory impairment. *Philosophical Transactions of the Royal Society of London, Series B* 352, 1711–1717.

Gallistel, C.R. (1980) *The Organization of Action – A New Synthesis*. Lawrence Erlbaum, Hillsdale, New Jersey.

Gibertini, M., Newton, C., Friedman, H. and Klein, T.W. (1995a) Spatial learning impairment in mice infected with *Legionella pneumophila* or administered exogenous interleukin-1-β. *Brain, Behavior, and Immunity* 9, 113–128.

Gilbertini, M., Newton, C., Klein, T.W. and Friedman, H. (1995b) *Legionella pneumophila*-induced visual learning impairment reversed by anti-interleukin-1β. *Proceedings of the Society for Experimental Behavioural Medicine* 210, 7–11.

Glassman, R. (1976) A neural systems theory of schizophrenia and tardive dyskinesia. *Behavioral Science* 21, 274–288.

Glavin, G.B. (1993) Vulnerability to stress ulcerogenesis in rats differing in anxiety: a dopaminergic correlate. *Journal of Physiology (Paris)* 87, 239–243.

Glavin, G.B., Murison, R., Overmier, J.B., Pare, W.P., Bakke, H.K., Henke, P.G. and Hernandez, D.E. (1991) The neurobiology of stress ulcers. *Brain Research Reviews* 16, 301–343.

Hall, F.S. (1998) Social deprivation of neonatal, adolescent, and adult rats has distinct neurochemical and behavioural consequences. *Critical Reviews in Neurobiology* 12, 129–162.

Harlow, H.F. and Harlow, M.K. (1962) Social deprivation in monkeys. *Scientific American* 207(5), 136–146.

Hart, B.L. (1988) Biological basis of the behaviour of sick animals. *Neuroscience and Biobehavioral Reviews* 12, 123–137.

Hediger, H. (1964) *Wild Animals in Captivity*. Dover Publications, New York.

Hirsh, R. (1974) The hippocampus and contextual retrieval of information from memory: a theory. *Behavioural Biology* 12, 421–444.

Hoyle, G. (1984) The scope of neuroethology. *Behavioral and Brain Sciences* 7, 367–412.

Jacobs, W.J. and Nadel, L. (1985) Stress-induced recovery of fears and phobias. *Psychological Review* 92, 512–531.

Jensen, P. and Toates, F.M. (1997) Stress as a state of motivational systems. *Applied Animal Behaviour Science* 53, 145–156.

Jones, G.H., Mittleman, G. and Robbins, T.W. (1989) Attenuation of amphetamine-stereotypy by mesostriatal dopamine depletion enhances plasma corticosterone: implications for stereotypy as a coping response. *Behavioral and Neural Biology* 51, 80–91.

Katz, R.J. (1982) Dopamine and the limits of behavioural reduction – or why aren't all schizophrenics fat and happy. *Behavioral and Brain Sciences* 5, 60–61.

Keefe, K.A., Salamone, J.D., Zigmond, M.J. and Stricker, E.M. (1989) Paradoxical kinesia in Parkinsonism is not caused by dopamine release. Studies in an animal model. *Archives of Neurology* 46, 1070–1075.

Kehoe, P., Shoemaker, W.J., Triano, L., Callahan, M. and Rappolt, G. (1998) Adult rats stressed as neonates show exaggerated behavioural responses to both pharmacological and environmental challenges. *Behavioural Neuroscience* 112, 116–125.

Kelleher, R.T. and Morse, W.H. (1968) Schedules using noxious stimuli. III. Responding maintained with response-produced electric shocks. *Journal of the Experimental Analysis of Behavior* 11, 819–838.

Kelley, A.E., Lang, C.G. and Gauthier, A.M. (1988) Induction of oral stereotypy following amphetamine microinjection into a discrete subregion of the striatum. *Psychopharmacology* 95, 556–559.

Kennes, D., Ödberg, F.O., Bouquet, Y. and De Rycke, P.H. (1988) Changes in naloxone and haloperidol effects during the development of captivity-induced jumping stereotypy in bank voles. *European Journal of Pharmacology* 153, 19–24.

Klemm, W.R. and Vertes, R.P. (1990) *Brainstem Mechanisms of Behaviour.* John Wiley & Sons, Chichester, UK.

Kraemer, G.W., Ebert, M.H., Lake, C.R. and McKinney, W.T. (1983) Amphetamine challenge: effects in previously isolated rhesus monkeys and implications for animal models of schizophrenia. In: Miczek, K.A. (ed.) *Ethopharmacology: Primate Models of Neuropsychiatric Disorders.* Alan R. Liss, New York, pp. 199–218.

Le Moal, M. and Simon, H. (1991) Mesocorticolimbic dopaminergic network: functional and regulatory roles. *Physiological Reviews* 71, 155–234.

Levy, D.M. (1944) On the problem of movement restraint – tics, stereotyped movements, hyperactivity. *American Journal of Orthopsychiatry* 14, 644–671.

Lewis, M.H., Gluck, J.P., Bodfish, J.W., Beauchamp, A.J. and Mailman, R.B. (1996) Neurobiological basis of stereotyped movement disorder. In: Sprague, R.L. and Newell, K.M. (eds) *Stereotyped Movements.* American Psychological Association, Washington, pp. 37–67.

Lorenz, K.Z. (1981) *The Foundations of Ethology.* Springer-Verlag, New York.

Lupien, S.J., de Leon, M., de Santi, S., Convit, A., Tarshish, C., Nair, N.P.V., Thakur, M., McEwen, B.S., Hauger, R.L. and Meaney, M.J. (1998) Cortisol levels during human aging predict hippocampal atrophy and memory deficits. *Nature Neuroscience* 1, 69–73.

McEwen, B.S., De Kloet, E.R. and Rostene, W. (1986) Adrenal steroid receptors and actions in the nervous system. *Physiological Reviews* 66, 1121–1188.

McFarland, D.J. (1966) On the causal and functional significance of displacement activities. *Zeitschrift für Tierpsychologie* 23, 217–235.

MacLennan, A.J. and Maier, S.F. (1983) Coping and the stress-induced potentiation of stimulant stereotypy in the rat. *Science* 219, 1091–1093.

Maier, S.F. and Watkins, L.R. (1998) Cytokines for psychologists: implications of bidirectional immune-to-brain communication for understanding behaviour, mood, and cognition. *Psychological Review* 105, 83–107.

Mason, G. and Mendl, M. (1997) Do the stereotypies of pigs, chickens and mink reflect adaptive species differences in the control of foraging? *Applied Animal Behaviour Science* 53, 45–58.

Mason, G.J. (1993) Forms of stereotypic behaviour. In: Lawrence, A.B. and Rushen, J. (eds) *Stereotypic Animal Behaviour – Fundamentals and Applications to Welfare.* CAB International, Wallingford, UK, pp. 7–40.

Meyer-Holzapfel, M. (1968) Abnormal behaviour in zoo animals. In: Fox, W.W. (ed.) *Abnormal Behaviour in Animals.* W.B.Saunders, Philadelphia, pp. 476–503.

Mittleman, G. and Valenstein, E.S. (1985) Individual differences in non-regulatory ingestive behaviour and catecholamine systems. *Brain Research* 348, 112–117.

Morris, R.G.M. (1983) Neural substrates of exploration in rats. In: Archer, J. and Birke, L.I.A. (eds) *Exploration in Animals and Humans.* Van Nostrand, Wokingham, UK, pp. 117–146.

Oitzl, M.S., de Kloet, R., Joëls, M., Schmid, W. and Cole, T.J. (1997) Spatial learning deficits in mice with a targeted glucocorticoid receptor gene disruption. *European Journal of Neuroscience* 9, 2284–2296.

O'Keefe, J. and Nadel, L. (1978) *The Hippocampus as a Cognitive Map.* The Clarendon Press, Oxford.

Olton, D.S., Becker, J.T. and Handelmann, G.E. (1979) Hippocampus, space and memory. *Behavioral and Brain Sciences* 2, 313–365.

Packard, M.G., Hirsh, R. and White, N.M. (1989) Differential effects of fornix and caudate nucleus lesions on two radial maze tasks: evidence for multiple memory systems. *Journal of Neuroscience* 9, 1465–1472.

Panksepp, J. (1998a) *Affective Neuroscience.* Oxford University Press, New York.

Panksepp, J. (1998b) Attention deficit hyperactivity disorders, psychostimulants, and intolerance of childhood playfulness: a tragedy in the making? *Current Directions in Psychological Science* 7, 91–98.

Piazza, P.V., Ferdico, M., Russo, D., Crescimanno, G., Benigno, A. and Amato, G. (1989a) Circling behaviour: ethological analysis and functional considerations. *Behavioural Brain Research* 31, 267–271.

Piazza, P.V., Ferdico, M., Russo, D., Crescimanno, G., Benigno, A. and Amato, G. (1989b) The influence of dopaminergic A10 neurons on the motor pattern evoked by substantia nigra (pars compacta) stimulation. *Behavioural Brain Research* 31, 273–278.

Piazza, P.V., Deminiere, J.M., LeMoal, M. and Simon, H. (1990) Stress- and pharmacologically-induced behavioural sensitization increases vulnerability to acquisition of amphetamine self-administration. *Brain Research* 514, 22–26.

Posluns, D. (1962) An analysis of chlorpromazine-induced suppression of the avoidance response. *Psychopharmacologia* 3, 361–373.

Ray, O.S. and Bivens, L.W. (1966) Performance as a function of drug, dose, and level of training. *Psychopharmacologia* 10, 103–109.

Ridley, R.M. (1994) The psychology of perseverative and stereotyped behaviour. *Progress in Neurobiology* 44, 221–231.

Ridley, R.M. and Baker, H.F. (1982) Stereotypy in monkeys and humans. *Psychological Medicine* 12, 61–72.

Ridley, R.M. and Baker, H.F. (1983) Is there a relationship between social isolation, cognitive inflexibility, and behavioural stereotypy? An analysis of the effects of amphetamine in the marmoset. In: Miczek, K.A. (ed.) *Ethopharmacology: Primate Models of Neuropsychiatric Disorders.* Alan R. Liss, New York, pp. 101–135.

Robbins, T.W. (1976) Relationship between reward-enhancing and stereotypical effects of psychomotor stimulant drugs. *Nature* 264, 57–59.

Robbins, T.W. (1978) A strange scientific tail. *New Scientist* 79, 764–766.

Robbins, T.W. and Koob, G.F. (1980) Selective disruption of displacement behaviour by lesions of the mesolimbic dopamine system. *Nature* 285, 409–412.

Robinson, P., Daley, M. and Wolff, P.C. (1967) Apomorphine induced reinforcement. *Psychonomic Science* 7, 117–118.

Robinson, T.E. and Berridge, K.C. (1993) The neural basis of drug craving: an incentive-sensitization theory of addiction. *Brain Research Reviews* 18, 247–291.

Rubovits, R. and Klawans, H.L. (1972) Implications of amphetamine-induced stereotyped behaviour as a model for tardive dyskinesias. *Archives of General Psychiatry* 27, 502–507.

Rushen, J., Lawrence, A.B. and Terlouw, E.M.C. (1993) The motivational basis of stereotypies. In: Lawrence, A.B. and Rushen, J. (eds) *Stereotypic Animal Behaviour – Fundamentals and Applications to Welfare.* CAB International, Wallingford, UK, pp. 41–64.

Sahakian, B.J. and Robbins, T.W. (1975) The effects of test environment and rearing condition on amphetamine-induced stereotypy in the guinea pig. *Psychopharmacologia* 45, 115–117.

Sahakian, B.J. and Robbins, T.W. (1977) Isolation-rearing enhances tail-pinch induced oral behaviour in rats. *Physiology and Behavior* 18, 53–58.

Sahakian, B.J., Robbins, T.W., Morgan, M.J. and Iversen, S.D. (1975) The effects of psychomotor stimulants on stereotypy and locomotor activity in socially-deprived and control rats. *Brain Research* 84, 195–205.

Sahakian, B.J., Robbins, T.W. and Iversen, S.D. (1977) The effects of isolation rearing on exploration in the rat. *Animal Learning and Behavior* 5, 193–198.

Salamone, J.D., Cousins, M.S. and Snyder, B.J. (1997) Behavioural functions of nucleus accumbens dopamine: empirical and conceptual problems with the anhedonia hypothesis. *Neuroscience and Biobehavioral Reviews* 21, 341–359.

Scatton, B., D'Angio, M., Driscoll, P. and Serrano, A. (1988) An *in vivo* voltammetric study of the response of mesocortical and mesoaccumbens dopaminergic neurons to environmental stimuli in strains of rats with differing levels of emotionality. *Annals of the New York Academy of Sciences* 537, 124–137.

Schiff, S.R. (1982) Conditioned dopaminergic activity. *Biological Psychiatry* 17, 135–154.

Schwartz, B. (1985) On the organization of stereotyped response sequences. *Animal Learning and Behavior* 13, 261–268.

Segal, E.F. (1990) Animal well-being: there are many paths to enlightenment. *Behavioral and Brain Sciences* 13, 36–37.

Serruya, D. and Eilam, D. (1996) Stereotypies, compulsions, and normal behaviour in the context of motor routines in the rock hyrax (*Procavia capensis*). *Psychobiology* 24, 235–246.

Silverin, B. (1997) The stress response and autumn dispersal behaviour in willow tits. *Animal Behaviour* 53, 451–459.

Singh, S.D. (1964) Habit strength and drug effects. *Journal of Comparative and Physiological Psychology* 58, 468–469.

Spealman, R.D. (1979) Behaviour maintained by termination of a schedule of self-administered cocaine. *Science* 204, 1231–1233.

Steiner, S.S., Beer, B. and Shaffer, M.M. (1969) Escape from self-produced rates of brain stimulation. *Science* 163, 90–91.

Stevens, J., Livermore, A. and Cronan, J, (1977) Effects of deafening and blindfolding on amphetamine induced stereotypy in the cat. *Physiology and Behavior* 18, 809–812.

Stewart, J., de Wit, H. and Eikelboom, R. (1984) Role of unconditioned and conditioned drug effects in the self-administration of opiates and stimulants. *Psychological Review* 91, 251–268.

Swerdlow, N.R. and Koob, G.F. (1984) Restrained rats learn amphetamine-conditioned locomotion, but not place preference. *Psychopharmacology* 84, 163–166.

Taché, Y. (1991) Effect of stress on gastric ulcer formation. In: Brown, M.R., Koob, G.F. and Rivier, C. (eds) *Stress – Neurobiology and Neuroendocrinology*. Marcel Dekker, New York, pp. 549–564.

Terlouw, E.M.C. (1993) Environmental and individual factors contributing to the occurrence of stereotypies in female pigs (*Sus scrofa*). PhD thesis, University of Groningen, The Netherlands.

Terlouw, E.M.C., Lawrence, A.B. and Illius, A.W. (1991) Relationship between agonistic behaviour and propensity to develop excessive drinking and chain manipulation in pigs. *Physiology and Behavior* 50, 493–498.

Thelen, E. (1979) Rhythmical stereotypies in normal human infants. *Animal Behaviour* 27, 699–715.

Thelen, E. (1996) Normal infant stereotypies: a dynamic systems approach. In: Sprague, R.L. and Newell, K.M. (eds) *Stereotyped Movements*. American Psychological Association, Washington, pp. 139–165.

Tinbergen, N. (1969) *The Study of Instinct*. Clarendon Press, Oxford.

Toates, F. (1995) *Stress – Conceptual and Biological Aspects*. John Wiley & Sons, Chichester, UK.

Toates, F. (1998a) The interaction of cognitive and stimulus-response processes in the control of behaviour. *Neuroscience and Biobehavioral Reviews* 22, 59–83.

Toates, F. (1998b) Biological bases of behaviour. In: Eysenck, M. (ed.) *Psychology – An Integrated Approach*. Addison Wesley Longman, Harlow, UK, pp. 23–67.

Toates, F. (1998c) Sensory systems. In: Eysenck, M. (ed.) *Psychology – An Integrated Approach*. Addison Wesley Longman, Harlow, UK, pp. 100–137.

Turkkan, J.S. (1990) Paradoxical experimental outcomes and animal suffering. *Behavioral and Brain Sciences* 13, 42–43.

Uno, H., Tarara, R., Else, J.G., Suleman, M.A. and Sapolsky, R.M. (1989) Hippocampal damage associated with prolonged and fatal stress in primates. *Journal of Neuroscience* 9, 1705–1711.

Ursin, H. (1988) Expectancy and activation: an attempt to systematize stress theory. In: Hellhammer, D., Florin, I. and Weiner, H. (eds) *Neurobiological Approaches to Human Disease*. Hans Huber, Toronto, pp. 313–334.

Valenstein, E.S. (1969) Behaviour elicited by hypothalamic stimulation – a prepotency hypothesis. *Brain Behaviour and Evolution* 2, 295–316.

Vestergaard, K. (1980) The regulation of dustbathing and other behaviour patterns in the laying hen: a Lorenzian approach. In: Moss, R. (ed.) *The Laying Hen and its Environment*. Martinus Nijhoff, The Hague, The Netherlands, pp. 101–113.

Von Holst, D. (1986) Vegetative and somatic components of tree shrews' behaviour. *Journal of the Autonomic Nervous System* (Suppl.) 657–670.

Walker, S. (1990) Natural and unnatural justice in animal care. *Behavioral and Brain Sciences* 13, 43.

Wayner, M.J., Barone, F.C. and Loullis, C.C. (1981) The lateral hypothalamus and adjunctive behaviour. In: Morgane, P.J. and Panksepp, J. (eds) *Handbook of the Hypothalamus* vol. 3, Part B, *Behavioural Studies of the Hypothalamus*. Marcel Dekker, New York, pp. 107–145.

Weiss, J.M. (1971) Effects of coping behaviour in different warning signal conditions on stress pathology in rats. *Journal of Comparative and Physiological Psychology* 77, 1–13.

Weiss, J.M., Pohorecky, L.A., Salman, S. and Gruenthal, M. (1976) Attenuation of gastric lesions by psychological aspects of aggression in rats. *Journal of Comparative and Physiological Psychology* 90, 252–259.

Wemelsfelder, F. (1993) The concept of animal boredom and its relationship to stereotyped behaviour. In: Lawrence, A.B. and Rushen, J. (eds) *Stereotypic*

Animal Behaviour – Fundamentals and Applications to Welfare. CAB International, Wallingford, UK, pp. 65–95.

Whishaw, I.Q. and Kolb, B. (1984) Decortication abolishes place but not cue learning in rats. *Behavioural Brain Research* 11, 123–134.

Whishaw, I.Q., Mittleman, G., Bunch, S.T. and Dunnett, S.B. (1987) Impairments in the acquisition, retention and selection of spatial navigation strategies after medial caudate–putamen lesions in rats. *Behavioural Brain Research* 24, 125–138.

White, N.M. (1989) Reward or reinforcement: what's the difference? *Neuroscience and Biobehavioral Reviews* 13, 181–186

Wiedenmayer, C. (1997) Causation of the ontogenetic development of stereotypic digging in gerbils. *Animal Behaviour* 53, 461–470.

Wiepkema, P.R. (1987) Behavioural aspects of stress. In: Wiepkema, P.R. and Van Adrichem, P.W.M. (eds) *Biology of Stress in Farm Animals: An Integrative Approach.* Martinus Nijhoff, Dordrecht, The Netherlands, pp. 113–133.

Wiepkema, P.R. and Schouten, W.G.P. (1992) Stereotypies in sows during chronic stress. *Psychotherapy and Psychosomatics* 57, 194–199.

Wise, R.A. (1982) Neuroleptics and operant behaviour: the anhedonia hypothesis. *Behavioral and Brain Sciences* 5, 39–87.

Wolff, P.H. (1968) Stereotypic behaviour and development. *Canadian Psychologist* 9, 474–484.

Würbel, H. and Stauffacher, M. (1997) Age and weight at weaning affect corticosterone level and development of stereotypies in ICR-mice. *Animal Behaviour* 53, 891–900.

Würbel, H., Freire, R. and Nicol, C.J. (1998) Prevention of stereotypic wire-gnawing in laboratory mice: effects on behaviour and implications for stereotypy as a coping response. *Behavioural Processes* 42, 61–72.

Young, A.M.J., Joseph, M.H. and Gray, J.A. (1993) Latent inhibition of conditioned dopamine release in rat nucleus accumbens. *Neuroscience* 54, 5–9.

Chronic Social Stress: Studies in Non-human Primates

11

S.P. Mendoza,[1] J.P. Capitanio[1,2] and W.A. Mason[2]

[1]Department of Psychology and [2]California Regional Primate Research Center, University of California, Davis, California, USA

Introduction

In describing the general adaptation syndrome, Hans Selye (1936, 1946) identified three stages of the stress response: the alarm reaction, the stage of resistance and the stage of exhaustion. Most experimental studies of stress have focused on the alarm reaction and the stage of resistance which together are referred to as the acute stress response or, simply, the stress response. Central to the alarm reaction is activation of the hypothalamic–pituitary–adrenal (HPA) system which in turn produces a variety of events that enable the organism to redirect cognitive, behavioural and physiological processes to respond effectively to emergency situations. The stage of resistance inevitably follows the alarm reaction and is the portion of the stress response in which homeostasis is restored and the organism is protected from the most deleterious consequences of stress. This is also the stage of the general adaptation syndrome in which glucocorticoids are elevated and, if resistance is complete, return to basal values. Although Selye noted that excessive elevation of glucocorticoids was detrimental, he also considered the primary role of glucocorticoids to be protective. This sentiment is echoed in the modern literature. Munck and colleagues (1984), for example, emphasize the important role that glucocorticoids play in terminating other aspects of the stress response, thereby preventing damage caused by excess stimulation of neuroendocrine and immunological agents activated early in the cascade of events comprising the acute stress response.

The focus of this chapter is chronic stress or, in Selye's terminology, the stage of exhaustion. Failure to restore homeostasis fully once the acute

 The Biology of Animal Stress (eds G.P. Moberg and J.A. Mench)

response to stress has been mounted is the defining characteristic of the stage of exhaustion. In other words, the stage of exhaustion is initiated if the stage of resistance is not completely successful. In terms of pituitary–adrenal activity this should be reflected in a failure to restore normal baseline levels of glucocorticoids. Although some attenuation of pituitary–adrenal activity may be expected as physiological processes initiated during the stage of resistance attempt to restore homeostasis, the stage of exhaustion is characterized by persistent elevations in glucocorticoids.

Embedded within this traditional view of stress are several assumptions regarding the cause and consequence of chronic stress. First, chronic stress begins with activation of the acute stress response. Second, failure to restore homeostasis following the acute response to stress is reflected in sustained elevations of adrenocortical activity. Third, chronic elevations in pituitary–adrenal activity are deleterious to health. This last assumption has led to the common practice of using glucocorticoids as a metric of stress. Moreover, since excess stimulation by glucocorticoids can be toxic, the notion that sustained elevations in pituitary–adrenal activity, in and of itself, can produce pathological outcomes provides a convincing and ready explanation of the diversity of adverse health outcomes associated with chronic stress (Sapolsky, 1992).

Repeated activation of the acute response to stress is particularly likely to lead ultimately to sustained elevations in glucocorticoids. The details regarding the nature of physiological changes attendant to repeated activation of the acute stress response have certainly been clarified and refined since Selye originally posited his model of stress. However, his model remains the predominant conceptual basis for understanding how stressful events can affect health and well-being. The important body of research regarding repeated stress is covered in several of the other chapters in this volume and will only be briefly touched upon here. The focus of this chapter is on social stress and is particularly concerned with the mechanisms and consequences of chronic social stress. Our studies of non-human primates suggest that social stress can produce either chronic elevations or chronic reductions in adrenocortical activity; that reductions in cortisol, like elevations, are associated with deleterious health outcomes; and that social stress produces chronic changes in pituitary–adrenal activity through mechanisms that differ from those associated with repeated activation of the acute stress response.

The traditional view

The HPA system plays a prominent role in orchestrating the diverse elements of the stress response. The hypothalamus, particularly the paraventricular nucleus, is activated during an emergency reaction probably through inputs from the amygdala and other parts of the central nervous system (CNS) involved in processing fearful or threatening stimuli. Corticotropin-releasing hormone (CRH) is released within seconds of recognition of the emergency and stimulates other elements of the stress response centrally (e.g. by activation of the sympathetic nervous system). It also acts peripherally by stimulating the anterior pituitary to produce and release adrenocorticotropic hormone (ACTH). ACTH acts peripherally to influence many aspects of immune activity and, most importantly, causes release of high levels of corticosteroids from the adrenal cortex. Corticosteroids are elevated in the circulation several minutes after the onset of the stress, too late to participate in the emergency reaction *per se*. However, corticosteroids do inhibit other aspects of the stress response, including further release of CRH and ACTH, and thus restore homeostasis (Munck *et al.*, 1984; Johnson *et al.*, 1992).

Pharmacological studies have amply demonstrated that excess glucocorticoids can have a variety of deleterious physiological consequences (Sapolsky, 1992). An immediate consequence of elevated glucocorticoids is a reduction in the number of glucocorticoid receptors (Young *et al.*, 1990; Song, 1991). Unless there are periods during which corticosteroids are low, these receptors are not replenished. This eventually leads to 'lowered responsiveness or frank resistance of target cells to glucocorticoid hormones' (Song, 1991, p. C172). Glucocorticoid resistance prevents the suppressive role that glucocorticoids ordinarily play during the stress response, allowing those systems activated by acute stressors to continue unchecked for prolonged periods. This includes, of course, continued secretion of the glucocorticoids themselves.

Multiple feedback mechanisms buffer the organism from experiencing many of the potentially harmful effects of glucorticoids (Keller-Wood and Dallman, 1984). For example, fast feedback operates via membrane receptors to reduce CRH and ACTH release immediately in response to CRH stimulation of the anterior pituitary (Widmaier and Dallman, 1984) and terminate the stress-induced activation of the system. Thus, the acute response to a stressor is typically short-lived, and homeostasis is restored relatively quickly, even when the eliciting conditions prevail.

Stimulation of the two main types of receptors for glucocorticoids accounts for both the complexity of feedback mechanisms and some of the deleterious effects of sustained elevations in glucocorticoids. In the CNS, occupation of high-affinity mineralocorticoid receptors (Type I, MRs) by glucocorticoids regulates basal pituitary–adrenal activity at the trough of the circadian cycle. There is increasing evidence that occupation of low-affinity glucocorticoid receptors (Type II, GRs) stimulates GR-responsive genes throughout the CNS and periphery (Bradbury *et al.*, 1991). Stimulation of GRs for a few hours once a day is sufficient to maintain steady-state levels of the gene product, whereas continuous stimulation results in excessive GR-mediated gene expression. In the normal course of events, GRs in the CNS are saturated at the circadian peak and immediately following the acute stress response (Bradbury *et al.*, 1994; Dallman *et al.*, 1994).

One function of GRs is to initiate a slow feedback mechanism that alters subsequent basal levels of ACTH and cortisol and maintains the mean daily output of glucocorticoids even on days when an acute stress response has been activated (Akana *et al.*, 1992). The impact of this type of feedback is, therefore, not evident immediately, but instead acts by modulating subsequent circadian rhythms. This means that under most circumstances the acute stress response may alter the timing of glucocorticoid stimulation of GRs (i.e. post-stress instead of circadian peak) but would not result in excessive stimulation by glucocorticoids. However, if the full panoply of physiological changes that occur with the activation of the stress response has not fully dissipated before the individual encounters another stressor, then mean daily output of glucocorticoids is exceeded, excessive GR stimulation occurs, and deleterious consequences follow.

The timing of repeated stressors is crucial (De Boer *et al.*, 1990). Multiple stressors encountered over prolonged periods of time may not have an impact on health and well-being. Encountering the same number and type of stressors within a more constricted time frame could result in sustained periods of vulnerability due to the catabolic and immunosuppressive actions of cortisol and could even result in irreversible neural damage due to cortisol's cytotoxic actions. Circadian variations in the slow feedback mechanisms suggest that 24 h is a critical time, and we may expect stressors encountered within 24 h of one another to have a greater impact on organismic integrity than stressors separated by more than 24 h. The paradigm employed by Dallman and colleagues of exposing rodents to daily physical restraint (Akana *et al.*, 1992) has proven to be a powerful model for the study of chronic stress and associated pathologies, including immunological suppression, metabolic dysfunction and cell death in the CNS (Sapolsky, 1992).

Studies of social stress in non-human primates

Social stress is particularly effective in producing chronic changes in HPA function (Sapolsky, 1987; 1993; Mendoza *et al.*, 1991; Levine, 1993; Brooke *et al.*, 1994). Our research has capitalized on a broad-based understanding of the social interactions of two species of New World monkeys, the monogamous titi monkeys (*Callicebus moloch*) and the polygynous squirrel monkeys (*Saimiri sciureus*), to investigate the sources of social stress. Alterations of critical aspects of the social situations of these monkeys provide naturalistic models of chronic stress.

Titi monkeys live in family groups consisting of an adult male and female and 1–3 offspring (Mason, 1968). The adults are monogamous and a strong bond develops between the adult male and female that closely resembles the filial attachment bond (Mason and Mendoza, 1998; Mendoza and Mason, 1999). Interactions between adults of the same sex are largely limited to ritualized encounters at territorial boundaries (Mason, 1968; Menzel, 1993). Infants are carried predominantly by the father and transfer to the mother for brief nursing bouts (Fragaszy *et al.*, 1982; Mendoza and Mason, 1986). Infants form an intense attachment bond with their father and a weaker bond with their mother, *neither* of which is reciprocated by the parents (Mendoza and Mason, 1986; Hoffman *et al.*, 1995).

In contrast, squirrel monkeys live in large groups and primarily associate with animals of the same sex and age categories. Relationships among adults can best be characterized as friendships. Intense bonds analogous to the filial attachment bond do not appear to exist among adults (Mendoza *et al.*, 1991). Within-sex dominance relationships are apparent among both males and females (Mendoza *et al.*, 1978a). Interactions between males and females are infrequent unless animals' same-sex partners are not available (Lyons *et al.*, 1992). Infants are cared for exclusively by females, predominantly the mother. Infants form an intense attachment bond with their mother, which is reciprocated (Mendoza *et al.*, 1980).

Given the different types of social organization displayed by these species it is not surprising that conditions that give rise to social stress are also different. For titi monkeys, studies of social stress have focused on disruption of the attachment bond by involuntary separation. For squirrel monkeys, the social environment is more complex in that it involves maintaining diverse relationships with multiple animals, suggesting that social instability produced by formation of new social groups would constitute a potent stressor. In both species, social stress produces prolonged alteration of pituitary–adrenal activity. In each species, however, the details of the response to social stress are not what we predicted based on prevalent models of chronic stress.

Adrenocortical response to social separation

In both squirrel monkeys and titi monkeys, involuntary disruption of an attachment bond by separation evokes an immediate and substantial increase in plasma cortisol levels. This was first demonstrated in a study of the response of 18 squirrel monkey mothers and infants to involuntary separation (Mendoza *et al.*, 1978b). We subsequently showed the same response in titi monkey mates separated from one another (Mendoza and Mason, 1986) and in titi infants separated from their fathers (Hoffman *et al.*, 1995). The adrenocortical response to social separation is only apparent for relationships that are characterized by an attachment bond, a finding that is general among all mammalian species studied thus far (Hennessy, 1997).

In titi monkeys, the adrenocortical response to separation from attachment figures persists through long periods of separation, with little or no attenuation. We examined the long-term response to separation in several contexts. The first was one in which six adult male titi monkeys were separated from their mates and placed in hospital-like surroundings (Mendoza, 1991). Blood samples were collected before animals were moved into the hospital and at weekly intervals for the 3 weeks after they were moved. Cortisol levels remained elevated (at approximately 40 $\mu g\ dl^{-1}$ above basal values) for the entire 3-week period the monkeys were housed in the hospital setting, and there was no difference between day 7 and day 21 levels. We cannot attribute the response solely to separation, however, inasmuch as these monkeys are also strongly reactive to novel environments.

To determine the extent to which separation and the degree of novelty influenced the response to the hospital situation, we monitored the adrenocortical response to separation in titi monkeys placed in three environments: the home cage, a new cage in the same room as the home cage, and a new cage in a new room (Hennessy *et al.*, 1995). All cages were identical in internal dimensions and perch arrangements. Blood samples were collected at 1, 25 and 49 h after moving the monkeys ($n = 10$, five pairs) to the test cage (or following capture and return to the home cage) when the monkeys were with or without their mate. Titi monkeys responded to both novelty and mate separation. The presence of the mate attenuated but did not eliminate the response to the novel environments. Separation effects were apparent in all environments for the 2-day test period. The independent response to a new cage in a familiar room dissipated within 24 h.

We have extended separation time in studies in which adult and juvenile offspring were separated from their parents. Results from one study established that even older offspring respond to removal of their parents for extended periods of time (Valeggia, 1996). Subjects were in three age groups: 6–12 months of age (infants, $n = 8$), 1–2 years of age (juveniles, $n = 7$), and 2–3 years of age (adults, $n = 6$). Each family group consisted of one to three offspring. Blood samples were collected 1 day and 5 days following removal of the parents. The results from the separation condition were compared with a separation–reunion condition in which the parents were separated from their offspring for 1–2 h the day before the first blood sample was obtained. All offspring responded to separation from their parents with an increase in plasma cortisol levels, even though all were tested in a familiar environment and 20 of the 21 subjects were housed with siblings throughout the separation period. As in the previous studies, the separation response did not attenuate during the 5-day period.

In another study, adult daughters ($n = 8$) were separated from their natal groups for 30 days and placed alone in a new cage in the same colony room and then returned to their family groups and monitored for the ensuing 30 days (Valeggia, 1996; Hoffman, 1998). Blood samples were collected before the relocation, at 1, 8, 15, 22 and 29 days following separation, and at 4 h (day 1), 2, 9, 16 and 23 days following reunion. Plasma cortisol levels were elevated on day 1 of separation relative to pre-separation levels and remained elevated throughout the separation period. Although cortisol elevations were evident 4 h after reunion, adrenocortical activity returned to pre-separation levels by the next sample collection time, 2 days after the females had been reunited with their family groups (see Fig. 11.1).

In a related study, we monitored the response of adult titi monkeys to formation of new heterosexual pairs (Hoffman, 1998). Ten new pairs were formed. Most of the subjects (six males and seven females) were living with their parents and siblings prior to pair formation, and thus were separated from their natal groups as in the previous study in order to form new pairs. Blood samples were collected twice a week for 3 weeks before and for 4 weeks following pair formation. Cortisol levels were elevated 24 h following pair formation but, in contrast to the previous study, the presence of a new mate apparently truncated the response to separation from the parents as indicated by the return to pre-separation cortisol levels by day 3 of separation. The response to formation of new pairs was the same for monkeys who were removed from their natal groups and for monkeys who were not currently living with either a mate or their parents at the time of pair formation.

In combination, the results of these studies indicate that disruption of attachment bonds by involuntary separation leads to elevation of plasma cortisol levels within a short time following the onset of separation. The

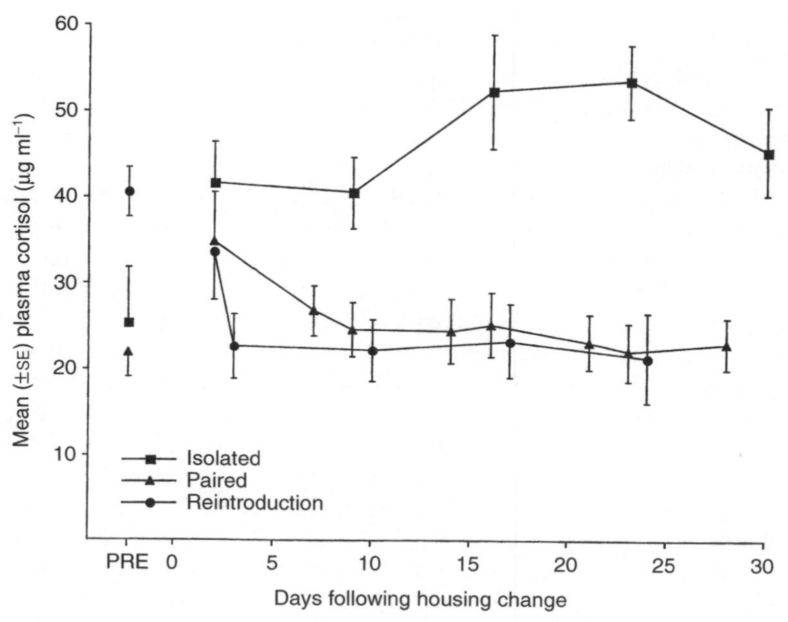

Fig. 11.1. Mean plasma cortisol levels for: female titi monkeys that were removed from their families (isolated) and placed in identical cages in the same colony room; males and female titi monkeys that were formed into new pairs (paired); and female titi monkeys that were returned to their family groups following a prolonged period of isolation from their natal groups (reintroduction).

heightened adrenocortical activity does not attenuate even if the separation period persists for several weeks. Heightened adrenocortical activity appears specific to lack of an appropriate social partner. Restoration of basal values occurs following return of the original attachment figure or provision of a potential attachment figure, as when new pairs are formed. The adrenocortical response is evident whether animals remain in the home cage during separation or are placed in a novel cage. Availability of social partners who are not appropriate attachment figures, such as siblings or offspring, does not alter cortisol levels during separation. Similarly, reduction in cortisol upon return or replacement of an attachment figure occurs in the familiar home cage or in a novel environment at approximately the same rate.

With respect to understanding mechanisms involved in chronic stress, the separation studies in titi monkeys present a particularly interesting outcome. There is no evidence that negative feedback mechanisms operate in any way during the entire separation period. Although the mechanisms accounting for heightened cortisol levels are not known, it is unlikely that loss of glucocorticoid receptors can account for the response. If the high levels of glucocorticoids evident during separation led to down-regulation of glucocorticoid receptors we might expect to see a gradual increase in glucocorticoid levels throughout the separation period as more receptors are lost. Moreover, it may be expected that the negative feedback mechanisms would be disabled for long periods post-separation as the receptor population is replenished. Neither effect is apparent in our data. Cortisol levels are stable throughout the separation period and return promptly to pre-separation levels upon return of attachment figures.

The ineffectiveness of negative feedback mechanisms during separation in titi monkeys is in sharp contrast to the efficacy of negative feedback mechanisms in this species in other situations. A sensitive measure of negative feedback sensitivity is provided by studies of dexamethasone suppression of pituitary–adrenal activity. This synthetic glucocorticoid stimulates glucocorticoid receptors (but not MRs) to suppress pituitary–adrenal activity for several hours following administration. Titi monkeys are more sensitive to lower doses of dexamethasone than are squirrel monkeys and show a much greater diminution of cortisol levels in response to its administration (Mendoza and Moberg, 1985). We have suggested that titi monkeys have very sensitive negative feedback mechanisms that effectively limit the duration of the acute stress response (Cubicciotti *et al.*, 1986; Mendoza and Mason, 1997). Apparently separation somehow inhibits, but does not permanently disable, the negative feedback mechanisms from operating during the period of separation.

There are no studies that compare the consequences of sustained elevation in pituitary–adrenal activity produced by repeated stress with those produced by separation. To the extent that the consequences of stress are due solely to elevated glucocorticoids, the route of activation should not matter in evaluating the consequences of chronic stress. We should not assume, however, that all effects of stress are due to corticosteroids. For example, there is increasing evidence that the amygdala is involved in activation of the pituitary–adrenal system in response to psychosocial stressors but not physical stressors. Amygdalectomized animals do not show the expected elevations in cortisol in response to fear-inducing stimuli, whereas the response to physical restraint is only slightly dampened (Prewitt and Herman, 1994). Since the amygdala has also been implicated in immune processes (Brooks *et al.*, 1982, Grijalva *et al.*, 1990, Nistico *et al.*, 1994), it is possible that psychosocial stressors pose different risks for the individual from physical stressors even though both can lead to excess circulating glucocorticoids.

These results also have implications for welfare issues in the management of these monkeys. It is common practice, for example, to remove sick or injured animals from the colony and place them in small hospital cages for treatment. Our studies suggest that separation from attachment figures in adult and juvenile monkeys induces a physiological state that is believed to induce further pathologies.

Group formation studies in squirrel monkeys

Social instability has long been supposed to induce chronic stress in group-living monkeys and is associated with high mortality (Zuckerman, 1932). In established groups, periods of social instability occur but are infrequent. We attempted to produce social instability in the laboratory by formation of new social groups consisting of individuals who were all unfamiliar with one another. Our initial study of group formation in squirrel monkeys was undertaken to examine the extent and duration of hypothesized increases in pituitary–adrenal activity during the initial stages of new relationship formation. To our surprise, formation of female triads produced a sustained reduction in cortisol levels in all subjects ($n = 15$; Mendoza and Mason, 1991, see Fig. 11.2). This reduction in cortisol levels took several days to become apparent, but persisted for the entire 4-week period during which females lived in isosexual triads. A single male was then introduced to the females. The immediate effect of introduction of the male was to stimulate gonadal hormone activity in females. Cortisol levels in the females remained low for an additional 3 weeks and then began to rise, reaching pre-formation levels by the end of 4 weeks following formation of heterosexual groups. During the month they lived with males, several females became pregnant. It is possible that the increase in cortisol following formation of heterosexual groups occurred as a result of the pregnancy in several of the females. However, even females who did not conceive showed the increase in cortisol when mating activities began in the groups.

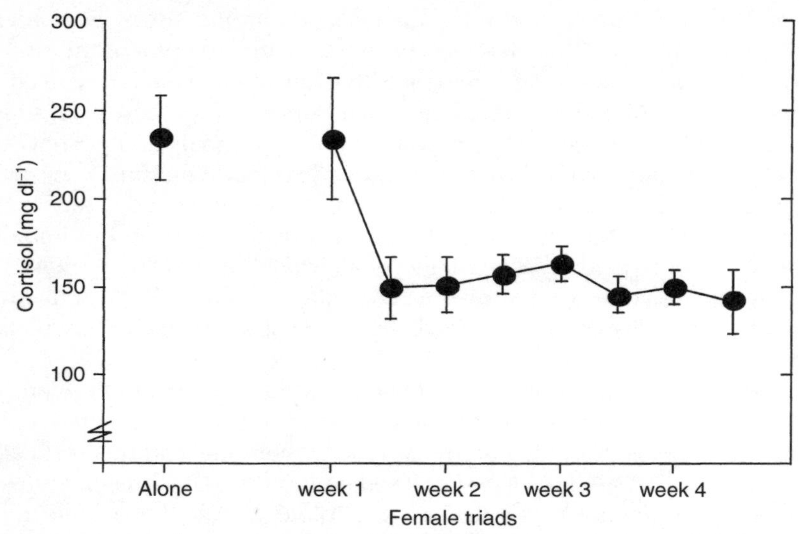

Fig. 11.2. Mean plasma cortisol level for female squirrel monkeys while living alone and for the first 4 weeks following formation of unisexual triads.

The initial reduction in cortisol upon formation of like-sex relationships has since been replicated in several studies of both male and female squirrel monkeys (Saltzman *et al.*, 1991; Mendoza *et al.*, 1992; Lyons *et al.*, 1994). Comparison of circadian rhythms in adrenocortical activity of monkeys living in newly formed social groups (15 males, 14 females) with monkeys housed individually (ten males, eight females) showed that hypocortisolaemia can persist for several months and is apparent in the socially housed animals at all points during the circadian cycle. The social groups for this study were formed in two stages. Isosexual triads were formed first (five male and ten female triads) and 6 months later heterosexual groups were formed by combining a male and a female triad or by adding a single male to a female triad. Circadian rhythms were evaluated at three points following formation of the heterosexual social groups: 5 months, 7 months and 10 months. Blood samples were collected at six time points representing 4-h intervals within the 24 h cycle. Cortisol levels increased in group-housed animals the longer they were together and as the breeding season approached. However, even by the final round of data collection, when the like-sex companions had been together for 16 months (see Fig. 11.3), cortisol levels were lower at each point in the circadian cycle for animals in social groups than for individually housed animals. With the onset of breeding some 18 months following formation of isosexual groups, cortisol levels increased in both group-housed and individually housed animals and the differences between group-housed and individually housed animals were no longer apparent (Schiml *et al.*, 1996, 1999). It is important to note that

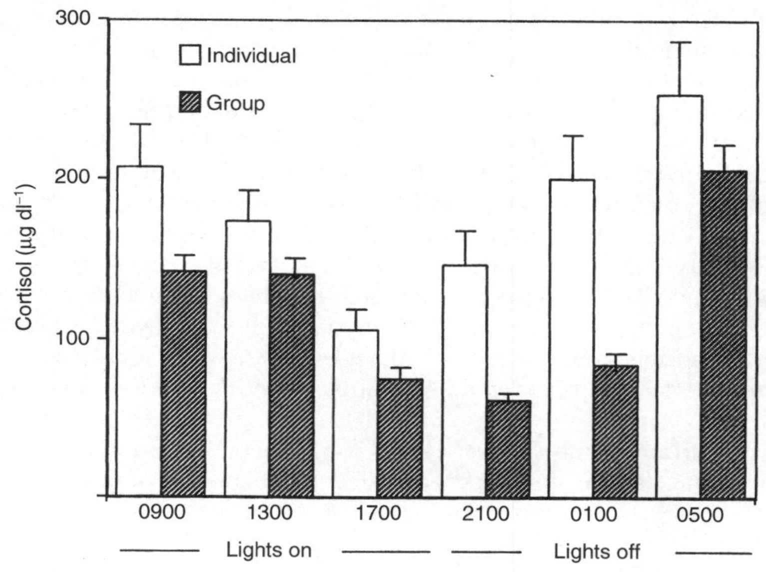

Fig. 11.3. Mean plasma cortisol levels for individually and group-housed squirrel monkeys at six time points in the circadian cycle. Data were collected approximately 2 months prior to breeding, 16 months following formation of isosexual triads and 10 months following formation of heterosexual groups.

cortisol levels for the group-housed animals increased to those exhibited by the individually housed animals during the previous breeding season, and differences between individually housed and group-housed animals did not re-emerge in the next 2 years. Nor did the group-housed animals ever again exhibit the extremely low levels that were evident in the first 18 months following group formation. It thus appears that hypocortisolaemia in group-housed animals is evident only during the period following formation of new social relationships and defines the period of social instability. Moreover, the induction of breeding and the social reorientation that accompanies it reverses the hypocortisolaemia induced by social instability.

Immunodeficiency disease progression in rhesus macaques

Our studies with squirrel monkeys led naturally to the question of whether or not reduced cortisol levels have implications for disease progression. To answer this question we turned to another species more commonly used as subjects in biomedical research, rhesus macaques (*Macaca mulatta*). Like squirrel monkeys, rhesus monkeys are polygynous and live in large social groups comprising several individuals in each age and sex category. We hypothesized that the rhesus monkeys would perceive formation of new social units as a stressful event. Since rhesus monkeys are the principal animal model for studies of immunodeficency virus disease, we tested the hypothesis that social stress would result in faster simian immunodeficiency virus (SIV) disease progression (Capitanio *et al.*, 1998). Monkeys ($n = 36$) were assigned equally to Stable or Unstable experimental conditions and half of the monkeys in each condition were then assigned to either Inoculated ($10^3 TCID_{50}$ $SIV_{mac}251$) or Control groups. All social exposures were conducted within an inoculation condition; that is, inoculated animals never interacted with control animals. Stable monkeys were placed with the same social partners in triads for 110 min day^{-1}, 3–5 days week^{-1}. Unstable subjects had the same amount of social experience, but the identity and number of interactants varied daily; groups ranged in size from two to four monkeys.

Comparison of the survival curves for SIV-inoculated monkeys in the Stable and Unstable conditions revealed that Unstable monkeys had a

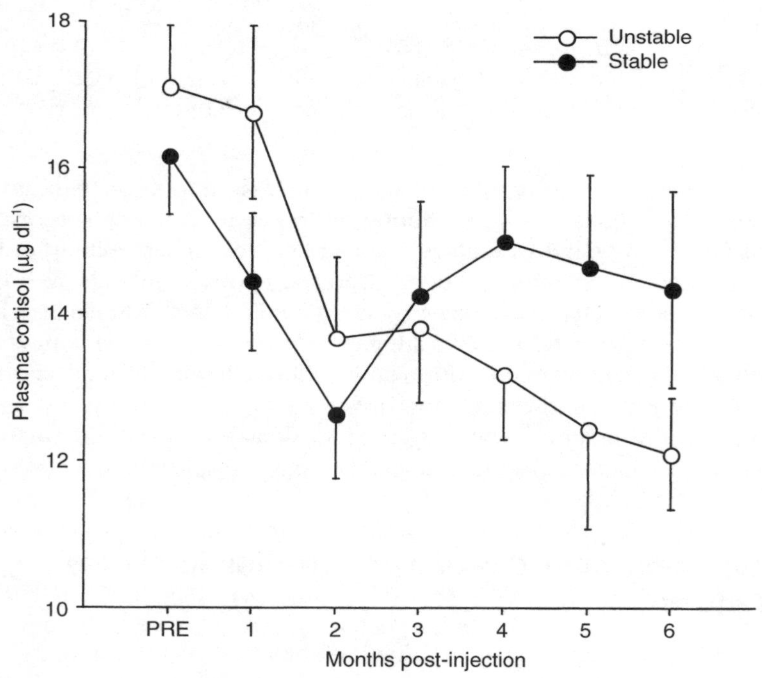

Fig. 11.4. Mean plasma cortisol levels for male rhesus monkeys before (pre) and following inoculation ($SIV_{mac}251$ or saline control). Monkeys in the Stable social condition were placed with the same social partners three to five times per week; monkeys in the Unstable social condition were placed in groups of two to four monkeys of varying identity three to five times per week.

significantly shorter survival (log-rank test, $P < 0.05$). Median survival for Stable monkeys was 588.5 days, with a range of 238 to 1191 days. For Unstable monkeys, median survival was 421 days, with a range of 141 to 576 days. Four Stable monkeys (44.4%) survived beyond the longest surviving Unstable animal. No mortality was observed among Stable or Unstable monkeys who were not inoculated with SIV. Although the differential ability to mount an antibody response to the virus was significantly related to survival, Stable and Unstable monkeys did not differ significantly in formation of SIV-specific antibody and therefore antibody formation to the virus could not account for social condition effects on survival.

A principal hypothesis of the research project, that social condition would result in differences in basal plasma cortisol concentrations, was confirmed. We assayed plasma cortisol from blood samples drawn on the monkeys at 4-week intervals (Fig. 11.4). All such samples were drawn while the monkeys were in their living cages on days when they did not experience their social conditions. Consequently hormonal concentrations do not reflect acute effects of social experience, but rather long-term socially induced alterations of pituitary–adrenal regulation. All blood samples were collected in the afternoon, reflecting values near the trough of the circadian cycle. Basal cortisol concentrations declined for monkeys in both the Stable and Unstable conditions for the first 8 weeks. Cortisol levels in the Stable monkeys subsequently rose, while levels continued to decline for 24 weeks among Unstable monkeys. After this time, cortisol levels increased for Unstable monkeys, and Stable and Unstable monkeys no longer differed consistently. Thus, monkeys in the Unstable social condition showed a more pronounced and longer lasting suppression of cortisol and disease progressed more rapidly.

Altered regulation of the pituitary–adrenal system in these monkeys is also revealed in the acute stress response. The adrenocortical response to physical restraint was evaluated with and without pretreatment with dexamethasone (50 $\mu g\ kg^{-1}$ i.m.) 6 h prior to restraint. These tests were performed at weeks 12 and 18 post-innoculation when differences in basal levels between Stable and Unstable animals first emerged in our data. Unstable monkeys showed a blunted cortisol response over 2 h of restraint and a greater response to dexamethasone. Pretreatment with dexamethasone attenuated, but did not eliminate the differences between Stable and Unstable monkeys in adrenocortical responsiveness to restraint stress. Whether or not the animals had been inoculated with SIV did not have an impact on the pituitary–adrenal response to restraint or dexamethasone.

The pattern of cortisol release found among Unstable monkeys – lower basal concentrations, blunted responsiveness to an acute stressor and greater suppression following dexamethasone challenge – suggests that social instability enhances negative feedback sensitivity. These data are reminiscent of those reported by Yehuda and colleagues (Yehuda *et al.*,

1993, 1995) for patients suffering from post-traumatic stress disorder (PTSD). Because the PTSD research suggested an inverse relationship between cortisol levels and glucocorticoid receptor number on lymphocytes, and because glucocorticoids appear to regulate cytokine production (Munck and Guyre, 1991), we examined whether Stable and Unstable monkeys differed on non-SIV-related measures of immune function at a time when group differences in cortisol were maximal. Monkeys received an intramuscular boost of tetanus toxoid during week 21 post-inoculation. Blood samples were drawn at the time of immunization, and 10 and 24 days later, and were analysed for anti-tetanus antibody using ELISA. Unstable monkeys, which had lower cortisol at the time of immunization, had a significantly higher antibody response than did Stable monkeys. Thus, the ability to form antibodies follows the patterns predicted by the traditional view of cortisol as an immunosuppressive agent and cannot account for adrenocortical influences on survival.

At this point we do not know how reduced cortisol or other aspects of social instability function to promote disease progression. It is clear, however, that hypocortisolaemia is a health risk and can also be used as a marker of social stress. This does not negate the large literature suggesting that cortisol is immunosuppressive, particularly with respect to antibody formation. Our data confirm this association. Clearly, the actions of cortisol on disease progression need to be examined in a broader context in order to elucidate the consequences of both elevations and reductions in cortisol on organismic integrity.

Summary and implications

A variety of conditions considered to be models of chronic stress have been shown to lead to prolonged changes in pituitary–adrenal function (e.g. alcoholism, Eskay *et al.*, 1995; severe environmental conditions, Janssens *et al.*, 1995; repeated encounters with acute stressors, Akana *et al.*, 1992; bereavement, Hofer, 1984; social subordination, De Goeij *et al.*, 1992; Saltzman *et al.*, 1994; Sapolsky, 1995; Spencer *et al.*, 1996; social instability, Sapolsky, 1983; 1993; Capitanio *et al.*, 1998; severe trauma with attendant changes in psychological and social functioning, Bauer *et al.*, 1994; Fleshner *et al.*, 1995). Our research and that of several other laboratories suggest that social stress is particularly effective in producing chronic changes in pituitary–adrenal function (Sapolsky, 1987, 1993; Mendoza *et al.*, 1991; Levine, 1993; Brooke *et al.*, 1994). Clearly, the presumption that chronic stress, like the response to acute stress, involves regulation (or dysregulation) of the HPA system is established.

One consequence of chronic stress is likely to be a change in negative feedback regulation of the pituitary–adrenal system. The way in which negative feedback regulation is altered and how such alteration is reflected in corticosteroids depends on how chronic stress is induced. Repeated activation of the acute stress response leads to reduction in the number of GR and/or neurons responsible for initiating negative feedback mechanisms (Sapolsky, 1992; Dallman *et al.*, 1994; Spencer *et al.*, 1996). Mean daily output of cortisol gradually increases with repeated stress, an effect that takes several days to manifest. In titi monkeys, disruption of attachment relationships also leads to long-term increases in cortisol, but the mechanism appears to be different in that the effect does not develop slowly and is apparently immediately reversible. This implies that with this type of chronic stress (disruption of attachment relationships) negative feedback mechanisms are inhibited but that neither receptor populations nor neurons mediating negative feedback are destroyed.

It is not known, but is important to discover, whether or not different means of producing hypercortisolaemia, by presumably different neurophysiological mechanisms, hold the same risks to health and well-being. The assumption that stress-induced elevations in glucocorticoids account for the health risks associated with chronic stress has rarely been tested directly. There is reason to believe that the means of producing hypercortisolaemia is an important determinant of the consequences of this condition. For example, alterations in pituitary–adrenal responsiveness are different when corticosteroids are elevated by stressors than when corticosteroids are manipulated pharmacologically (Wilkenson *et al.*, 1979; Akana *et al.*, 1992; Strack *et al.*, 1995).

In two species of non-human primates, social instability has been found to produce hypocortisolaemia. The studies of patients with post-traumatic stress disorder (PTSD) suggest that humans are a third species for which chronic stress may lead to reduced cortisol. The variety of conditions that have been implicated in producing PTSD all include disruption of the social environment, suggesting that there is a possibility that the causes of hypocortisolaemia may be similar. More research is necessary, however, before a firm connection between the causes of hypocortisolaemia in human and non-human primates can be established. Reductions in cortisol associated with social instability and PTSD are accompanied by enhanced negative feedback (Yehuda *et al.*, 1993, 1995; Capitanio *et al.*, 1998) and the time course for the reduction in cortisol (days to weeks) suggests that changes in receptor populations may contribute to hypocortisolaemia.

Most of what is known about the potential consequences of a reduction in glucocorticoids comes from studies in which corticoids are completely or nearly eliminated from the circulation, as with adrenalectomy or Addison's disease. Relatively little is known about the potentially deleterious effects of reduced glucocorticoid activity. Studies of PTSD offer the most promising avenue for investigation of stress-induced hypocortisolaemia in human populations. However, to date the research has concentrated on mechanims of reduced cortisol rather than consequences. The need for animal models of hypocortisolaemia is highlighted by criticisms of PTSD as a model of chronic stress since only a portion of the population exposed to a particular traumatic event displays the disorder, leaving the possibility that lowered cortisol creates susceptibility to the disorder rather than resulting from it (Sapolsky, 1996). Nonetheless, studies of various disease processes make it clear that glucocorticoid insufficiency produced by either adrenalectomy or disease can have serious implications for organismic integrity (Tsigos and Chrousos, 1994).

The only study to our knowledge of disease progression in monkeys with hypocortisolaemia comes from our study of rhesus monkeys infected with the simian AIDS virus (Capitanio *et al.*, 1998). We found that social instability was associated with reduced cortisol for several weeks and that disease progression was more rapid in monkeys with greater and more sustained hypocortisolaemia. The mechanism by which lowered cortisol affects progression to simian AIDS is unclear. Low cortisol does not impair antibody production; in fact, antibody production is greater in monkeys with low cortisol. A possible mechanism is via glucocorticoid influences on cytokine production. IL-1 activity, for example, is suppressed by glucocorticoids, but production of the receptor for IL-1 is stimulated by glucocorticoids, creating the need for an optimal level of glucocorticoids for efficient IL-1 activity (Munck and Náray-Fejes-Tóth, 1992).

The most general implication of the research reviewed here is that social stress produces long-term changes in HPA function through mechanisms that are not predicted by traditional models of stress. The concept of stress subsumes multiple distinct syndromes each associated with its own symptoms and risks. A basic distinction is between acute and chronic stress. The most commonly evoked metaphor for understanding the physiological response to stress is a prey's encounter with a predator. Nothing is more important than escape from the certain death that awaits. All physiological systems involved with mobilizing energetic resources are activated while those concerned with immediately inessential processes are shut down. The studies of chronic stress presented here suggest that the acute response to stress may or may not have a direct relationship with the conditions that produce chronic stress. Moreover, the use of cortisol as a metric of stress may have some validity in studies of acute stress but not in evaluating conditions of chronic stress.

Given the complexity of the pituitary–adrenal response to chronic stress, can glucocorticoids continue to be used to monitor stress and well-being? We believe the answer is yes. The methods employed require attention, however, to the potential range of possible responses to chronic stress and recognition that either excess or insufficient glucocorticoids may be indicative of stress.

Acknowledgements

Preparation of this manuscript was supported by grants RR00169 and MH49033. We would like to thank Joy Mench, DeeAnn Reeder and John Ruys for insightful comments on the manuscript and Kurt Hoffman, Patricia Schiml and Claudia Valeggia for the quality dissertations that made the contribution possible.

References

Akana, S.F., Scribner, K.A., Bradbury, M.J., Strack, A.M., Walker, C.D. and Dallman, M.F. (1992) Feedback sensitivity of the rat hypothalamo-pituitary-adrenal axis and its capacity to adjust to exogenous corticosterone. *Endocrinology* 131, 585–594.

Bauer, M., Priebe, S., Graf, K.J. and Kurten, I.B.A. (1994) Psychological and endocrine abnormalities in refugees from East Germany: Part II. Serum levels of cortisol, prolactin, luteinizing hormone, follicle stimulating hormone, and testosterone. *Psychiatry Research* 51, 75–85.

Bradbury, M.J., Akana, S.F., Cascio, C.S., Levin, N., Jacobson, L. and Dallman, M.F. (1991) Regulation of basal ACTH secretion by corticosterone is mediated by both type I (MR) and type II (GR) receptors in rat brain. *Journal of Steroid Biochemistry and Molecular Biology* 40, 133–142.

Bradbury, M.J., Akana, S.F. and Dallman, M.F. (1994) Roles of type I and II corticosteroid receptors in regulation of basal activity in the hypothalamo-pituitary-adrenal axis during the diurnal trough and the peak: evidence for a nonadditive effect of combined receptor occupation. *Endocrinology* 134, 1286–1296.

Brooke, S.M., de Haas-Johnson, A.M., Kaplan, J.R., Manuck, S.B. and Sapolsky, R.M. (1994) Dexamethasone resistance among nonhuman primates associated with a selective decrease of glucocorticoid receptors in the hippocampus and a history of social instability. *Neuroendocrinology* 60, 134–140.

Brooks, W.H., Cross, R.J., Roszman, T.L. and Marksbery, W.R. (1982) Neuroimmunomodulation: neural anatomical basis for impairment and facilitation. *Annals of Neurology* 12, 56–61.

Capitanio, J.P., Mendoza, S.P., Lerche, N.W. and Mason, W.A. (1998) Social stress results in altered glucocorticoid regulation and shorter survival in simian acquired immune deficiency syndrome. *Proceedings of the National Academy of Sciences of the United States of America* 95, 4714–4719.

Cubicciotti, D.D., III, Mendoza, S.P., Mason, W.A. and Sassenrath, E.N. (1986) Differences between *Saimiri sciureus* and *Callicebus moloch* in physiological responsiveness: implications for behavior. *Journal of Comparative Psychology* 100, 385–391.

Dallman, M.F., Akana, S.F., Levin, N., Walker, C.D., Bradbury, M.J., Suemaru, S. and Scribner, K.S. (1994) Corticosteroids and the control of function in the hypothalamo-pituitary–adrenal (HPA) axis. *Annals of the New York Academy of Sciences* 746, 22–31; discussion 31–32, 64–67.

De Boer, S.F., Koopman, S.J., Slangen, J.L. and Van der Gugten, J. (1990) Catecholamine, corticosterone and glucose responses to repeated stress in rats: effect of interstressor interval length. *Physiology and Behavior* 47, 1117–1124.

De Goeij, D.C.E., Dijkstra, H. and Tilders, F.J.H. (1992) Chronic psychosocial stress enchances vasopressin, but not corticotropin-releasing factor, in the external zone of the median eminance of male rats: relationship to subordinate status. *Endocrinology* 131, 847–852.

Eskay, R.L., Chautard, T., Torda, T., Daoud, R.I. and Hamelink, C. (1995) Alcohol, corticosteroids, energy utilization, and hippocampal endangerment. *Annals of the New York Academy of Sciences* 771, 105–114.

Fleshner, M., Deak, T., Spencer, R.L., Laudenslager, M.L., Watkins, L.R. and Maier, S.F. (1995) A long-term increase in basal levels of corticosterone and a decrease in corticosteroid-binding globulin after acute stressor exposure. *Endocrinology* 136, 5336–5342.

Fragaszy, D.M., Schwarz, S. and Shimosaka, D. (1982) Longitudinal observations of care and development of infant titi monkeys (*Callicebus moloch*). *American Journal of Primatology* 2, 191–200.

Grijalva, C.V., Levin, E.D., Morgan, M., Roland, B. and Martin, F.C. (1990) Contrasting effects of centromedial and basolateral amygdaloid lesions on stress-related responses in the rat. *Physiology and Behavior* 48, 495–500.

Hennessy, M.B. (1997) Hypothalamic–pituitary–adrenal responses to brief social separation. *Neuroscience and Biobehavioral Reviews* 21, 11–29.

Hennessy, M.B., Mendoza, S.P., Mason, W.A. and Moberg, G.P. (1995) Endocrine sensitivity to novelty in squirrel monkeys and titi monkeys: species differences in characteristic modes of responding to the environment. *Physiology and Behavior* 57, 331–338.

Hofer, M.A. (1984) Relationships as regulators: a psychobiologic perspective on bereavement. *Psychosomatic Medicine* 46, 183–197.

Hoffman, K.A. (1998) Transition from juvenile to adult stages of development in titi monkeys (*Callicebus moloch*). PhD dissertation, University of California, Davis, California.

Hoffman, K.A., Mendoza, S.P., Hennessy, M.B. and Mason, W.A. (1995) Responses of infant titi monkeys, *Callicebus moloch*, to removal of one or both parents: evidence for paternal attachment. *Developmental Psychobiology* 28, 399–407.

Janssens, C.J., Helmond, F.A., Loyens, L.W., Schouten, W.G. and Wiegant, V.M. (1995) Chronic stress increases the opioid-mediated inhibition of the pituitary-adrenocortical response to acute stress in pigs. *Endocrinology* 136, 1468–1473.

Johnson, E.O., Kamilaris, T.C., Chrousos, G.P. and Gold, P.W. (1992) Mechanisms of stress: a dynamic overview of hormonal and behavioral homeostasis. *Neuroscience and Biobehavioral Reviews* 16, 115–130.

Keller-Wood, M.E. and Dallman, M.F. (1984) Corticosteroid inhibition of ACTH secretion. *Endocrine Reviews* 5, 1–24.

Levine, S. (1993) The influence of social factors on the response to stress. *Psychotherapy and Psychosomatics* 60, 33–38.

Lyons, D.M., Mendoza, S.P. and Mason, W.A. (1992) Sexual segregation in squirrel monkeys (*Saimiri sciureus*): a transactional analysis of adult social dynamics. *Journal of Comparative Psychology* 106, 323–330.

Lyons, D.M., Mendoza, S.P. and Mason, W.A. (1994) Psychosocial and hormonal aspects of hierarchy formation in groups of male squirrel monkeys. *American Journal of Primatology* 32, 109–122.

Mason, W.A. (1968) Use of space by *Callicebus* groups. In: Jay, P.C. (ed.) *Primates: Studies in Adaptation and Variability.* Holt, Rinehart & Winston, New York, pp. 200–216.

Mason, W.A. and Mendoza, S.P. (1998) Generic aspects of attachment: parents, offspring and mates. *Psychoneuroendocrinology* 23, 765–778.

Mendoza, S.P. (1991) Behavioural and physiological indices of social relationships: comparative studies of new world monkeys. In: Box, H.O. (ed.) *Primate Responses to Environmental Change.* Chapman and Hall, London, pp. 311–335.

Mendoza, S.P. and Mason, W.A. (1986) Parental division of labour and differentiation of attachments in a monogamous primate (*Callicebus moloch*). *Animal Behaviour* 34, 1336–1347.

Mendoza, S.P. and Mason, W.A. (1991) Breeding readiness in squirrel monkeys: female-primed females are triggered by males. *Physiology and Behavior* 49, 471–479.

Mendoza, S.P. and Mason, W.A. (1997) Autonomic balance in *Saimiri sciureus* and *Callicebus moloch*: relation to life-style. *Folia Primatologica* 68, 307–318.

Mendoza, S.P. and Mason, W.A. (1999) Attachment relationships in New World primates. In: Carter, C.S., Lederhendler, I.I. and Kirkpatrick, B. (eds) *The Integrative Neurobiology of Affiliation.* MIT Press, Cambridge, Massachusetts, pp. 83–93.

Mendoza, S.P. and Moberg, G.P. (1985) Species differences in adrenocortical activity of new world primates: responses to dexamethasone suppression. *American Journal of Primatology* 8, 215–224.

Mendoza, S.P., Lowe, E.L. and Levine, S. (1978a) Social organization and social behavior in two subspecies of squirrel monkeys (*Saimiri sciureus*). *Folia Primatologica* 30, 126–144.

Mendoza, S.P., Smotherman, W.P., Miner, M., Kaplan, J. and Levine, S. (1978b). Pituitary-adrenal response to separation in mother and infant squirrel monkeys. *Developmental Psychobiology* 11, 169–175.

Mendoza, S.P., Coe, C.L., Smotherman, W.P., Kaplan, J. and Levine, S. (1980) Functional consequences of attachment: a comparison of two species. In: Bell, R.W. and Smotherman, W.P. (eds) *Maternal Influences and Early Behavior.* Spectrum Publications, New York, pp. 235–252.

Mendoza, S.P., Lyons, D.M. and Saltzman, W. (1991) Sociophysiology of squirrel monkeys. *American Journal of Primatology* 23, 37–54.

Mendoza, S.P., Hennessy, M.B. and Lyons, D.M. (1992) Distinct immediate and prolonged effects of separation on plasma cortisol in adult female squirrel monkeys. *Psychobiology* 20, 300–306.

Menzel, C.R. (1993) Coordination and conflict in *Callicebus* social groups. In: Mason, W.A. and Mendoza, S.P. (eds) *Primate Social Conflict.* State University of New York Press, Albany, New York, pp. 253–290.

Munck, A. and Guyre, P.M. (1991) Glucocorticoids and immune function. In: Ader, R., Felten, D.L. and Cohen, N. (eds) *Psychoneuroimmunology*, 2nd edn. Academic Press, New York, pp. 447–474.

Munck, A. and Náray-Fejes-Tóth, A. (1992) The ups and downs of glucocorticoid physiology. Permissive and suppressive effects revisited. *Molecular and Cellular Endocrinology* 90, C1–C4.

Munck, A., Guyre, P.M. and Holbrook, N.J. (1984) Physiological functions of glucocorticoids in stress and their relation to pharmacological actions. *Endocrine Reviews* 5, 25–44.

Nistico, G., Caroleo, M.C., Arbitrio, M. and Pulvirenti, L. (1994) Evidence for an involvement of dopamine D1 receptors in the limbic system in the control of immune mechanisms. *Neuroimmunomodulation* 1, 174–180.

Prewitt, C.M. and Herman, J.P. (1994) Lesions of the central nucleus of the amygdala decreases basal CRH mRNA expression and stress-induced ACTH release. *Annals of the New York Academy of Sciences* 746, 438–440.

Saltzman, W., Mendoza, S.P. and Mason, W.A. (1991) Sociophysiology of relationships in squirrel monkeys. I. Formation of female dyads. *Physiology and Behavior* 50, 271–280.

Saltzman, W., Schultz-Darken, N., Scheffler, G., Wegner, F.H. and Abbott, D.H. (1994) Social and reproductive influences on plasma cortisol in female marmoset monkeys. *Physiology and Behavior* 56, 801–810.

Sapolsky, R.M. (1983) Endocrine aspects of social instability in the olive baboon (*Papio anubis*). *American Journal of Primatology* 5, 365–379.

Sapolsky, R.M. (1987) Stress, social status, and reproductive physiology in free-living baboons. In: Crews, D. (ed.) *Psychobiology of Reproductive Behavior. An Evolutionary Perspective.* Prentice-Hall, New Jersey, pp. 291–322.

Sapolsky, R.M. (1992) *Stress, the Aging Brain, and the Mechanisms of Neuron Death.* MIT Press, Cambridge, Massachusetts.

Sapolsky, R.M. (1993) The physiology of dominance in stable and unstable social hierarchies. In: Mason, W.A. and Mendoza, S.P. (eds) *Primate Social Conflict.* State University of New York Press, Albany, New York, pp.171–204.

Sapolsky, R.M. (1995) Social subordinance as a marker of hypocortisolism: some unexpected subtleties. *Annals of the New York Academy of Sciences* 771, 626–639.

Sapolsky, R.M. (1996) Why stress is bad for your brain. *Science* 273, 749–750.

Schiml, P.A., Mendoza, S.P., Saltzman, W., Lyons, D.M. and Mason, W.A. (1996) Seasonality in squirrel monkeys (*Saimiri sciureus*): social facilitation by females. *Physiology and Behavior* 60, 1105–1113.

Schiml, P.A., Mendoza, S.P., Lyons, D.M. and Mason, W.A. (1999) Annual physiological changes in individually housed squirrel monkeys (*Saimiri sciureus*). *American Journal of Primatology* 47, 93–103.

Selye, H. (1936) The alarm reaction. *Canadian Medical Association Journal* 34, 706.

Selye, H. (1946) The general adaptation syndrome and the diseases of adaptation. *Journal of Clinical Endocrinology* 6, 117–230.

Song, L.N. (1991) Stress-induced changes in glucocorticoid receptors: molecular mechanisms and clinical implications. *Molecular and Cellular Endocrinology* 80, C171–174.

Spencer, R.L., Miller, A.H., Moday, H., McEwen, B.S., Blanchard, R.J., Blanchard, D.C. and Sakai, R.R. (1996) Chronic social stress produces reductions in available splenic type II corticosteroid receptor binding and plasma corticosteroid binding globulin levels. *Psychoneuroendocrinology* 21, 95–109.

Strack, A.M., Bradbury, M.J. and Dallman, M.F. (1995) Corticosterone decreases nonshivering thermogenesis and increases lipid storage in brown adipose tissue. *American Journal of Physiology* 268, R183-R191.

Tsigos, C. and Chrousos, G.P. (1994) Physiology of the hypothalamic–pituitary–adrenal axis in health and dysregulation in psychiatric and autoimmune disorders. *Endocrinology and Metabolism Clinics of North America* 23, 451–466.

Valeggia, C.R. (1996) Social influences on the development of sexual physiology and behavior in titi monkey females (*Callicebus moloch*). PhD dissertation, University of California, Davis, California.

Widmaier, E.P. and Dallman, M.F. (1984) The effects of corticotropin-releasing factor on adrenocorticotropin secretion from perfused pituitaries *in vitro*: rapid inhibition by glucocorticoids. *Endocrinology* 115, 2368–2374.

Wilkenson, C.W., Shinsako, J. and Dallman, M.F. (1979) Daily rhythms in adrenal responsiveness to adrenocorticotropin are determined primarily by the time of feeding in the rat. *Endocrinology* 104, 350–359.

Yehuda, R., Resnick, H., Kahan, B. and Giller, E.L. (1993) Long-lasting hormonal alterations to extreme stress in humans: normative or maladaptive? *Psychosomatic Medicine* 55, 287–297

Yehuda, R., Kahana, B., Binder-Brynes, K., Southwick, S.M., Mason, J.W. and Giller, E.L. (1995) Low urinary cortisol excretion in Holocaust survivors with post-traumatic stress disorder. *American Journal of Psychiatry* 152, 982–986.

Young, E.A., Akana, S. and Dallman, M.F. (1990) Decreased sensitivity to glucocorticoid fast feedback in chronically stressed rats. *Neuroendocrinology* 51, 536–542.

Zuckerman, S. (1932) *The Social Life of Monkeys and Apes*. Routledge, London.

12 Consequences of Stress During Development

D.C. Lay Jr

Department of Animal Science, Iowa State University, Ames, Iowa, USA

Introduction

The deleterious effects of stress on animals are well known. Impaired immune function, gastric ulcers, impaired growth and reproduction, abnormal behaviour and disease are all known to occur when animals are exposed to potent stressors (such as disease, pain, aggression, restraint) and to chronic stressors (such as isolation, social instability, loss of predictability, loss of control). Animals can often cope with stressors if the exposure is of limited duration. However, if the animal cannot escape from the stressor, the classic symptoms of stress appear. Although stress and its effects are commonly discussed, many of the mechanisms by which stress affects the individual are still poorly understood. Psychological stressors are accepted as much more potent challenges than physiological stressors, and chronic stressors are thought to be much more detrimental than acute ones. In addition, not all stress can be considered detrimental. The term 'eustress' is often used to indicate a stressor that may be perceived as pleasurable, such as animals running during play and the act of copulation. In both of these instances, classic responses to stress are invoked: increased heart rate, glucocorticoids, catecholamines, etc. We would not consider these acts necessarily detrimental or harmful to the animal. The term 'distress' is often used to indicate specifically that the stressor may cause harm or decrease the welfare and (or) fitness of the organism. The ideas of eustress and distress and the different mechanisms by which they act to alter the physiology of the animal remain important areas of research today.

Even less understood is the effect that stress may have on the developing individual. The timing of events during fetal and neonatal development is critical to the integrity and normal function of all animals. Slight changes in the typical developmental scheme can have profound effects on the developing fetus, such as the effect found in sheep where accidental ingestion of skunk cabbage (*Veratum californicum*) by the pregnant ewe results in cyclopean offspring that are lacking, among other things, a portion of their pituitary gland. Similarly, in some human males in the Dominican Republic there is a lack of exposure to dihydrotestosterone, due to a deficiency or instability of 5α-reductase during neonatal development and this causes characteristic female phenotypic expression (Griffin and Ojeda, 1992). It is not until plasma testosterone concentrations reach elevated concentrations at puberty, thereby becoming converted to significantly more dihydrotestosterone, that the person becomes phenotypically male. Because stress removes animals from homeostasis, many endocrine and neuroendocrine systems are disturbed. Thus exposure to a stressor during fetal and neonatal life may also have significant effects on the physiology and anatomy of the offspring in question.

The concern as to whether exposing animals to stress during pregnancy will affect their developing offspring is important because many livestock are exposed to stress during gestation, even under conditions of the highest quality management. Common management procedures such as handling (Mitchell *et al.*, 1988; Lay *et al.*, 1992), transportation (Fell and Shutt, 1986; Mitchell *et al.*, 1988; Clark *et al.*, 1993) and changes in housing (Friend *et al.*, 1985, 1988) cause activation of the hypothalamic–pituitary–adrenal (HPA) axis indicative of a stress reaction. A model proposed by Gluckman (1985) suggests that endocrine control between the central and peripheral nervous system is affected by pulsatile release of hypothalamic hormones during fetal development. These hormones act on the pituitary gland to produce an associated peripheral hormone effect, which then negatively feeds back on the hypothalamus to create a 'set point' in organ development. Theoretically, the mechanism controlling this set point is created by increasing or decreasing the number of responsive cells in the organ, the number of receptors on the cells, and (or) the ability of the cell to produce the specified hormone. If the developing fetus is exposed to 'stress' hormones, then its HPA axis could be permanently altered by such a mechanism. In addition, many species, like rodents and humans, are physiologically immature at birth and their HPA axis continues to mature during the neonatal period. Thus, exposure to 'stress' hormones during neonatal development could also permanently alter animals physiologically.

The welfare implications of altering livestock's ability to cope with stress are of extreme importance to livestock producers and the public as a whole. Most current livestock production systems require livestock to adjust to: (i) elevated social pressures due to high stocking densities; (ii) physical and mental stress due to a lack of ability to carry out a range of behaviours; and

(iii) the close proximity of humans. If livestock have an impaired ability to cope with stress then these everyday challenges will negatively affect their welfare. For instance, Broom *et al.* (1995) found that sows housed in stalls, which effectively prevent social interactions and can cause stress due to the inability to perform highly motivated behaviours, exhibit more welfare problems than group-housed sows. The welfare problem is characterized by a greater performance of stereotypies and aggression by stall-housed sows. In addition, an extensive review (Hemsworth and Coleman, 1998) of research investigating the fear induced by close proximity to humans that is experienced by cattle, swine and poultry indicates that the animals experiencing fear due to human–animal interactions can have both reduced production (i.e. growth and reproduction) and reduced welfare (i.e. injuries due to escape responses, increased stereotypies and immunosuppression).

Much of the research on stress during development indicates that animals (mainly rats) neonatally exposed to stressors have an increased ability to cope with stress when they mature. Increased coping and adaptation to stress are evidenced by less time immobile in Porsolt's swim test (Hilakivi-Clarke *et al.*, 1991), less emotionality in an open-field test (Levine *et al.*, 1967), reduced reactions to electric shock (Haltmeyer *et al.*, 1967), lower incidences of ulceration in stressful situations (Lambert *et al.*, 1995) and an increased antibody response to the mitogen keyhole limpet haemocyanin (Klein and Rager, 1995). However, there is also a great deal of research to indicate that stress during development is detrimental to the animal. Prenatal stress causes increased emotionality and HPA response in rhesus monkeys (Clarke and Schneider, 1993; Clarke *et al.*, 1994) and rats (e.g. Henry *et al.*, 1994; McCormick *et al.*, 1995; Vallée *et al.*, 1997; Weinstock, 1997). A great deal more research is required for us to understand this phenomenon. However, if exposing livestock to stress during development can increase their ability to cope with stressful situations, then manipulation of this phenomenon in a controlled manner will allow producers to 'programme' stock for specific management systems. By matching the animal's stress response to its environment, animal welfare could be improved.

This chapter will briefly review work that examines the effects of pre- and postnatal stress on rodents, followed by a more in-depth review of the research using livestock, and conclude with a discussion of possible physiological mechanisms that may be responsible for altering the animal's behavioural and physiological response to stress. The use of the phrase 'prenatal stress' in this chapter refers to the imposition of stress on the dam, not the fetus. This term may be erroneous in that the fetus is not actually stressed, but it will be used by convention.

During this review there are several points to keep in mind. The research protocols used to study either neonatal or prenatal stress have not used the same stressors and the developmental pattern of species is not identical. The 'stressor' could mean that the subjects were subjected to 'manipulation in infancy', electric shock or restraint. Also, 'manipulation in infancy' could mean that the subject was simply picked up and replaced or that it was picked up, isolated for several minutes, stroked and then replaced. Because the stressor varies in each of the experiments, I will usually refer to pre- or neonatal stress, but will also inform the reader of the source of the stress. Another significant difference to consider is the degree of maturity at birth of the species in question. Many studies have used rodents, which are extremely altricial at birth. In contrast, work with guinea pigs and livestock use an animal that is much more precocial.

Although not a great deal is known about the level of development of the HPA axis between species, it is apparent that the fetal adrenal is functional at some level prior to birth as it is critical for parturition in most species. This does not preclude the possibility, however, of maturation of the HPA axis postnatally. Indeed, there is evidence in many animals that adrenal maturation does occur during neonatal life. For example, human infants have a fetal zone at birth that regresses by their first year of life (Milgrom, 1990). Swine have a progressive decrease in adrenal glucocorticoid production from birth to approximately 30 days (Dvorak, 1972; Kattesh *et al.*, 1990), indicating that the function of the adrenal gland is undergoing developmental changes. Development of the hypothalamic–pituitary–gonadal axis also differs between species. In rats it is the surge of testosterone that occurs on days 18–19 of gestation that masculinizes the male pup (Ward and Weisz, 1980). In contrast, masculinization of the swine fetus occurs for months after parturition, and is therefore less apt to be impaired by a bout of an acute stress as compared with the rodent. Although there are many challenges in reviewing this literature there is also a great deal to learn.

Effects of neo- and prenatal stress in laboratory animals

Research on stress during development started in the mid-1950s with work by researchers like Seymour Levine, Gary Haltmeyer, George Karas and Victor Denenberg. The majority of the work at that time was designed to examine the effect of handling on neonatal rats. Research on the effects of prenatal stress came later, in the early 1970s. Therefore, I will first review neonatal stress and then end with prenatal stress.

Neonatal stress

In the mid-1950s Seymour Levine began to study rats that had been 'manipulated in infancy'. This manipulation consisted simply of placing the newborn rat into a box away from its dam for 3 min, after which time it was returned to her care (Levine *et al.*, 1958). This procedure occurred daily for varying numbers of days after birth. Rats handled as newborns were later subjected to cold stress (5°C) for 90 min. They were then sacrificed at 10, 12, 14 and 16 days of age to collect their adrenal glands, which were assayed for ascorbic acid depletion. It has been known since 1944 that adrenocorticotrophic hormone (ACTH) depletes the adrenal gland of ascorbic acid (Sayers *et al.*, 1944); therefore, its measurement was used as one of the first assays for ACTH, often referred to as the Sayers assay. These authors found that 10-day-old rats were unable to respond to the stressor. However, by day 12 the 'handled' infants did respond to the cold stress, and they were more responsive than the control rats on the following 2 days of sacrifice. These data indicate that the handling procedure caused an early maturation of the rats' HPA axis.

Researchers were fascinated by these findings and continued to explore this area. Levine (1959) noted that many researchers were using diverse 'early stimulation' such as stroking, partial starvation and electric shock, but that their results were similar: improved learning, reduced emotionality and resistance to stress. However, all of these stressors had handling in common. In addition, one component that could not be separated from the effects of handling was the role of the dam. Many researchers noted that when the rat pup was returned to its dam, she groomed the pup at a much higher rate than normal. One possibility was that it was not the handling that caused the effects associated with neonatal stress, but the increased attention of the rat dam to its offspring.

To determine if direct handling by humans was causing the effects associated with neonatal stress, Levine (1959) stimulated rats without handling. To do this he used a shaker on which the entire cage with the rat pup and dam were placed and shaken each day. Offspring that were treated this way defaecated less in an open-field test, indicating that they were less emotional than rats that were not shaken. Therefore, Levine concluded that it was not the removal of the pup that created less emotionality, increased learning and resistance to stress in individuals, but that it was a direct effect of stimulation (in this case shaking the cage) of the infant. Later, after several other studies (Barnett and Burn, 1967; Levine, 1975), these researchers noted that the responses of the dams appeared to be more important in causing the enhanced ability of their pups to cope with stress. Conducting several studies using different manipulations, including handling, ear punching and chilling, these authors showed that the dam responded with increased grooming behaviour toward the offspring and that this appeared to be responsible for their altered behavioural and physiological responses.

Denenberg and Karas (1960) attempted to discover if there was a critical time during which the effects of early stimulation were more pronounced. They found that early stimulation (3 min of isolation) worked best if the rat pup was exposed before 10 days of age. When these rats were tested at 62 days of age they proved to be less emotional and more capable of learning in a shock avoidance learning paradigm. Additionally, when challenged in an open field, neonatally handled rats had lower corticosterone concentrations, which these authors suggested conferred an increased resistance to disease (Levine *et al.*, 1967). Both experiments showed that neonatal stress was advantageous to the rat, and that stimulation had to occur during a critical period shortly after birth in order effectively to change neonatal rats' HPA axis.

Thus this early work (reviewed by Levine and Mullins, 1968) generally showed that individuals stimulated early in life were able to respond to stressful environments more appropriately when tested at an older age (varying from 10 to 60 days of age). Furthermore, it appeared that intense stimuli, such as electric shock, which immediately challenged the animals' homeostasis caused a greater behavioural and physiological reaction (40 μg 100 ml^{-1} of plasma corticosteroid), whereas relatively harmless situations (like the open-field test) caused less of a reaction (20 μg 100 ml^{-1} of plasma corticosteroid). Interestingly, in these studies the neonatally stressed rats actually had lower concentrations than the non-neonatally stressed rats (27 μg 100 ml^{-1} of plasma corticosteroid) when exposed to the open-field test. This phenomenon was believed to be a form of 'emotional immunization' in that stress early in life protected against stress later in life. The mechanism was believed to be the creation of differing thresholds at which rodents would respond to a stressor, such that neonatally stressed rodents had a higher threshold which the stressor had to surpass before the animals' protective stress response was activated. It was suggested that the importance of the stressor occurring during neonatal life was that these thresholds could only be altered during a critical period.

Research on neonatal handling continues today, and recent work suggests that it is the actual grooming that the rat dam does to the pup that causes the effects observed, not the handling procedure itself. To tease these effects apart, Liu *et al.* (1997) conducted a study in which they selected dams that naturally showed a high level of grooming of their offspring and ones that did not. No neonatal manipulations of the offspring were conducted, but they were subjected to restraint stress as adults. Infants from dams that performed a lot of maternal care (licking, grooming, etc.) had an increase in hippocampal glucocorticoid receptor concentrations and lower concentrations of plasma corticosterone, ACTH and corticotropin-releasing hormone (CRH).

Prenatal stress

Approximately 10 years after the field of neonatal stress research gained international importance, researchers found that prenatal stress could also change many aspects of the developing fetus. I will not give an exhaustive review of the literature here, but will discuss key references that led to a general understanding of these phenomena. For a more complete review, there is an excellent article by Weinstock (1997) about the effects of prenatal stress on the HPA axis.

Ward (1972) subjected pregnant mice to restraint during days 14–21 of gestation and found that their male offspring exhibited a low level of copulatory behaviour and a high rate of lordosis. Much later, Kerchner *et al.* (1995) found that prenatal stress attenuated the surge of testosterone in rats and that this brief suppression caused feminization due to sensitivity of the sexually dimorphic nucleus of the preoptic area. Sinha *et al.* (1997) identified the maternal adrenal as being important in this effect, since the testosterone surge on gestational days 18 and 19 was greatly attenuated in male offspring of adrenalectomized rat dams.

Prenatal stress also affects sexual characteristics in guinea pigs. Sachser and Kaiser (1996) subjected pregnant guinea pigs to an unstable social environment in order to induce stress and found that the female offspring of these dams displayed male-typical courtship when they reached puberty. Based on previous research showing that administration of androgens to gestating guinea pigs caused masculinization of their female fetuses, these authors hypothesized that maternal adrenal androgens were responsible for this effect of prenatal stress.

Like neonatal stress, prenatal stress is also known to affect the HPA axis of rodents. However, the changes observed thus far appear to be in the opposite direction. Henry *et al.* (1994) found that the offspring of rat dams restrained during the third week of gestation had greater corticosterone concentrations at 3, 21 and 90 days of age when exposed to a novel environment than did non-prenatally stressed offspring. In addition, there were fewer hippocampal type I and II corticosteroid receptors in prenatally stressed individuals at 21 and 90 days of age, but not at 3 days of age. The observation of differential HPA regulation over time demonstrates a maturation of the axis, and this was one of the first studies to show that the effects of prenatal stress were maintained into maturity.

McCormick *et al.* (1995) also restrained pregnant rats at the end of gestation and found that their offspring exhibited a sex-dependent alteration in their response to subsequent stressors. These authors challenged the prenatally stressed offspring at 5 months of age with restraint stress and found that, while the males did not respond differently from the control pups, the females had greater plasma ACTH, corticosterone and corticosteroid binding globulin concentrations than did control pups. These authors also noted an increase in glucocorticoid receptor numbers in the septum, frontal cortex and amygdala of prenatally stressed pups compared with control pups.

Clarke and Schneider (1997) evaluated the long-term effects of prenatal stress in rhesus monkeys. They subjected pregnant monkeys to unpredictable noise stress for 5 days a week from mid- to late gestation. They then allowed the offspring to mature to 4 years of age and challenged them with new group formations and the stress of novelty (unfamiliar playroom). The prenatally stressed monkeys showed greater locomotion, stereotypic behaviour, self-clasping and generalized disturbance behaviour. In sharp contrast, the control monkeys, whose dams were not exposed to noise stress, exhibited play behaviour at six times the rate of the prenatally stressed monkeys.

It is important at this point to ask which physiological and behavioural responses are considered 'normal'. Because we impose the stress during development, it is often taken for granted that this change in the animal's physiology is the 'treatment' as opposed to the 'control'. However, it is quite possible that we instead have recreated the 'normal' behaviour and physiology of a rodent in the wild. A rat in the wild is constantly being challenged by its environment: predators, thermal challenges and food challenges. It lives in a complex environment which requires it to meet its own demands. In sharp contrast, a rat in a laboratory is provided with a safe cage to live in, abundant food and water, and a thermoneutral temperature. Research on environmental enrichment has illustrated the many deleterious effects that a sterile, non-complex environment can have on laboratory animals (for a comprehensive list of articles on environmental enrichment refer to AWIC, 1995). Stress during development might actually produce an animal that has physiological and behavioural responses much more similar to its wild counterparts. It may also prove that this early stimulation is necessary for normal development.

Effects of neo- and prenatal stress in livestock

Neonatal stress in livestock

Compared with the vast amount of research on prenatal and neonatal stress in rodents, there is very little work investigating this phenomenon in livestock. However, neonatal stress in livestock is fairly common due to our methods of production. Dairy calves are taken from their dams the day they are born; veal calves are raised in stalls in which they cannot turn or graze, causing many to exhibit stereotypic behaviour and develop gastric ulcers (Wiepkema, 1987); calves, lambs and foals can become orphaned for various reasons; young animals may be transported, restrained and subjected to other manipulations like castration or dehorning.

To examine if such stressors alter the behaviour of these individuals, Moberg and Wood (1982) challenged lambs with open-field and novelty tests after they had been raised in isolation, peer raised, or left with their dam. They found that isolation-reared lambs were more withdrawn and performed more stereotypic behaviour at 42 days of age than either peer- or dam-raised controls. However, although the behaviour of the isolation-reared lambs indicated that they were more stressed by the open-field test procedure, the authors did not find treatment differences for plasma cortisol concentrations during this challenge. In a similar vein, Creel and Albright (1988) found that isolation-reared calves showed more distress, but their calves also exhibited greater serum cortisol concentrations in response to a short-term stress at 10 weeks of age compared with peer-raised calves. Likewise, Lay *et al.* (1998) group raised calves without their dams and only allowed them to suckle once a day from approximately 20 to 40 days of age. After this time the calves were returned to live with their dams until weaning. When these calves were challenged with restraint stress at 6.5 months of age they had greater plasma cortisol and heart rates than control calves that had never been separated from their dams. Apparently this brief 20-day maternal deprivation altered the calves' HPA axis or psyche such that the calves had a more robust response to restraint stress.

In contrast to these findings for calves and sheep, Houpt and Hintz (1982/83) found that orphan foals were less emotional in a novel environment. However, one methodological difference to be mindful of is that foals were peer raised without their dam, not raised in isolation; but then so were the calves in the Lay *et al.* (1998) study. This raises a valid question of interpretation. Does the increase in glucocorticoids found in many studies of prenatal stress indicate that the animal is not coping well with the stress, *or* that it is coping well with the stress? When there are challenges to homeostasis, activation of the HPA axis is the appropriate response to aid in the animal's survival. Only when the animal is subjected to stress of long duration do the negative aspects of stress occur. Therefore, I suggest that the behavioural responses to a stressor must be evaluated in tandem with the animal's endocrine responses to that stressor in order to make an informative decision as to whether stress during development may be deleterious to the individual. If an animal challenged with a stressor exhibits elevated plasma glucocorticoids but its behaviour is indicative of an animal that is coping well, it is possible that the endocrine response to the stress is allowing that animal to cope well.

In general, it appears that rodents and livestock do not respond similarly to neonatal stress. Neonatal stress in livestock, in the form of maternal deprivation and social isolation, causes an increase in emotionality (except for foals) and an increase in HPA activity in response to novelty and stress. However, in rodents, neonatal stress causes a decrease in emotional responses to novelty and a decrease in the HPA axis response to stress driven by decreases in hippocampal glucocorticoid receptors. These are generalizations, and the reader should keep in mind that along with the wide variation in stressors (handling, social isolation, electric shock, etc.) comes variation in experimental results. A point to consider when comparing rodent and livestock data is that the developmental programmes of the two are quite different. A newborn calf is much more developed, psychologically and physically, than a newborn rat. Unlike the calf, newborn rats are unable to see, thermoregulate or ambulate. Endocrine differences are also apparent during the neonatal stage as the rat goes through a 'stress-hyporesponsive period' from days 4 to 14 of age, during which it cannot respond to stress and exhibits low concentrations of glucocorticoids (Levine, 1994). Calves, on the other hand, maintain elevated glucocorticoids until at least 14 days of age and are responsive to stress during this period (Lay *et al.*, 1997a). Therefore, the critical period during development in which stress may affect animals is likely to occur at different times for each species.

Prenatal stress in livestock

One of the first studies to examine prenatal stress in livestock was that of Kattesh *et al.* (1979) in which sows were subjected to elevated temperature and crowding. These authors chose to examine the effects of prenatal stress on reproductive behaviour and physiology of the male offspring of these sows because of the previously described work of Ward (1972), who found that prenatal stress feminized and demasculinized the behaviour of male rats. However, Kattesh *et al.* (1979) did not find differences in either plasma testosterone or libido in boars from these stressed sows. Since that time, it has been shown that prenatal stress in rodents attenuates the fetal surge of testosterone on days 18 and 19 of gestation thereby preventing masculinization of the offspring (Kerchner *et al.*, 1995). In contrast, in male swine the neural pathways that affect reproductive behaviour are still being modified as late as 3 months of age (Ford and Christenson, 1987). Therefore, some aspects of reproduction in swine are more resistant to prenatal stress.

Although Kattesh *et al.* (1979) found that prenatal stress did not affect reproductive characteristics of swine, Lay *et al.* (1997a, b) did find that prenatal stress altered the HPA axis of cattle. In their study, pregnant Brahman cows were challenged with transportation stress (transported 24 km, unloaded into a holding pen for 1 h, then returned to the original ranch) at 60, 80, 100, 120 and 140 days of gestation. A second treatment involved injection of 1 IU kg^{-1} BW of ACTH into the cows at these same times, and a third treatment acted as a control in which cows were merely herded through the working chute with the rest of the cows. The offspring of the cows from both the ACTH-injected group and the transported group were affected in several ways. Calves of transported dams were heavier at birth and had enlarged pituitary glands (Lay *et al.*, 1997a, b). When calves were 10 and 150 days of age, they were challenged with a 180-min restraint test. At 10 days of age calves had a minimal plasma cortisol response (maximum twofold increase) that did not differ between calves from control dams or either ACTH or transported dams. However, at 150 days of age calves from both groups of treated dams exhibited significant increases in plasma cortisol, to four times above baseline values, whereas plasma cortisol in the control calves only doubled.

Additionally, calves from both the transported and the ACTH-injected cows maintained plasma cortisol concentrations above the levels of control calves throughout the 180-min restraint period. Another finding in this study was that the cortisol clearance rate of 150-day-old calves from transportation-stressed dams was much slower than that of calves from control dams (261 versus 473 ml min^{-1}, respectively). Also, while the calves were restrained to determine clearance rate, calves from the transported dams had higher heart rates than did the calves from either ACTH-treated or control dams (91, 79 and 77 ± 3 b.p.m.$^{-1}$, respectively). Because these authors found a faster clearance rate of cortisol and higher plasma cortisol concentrations in 150-day-old calves compared with 10-day-old calves, they hypothesized that the difference in HPA activity between 10- and 150-day-old calves may be due to an increased number of ACTH receptors on the adrenal gland cells, effectively decreasing the circulating plasma concentrations of ACTH. Indeed, this is exactly what Lay and Haussmann (unpublished data) found in prenatally stressed swine.

In order to examine if prenatal stress affected the HPA axis in swine, sows were either injected ($n = 9$) with ACTH (1 IU kg^{-1} BW, considered the 'prenatal stress' treatment) at 5, 6, 7, 8, 9, 10 and 11 weeks of gestation or ($n = 9$) were allowed to remain in gestation stalls undisturbed (Lay and Haussmann, unpublished data). Within 12 h of birth one pig from each litter was sacrificed to collect the hypothalamus, pituitary gland and adrenal gland. An additional two littermates were then sacrificed at 1 month and 2 months of age. The prenatally stressed pigs had smaller birth weights, smaller anogenital distances (for females only), more CRH and less β-endorphin in the hypothalamus at 1 month of age, more mRNA for the adrenal ACTH receptor at 1 month of age, and larger pituitary glands at 2 months of age. The physiology of these prenatally stressed pigs indicates that they would have a vigorous response to a stressor when challenged because a greater number of CRH and ACTH receptors would cause a profound increase in plasma glucocorticoids. In addition, the lower β-endorphin may indicate that they were more sensitive to painful stimuli and thus they may have poorer welfare. However, these are preliminary data that need further investigation to understand fully their implications for swine welfare.

Possible mechanisms

Neonatal stress

There is a vast literature on both neonatal and prenatal stress (a handful of which has been mentioned here) that clearly shows that stress during development can have profound effects on the physiology and behaviour of adult animals. This begs the question, however, as to how stress can permanently alter an animal's biology. Several hypotheses exist for both neonatal and prenatal stress. One hypothesis implicates thyroid hormones in these effects. Although Kuhn *et al.* (1978) found that rats deprived of their dam for several hours had a selective depression of serum growth hormone without affecting either prolactin, corticosterone or thyroid-stimulating hormone, Meaney *et al.* (1987) suggest that it is thyroid hormones that mediate the increase in hippocampal glucocorticoid receptors. One of the recurring effects of stress during development is that central nervous system glucocorticoid receptors are either increased or decreased, and the mechanism(s) for these changes has thus been of interest. Meaney and co-workers' hypothesis is based on findings that rats that were handled postnatally but had their endogenous thyroid hormone synthesis blocked (by propylthiouracil) did not exhibit the increased number of hippocampal glucocorticoid receptors shown by the postnatally handled control rats. In addition, administration of corticosterone during the first days of a rat's life did not cause an increase in hippocampal glucocorticoid receptors as is typically seen in postnatally handled (neonatal stress) rats. Therefore, they concluded that thyroid hormones, not glucocorticoids, were involved in the associated changes in the HPA axis of postnatally handled rats.

Other authors have suggested that glucocorticoids are the mediators that alter the neonatal rat's response to stress. De Kloet *et al.* (1996) discuss work that shows a stress-hyporesponsive period in the rat that occurs during postnatal days 4–14. During this period the neonatal rat does not respond to maternal separation with activation of the HPA axis and indeed exhibits low concentrations of circulating corticosterone and ACTH. A glucocorticoid response can be demonstrated in these animals if maternal separation is prolonged for up to 8–24 h. When the dam does give attention and tactile stimulation to the infant, neural paths that suppress the release of ACTH during this hyporesponsive period are suppressed, causing increased ACTH and glucocorticoid secretion. Thus the development of the neonatal HPA axis is dependent on the dam leaving the offspring for short periods during the day and returning later to groom and provide them with maternal care. It is easy to understand that this may be the 'natural' way in which HPA axis development occurs in wild rodents.

Prenatal stress

Effects similar to those found after prenatal stress can be induced by treating the animal with exogenous hormones associated with the HPA axis (e.g. Williams *et al.*, 1995; Lay *et al.*, 1997a, b). Research of this type provides some information as to how the HPA axis of the developing fetus is altered. Williams *et al.* (1995) administered subcutaneous injections of 20 μg CRH to pregnant rats on days 14–21 of gestation. Their resultant offspring vocalized more during an open-field test and showed reduced weight gain during the 2 weeks after birth. In addition, male offspring had a reduced anogenital distance. These findings are consistent with much of the literature on prenatal stress in rats, but do not provide information as to whether these effects are a direct cause of CRH administration or the resultant increase in ACTH, glucocorticoids and β-endorphin.

Similarly, the previously described research by Lay *et al.* (1997a, b) in which ACTH was injected into pregnant cattle to produce offspring with an altered HPA axis provides further information but still does not fully answer the question as to the mechanism by which prenatal stress exerts its effects. However, Barbazanges *et al.* (1996) conducted a series of experiments that provide good evidence for glucocorticoids as the mediator of prenatal stress. In these experiments they subjected both adrenalectomized and non-adrenalectomized rats to restraint stress during gestation. In addition, they supplemented a second group of pregnant adrenalectomized rats with glucocorticoids to simulate a stress. Those individuals whose glucocorticoid response to restraint stress was blocked by adrenalectomy did not produce offspring that demonstrated prenatal stress effects. However, those rats that were either supplemented with glucocorticoids or that were non-adrenalectomized produced offspring that had fewer hippocampal type I corticosteroid receptors and a greater corticosterone response to restraint stress as adults. This was one of the first projects to suggest that maternally derived glucocorticoids are at least one of the mechanisms by which prenatal stress affects the developing fetus. This finding is supported by the fact that radioactive corticosterone injected into pregnant rats can cross the placenta and bind in the fetal hypothalamus (Zarrow *et al.*, 1970). This mechanism may also play a role in the prenatal stress effects observed in swine, as Klemcke (1995) found that approximately 20% of fetal cortisol is derived from the sow at 50 days of gestation although this amount decreases as gestation continues.

However, placentation in rodents, cattle and swine is quite different (for a review, see Perry, 1981). The placenta of the rodent is haemoendothelial, comprising one layer of fetal endothelial cells bathed directly in the blood of the dam. In contrast, the placenta of cattle and swine is epithelialchorial, formed by three layers of maternal tissue (endothelium, stroma and uterine lumenal epithelium) as well as three layers of fetal tissue (placental epithelium, stroma and endothelium). The implications of these different placental types are important when discussing possible mechanisms by which prenatal stress may affect the developing fetus. The rodent fetus, which is separated from the maternal blood by only one layer of cells, is obviously much more susceptible to the possibility of influence of steroids that may pass across from the maternal blood. The bovine and swine placentae present a more formidable challenge with their six layers of cells for protection. In this light the work by Klemcke (1995) is interesting. Although he did find that maternal cortisol enters the fetal circulation, when the relatively low level of cortisol in fetal circulation (approximately 10 ng ml^{-1}) is compared with the relatively high level in the dam (70 ng ml^{-1}), her contribution is minimal.

Not all research on prenatal stress implicates metabolic hormones. Keshet and Weinstock (1995) subjected pregnant rats to noise and light stress three times a week throughout gestation. They used naltrexone

(10 mg kg^{-1} day^{-1}) to block the opioid receptors and found that offspring from naltrexone-treated dams were normal compared with prenatally stressed offspring, which exhibited a reduced anogenital distance (males), reduced growth rate, and increased anxiety in a novel environment. Thus, this research clearly indicates that opioids may be a strong player in causing the effects of prenatal stress.

The observation that prenatal stress in rodents increases anxiety and corticosterone response to novelty stress whereas neonatal stress (increased maternal grooming and handling) causes offspring to have low anxiety is intriguing. Even more interesting is the finding that adoption on the day of birth can reverse the effects of prenatal stress. Maccari *et al.* (1995) subjected dams to restraint stress during gestation and then either placed the offspring with a foster dam upon birth or left them with their own dam. Consistent with the literature, the authors found that prenatally stressed offspring had an elevated and prolonged corticosterone response to stress and that this was associated with a decrease in hippocampal glucocorticoid receptors. However, those offspring that had been fostered had the same stress response as the control rats. This reversal was attributed to the increase in maternal care that foster mothers gave to their infants as compared with biological mothers. Unquestionably, the phenomenon of stress during development demands more research.

Implications for welfare

Clearly we, as all animals, are challenged with stress in our daily lives. The stressors to which we are exposed are usually innocuous; however, they can at times be detrimental to our well-being. If possible, our predominant response to these stressors is to change our behaviour such that we are no longer affected by such stress. Unfortunately, our livestock do not have this luxury. Confinement agriculture cares for livestock by providing them with water, food and shelter. At the same time, however, it challenges them with stress due to social interactions, restriction of movement, thwarted drives and pain due to some standard agricultural practices. One major factor that would allow livestock to cope better with some of these stressors is forbidden: control. Because they cannot change their behaviour to solve the problem it is imperative that they be biologically prepared to cope with these stressors.

Exposure to a stressor, either during the neonatal or prenatal period, has the potential to alter many facets of an animal's physiology and behaviour. However, these alterations cannot be ascribed solely to 'stress' because a high level of grooming activity of the infant rat has also been shown to cause the alterations in physiology and behaviour as seen in neonatally handled and (or) shocked rats. Although the periods of pliability probably vary between species, what is clear is that the function of the HPA axis is open to alterations during development. This is most evident in the work of Maccari *et al.* (1995) which showed that the effects of prenatal stress could be reversed with neonatal stress in the rat. Unfortunately, the implications of these alterations are not yet fully understood.

A great deal of the literature on rodents indicates that neonatal stress promotes a minimal behavioural and physiological response to some stressors, while prenatal stress causes a heightened behavioural and physiological response to stress. The relatively minor amount of work on livestock does not allow us to draw such conclusions because both neonatal and prenatal stress result in increased plasma glucocorticoids in response to subsequent stressors. It is as yet unclear if exposure to prenatal and (or) neonatal stress allows animals ultimately to cope better with the situation or if it hinders their ability to cope. Increases in glucocorticoids either can be advantageous, meeting the challenge of such stressors, or they can be deleterious, causing wasting and disease.

Therefore, it is important that we determine the implications of both neonatal and prenatal stress in livestock. Research thus far indicates that both cause an increase in circulating glucocorticoids, heart rate and a decrease in hypothalamic β-endorphin. Livestock are often chronically challenged by potential stressors, which necessitates that they be able to cope well with these challenges. Chronically elevated glucocorticoids will impair the animal's ability to mount immune responses, utilize nutrients for growth and successfully synchronize events critical for reproduction. Predisposing livestock to become unthrifty and diseased is unacceptable for their welfare. The added observation that these animals may be deficient in β-endorphin suggests that they may be more prone to feeling pain, which is of great concern for their welfare as they are often housed on hard flooring and subjected to painful procedures (branding, castration, docking).

Coordinated efforts on the part of researchers will greatly accelerate our understanding of the implications of stress during development. Much of the current confusion is due to the application of very different stressors (restraint, electric shock, ear punching, handling, shaking, etc.) during either neonatal or prenatal development and then subsequent challenges with again very different stressors (restraint, electric shock, open-field test, etc.). Standardization would allow researchers to compare their data more effectively with those of other laboratories. The challenge of between-species differences will probably remain. The developmental stages among species are not synchronous, as previously explained, nor are the species' behavioural or physiological responses to stress similar. For instance, in an open-field test a frightened rodent will freeze and stay immobile until comfortable enough to ambulate. In contrast, a horse will immediately run from danger when frightened. As research progresses on the consequences of stress during development, specifically studying livestock, we will gain a better understanding of these phenomena to enable us to address the welfare concerns that it creates.

References

Animal Welfare Information Center (1995) *AWIC Resource Series No. 2 Environmental Enrichment Information Resources for Laboratory Animals.* Animal Welfare Information Center, Washington, DC and the Universities Federation for Animal Welfare, UK.

Barbazanges, A., Piazza, P.V., Le Moal, M. and Maccari, S. (1996) Maternal glucocorticoid secretion mediates long-term effects of prenatal stress. *Journal of Neuroscience* 16, 3943–3949.

Barnett, S.A. and Burn, J. (1967) Early stimulation and maternal behaviour. *Nature* 213, 150–152.

Broom, D.M., Mendl, M.T. and Zanella, A.J. (1995) A comparison of the welfare of sows in different housing conditions. *Animal Science* 61, 369–385.

Clark, D.K., Friend, T.H. and Dellmeier, G. (1993) The effect of orientation during trailer transport on heart rate, cortisol and balance in horses. *Applied Animal Behaviour Science* 38, 179–189.

Clarke, A.S. and Schneider, M.L. (1993) Prenatal stress has long-term effects on behavioral responses to stress in juvenile Rhesus monkeys. *Developmental Psychobiology* 26, 293–304.

Clarke, A.S. and Schneider, M.L. (1997) Effects of prenatal stress on behavior in adolescent Rhesus monkeys. *Annals of the New York Academy of Sciences* 807, 490–491.

Clarke, A.S., Wittwer, D.J., Abbott, D.H. and Schneider, M.L. (1994) Long-term effects of prenatal stress on HPA axis activity in juvenile Rhesus monkeys. *Developmental Psychobiology* 27, 257–269.

Creel, S.R. and Albright, J.L. (1988) The effects of neonatal social isolation on the behavior and endocrine function of Holstein calves. *Applied Animal Behaviour Science* 21, 293–306.

de Kloet, E.R., Rots, N.Y. and Cools, A.R. (1996) Brain-corticosteroid hormone dialogue: slow and persistent. *Cellular and Molecular Neurobiology* 16, 345–356.

Denenberg, V.H. and Karas, G.G. (1960) Interactive effects of age and duration of infantile experience on adult learning. *Psychological Reports* 7, 313–322.

Dvorak, M. (1972) Adrenocortical function in foetal, neonatal and young pigs. *Journal of Endocrinology* 54, 473–481.

Fell, L.R. and Shutt, D.A. (1986) Adrenocortical response of calves to transport stress as measured by salivary cortisol. *Canadian Journal of Animal Science* 66, 637–641.

Ford, J.J. and Christenson, R.K. (1987) Influences of pre- and postnatal testosterone treatment on defeminization of sexual receptivity in pigs. *Biology of Reproduction* 36, 581–587.

Friend, T.H., Dellmeier, G.R. and Gbur, E.E. (1985) Comparison of four methods of calf confinement. I. Physiology. *Journal of Animal Science* 60, 1095–1101.

Friend, T.H., Taylor, L., Dellmeier, G.R., Knabe, D.A. and Smith, L.A. (1988) Effect of confinement method on physiology and production of gestating gilts. *Journal of Animal Science* 66, 2906–2915.

Gluckman, P.D. (1985) The onset and organization of hypothalamic control in the fetus. In: Jones, C.T. and Nathanielsz, P.W. (eds) *The Physiological Development of the Fetus and Newborn.* Academic Press, London, pp. 103–111.

Griffin, J.E. and Ojeda, S.R. (1992) Sexual differentiation. In: Griffin, J.E. and Ojeda, S.R. (eds) *Textbook of Endocrine Physiology*, 2nd edn. Oxford University Press, New York, pp. 118–133.

Haltmeyer, G.C., Denenberg, V.H. and Zarrow, M.X. (1967) Modification of the plasma corticosterone response as a function of infantile stimulation and electric shock parameters. *Physiology and Behavior* 2, 61–65.

Hemsworth, P.H. and Coleman, G.J. (1998) *Human–Livestock Interactions. The Stockperson and the Productivity and Welfare of Intensively Farmed Animals.* CAB International, Wallingford, UK.

Henry, C., Kabbaj, M., Simon, H., Le Moal, M. and Maccari, S. (1994) Prenatal stress increases the hypothalamo-pituitary-adrenal axis response in young and adult rats. *Journal of Neuroendocrinology* 6, 341–345.

Hilakivi-Clarke, L.A., Turkka, J., Lister, R.G. and Linnoila, M. (1991) Effects of early postnatal handling on brain β-adrenoceptors and behavior in tests related to stress. *Brain Research* 542, 286–292.

Houpt, K.A. and Hintz, H.F. (1982/83) Some effects of maternal deprivation on maintenance behavior, spatial relationships and responses to environmental novelty in foals. *Applied Animal Ethology* 9, 221–230.

Kattesh, H.G., Kornegay, E.T., Gwazdauskas, F.C., Knight, J.W. and Thomas, H.R. (1979) Peripheral plasma testosterone concentration and sexual behavior in young prenatally stressed boars. *Theriogenology* 12, 289–305.

Kattesh, H.G., Charles, S.F., Baumbach, G.A. and Gillespie, B.E. (1990) Plasma cortisol distribution in the pig from birth to six weeks of age. *Biology of the Neonate* 58, 220–226.

Kerchner, M., Malsbury, C.W., Ward, O.B. and Ward, I.L. (1995) Sexually dimorphic areas in the rat medial amygdala: resistance to the demasculinizing effect of prenatal stress. *Brain Research* 672, 251–260.

Keshet, G.I. and Weinstock, M. (1995) Maternal naltrexone prevents morphological and behavioral alterations induced in rats by prenatal stress. *Pharmacology, Biochemistry and Behavior* 50, 413–419.

Klein, S.L. and Rager, D.R. (1995) Prenatal stress alters immune function in the offspring of rats. *Developmental Psychobiology* 28, 321–336.

Klemcke, H.G. (1995) Placental metabolism of cortisol at mid- and late gestation in swine. *Biology of Reproduction* 53, 1293–1301.

Kuhn, C.M., Butler, S.R. and Schanberg, S.M. (1978) Selective depression of serum growth hormone during maternal deprivation in rat pups. *Science* 201, 1034–1036.

Lambert, K.G., Kinsley, C.H., Jones, H.E., Klein, S.L., Peretti, S.N. and Stewart, K.M. (1995) Prenatal stress attenuates ulceration in the activity stress paradigm. *Physiology and Behavior* 57, 989–994.

Lay, D.C., Jr, Friend, T.H., Randel, R.D., Bowers, C.L., Grissom, K.K. and Jenkins, O.C. (1992) Behavioral and physiological effects of freeze or hot-iron branding on crossbred cattle. *Journal of Animal Science* 70, 330–336.

Lay, D.C., Jr, Randel, R.D., Friend, T.H., Jenkins, O.C., Neuendorff, D.A., Bushong, D.M., Lanier, E.K. and Bjorge, M.K. (1997a) Effects of prenatal stress on suckling calves. *Journal of Animal Science* 75, 3143–3151.

Lay, D.C., Jr, Randel, R.D., Friend, T.H., Carroll, J.A., Welsh, T.H.J., Jenkins, O.C., Neuendorff, D.A., Bushong, D.M. and Kapp, G.M. (1997b) Effects of prenatal stress on the fetal calf. *Domestic Animal Endocrinology* 14, 73–80.

Lay, D.C., Jr, Friend, T.H., Randel, R.D., Bowers, C.L., Grissom, K.K., Neuendorff, D.A. and Jenkins, O.C. (1998) Effects of restricted nursing on physiological and behavioral reactions of Brahman calves to subsequent restraint and weaning. *Applied Animal Behaviour Science* 56, 109–119.

Levine, S. (1959) The effects of differential infantile stimulation on emotionality at weaning. *Canadian Journal of Psychology* 13, 243–247.

Levine, S. (1975) Psychosocial factors in growth and development. In: Levi, L. (ed.) *Society, Stress and Disease.* Oxford University Press, London, pp. 43–50.

Levine, S. (1994) The ontogeny of the hypothalamic–pituitary–adrenal axis: the influence of maternal factors. *Annals of the New York Academy of Sciences* 746, 275–293.

Levine, S. and Mullins, R.F.J. (1968) Hormones in infancy. In: Newton, G. and Levine, S. (eds) *Early Experience and Behavior: the Psychobiology of Development.* Charles C. Thomas, Springfield, Ohio, pp. 168–196.

Levine, S., Alpert, M. and Lewis, G.W. (1958) Differential maturation of an adrenal response to cold stress in rats manipulated in infancy. *Journal of Comparative and Physiological Psychology* 51, 774–777.

Levine, S., Haltmeyer, G., Karas, G. and Denenberg, V. (1967) Physiological and behavioral effects of infantile stimulation. *Physiology and Behaviour* 2, 55–59.

Liu, D., Diorio, J., Tannenbaum, B., Caldji, C., Francis, D., Freedman, A., Sharma, S., Pearson, D., Plotsky, P.M. and Meaney, M.J. (1997) Maternal care, hippocampal glucocorticoid receptors, and hypothalamic–pituitary–adrenal responses to stress. *Science* 277, 1659–1662.

Maccari, S., Piazza, P.V., Kabbaj, M., Barbazanges, A., Simon, H. and Le Moal, M. (1995) Adoption reverses the long-term impairment in glucocorticoid feedback induced by prenatal stress. *Journal of Neuroscience* 15, 110–116.

McCormick, C.M., Smythe, J.W., Sharma, S. and Meaney, M.J. (1995) Sex-specific effects of prenatal stress on hypothalamic–pituitary–adrenal responses to stress and brain glucocorticoid receptor density in adult rats. *Developmental Brain Research* 84, 55–61.

Meaney, M.J., Aitken, D.H. and Sapolsky, R.M. (1987) Thyroid hormones influence the development of hippocampal glucocorticoid receptors in the rat: a mechanism for the effects of postnatal handling on the development of the adrenocortical stress response. *Neuroendocrinology* 45, 278–283.

Milgrom, E. (1990) Steroid hormones. In: Baulieu, E.-E. and Kelly, P.A. (eds) *Hormones: From Molecules to Disease.* Hermann Publishers, London and Chapman and Hall, New York, pp. 387–442.

Mitchell, G., Hattingh, J. and Ganhao, M. (1988) Stress in cattle assessed after handling, after transport and after slaughter. *Veterinary Record* 123, 201–205.

Moberg, G.P. and Wood, V.A. (1982) Effect of differential rearing on the behavioral and adrenocortical response of lambs to a novel environment. *Applied Animal Ethology* 8, 269–279.

Perry, J.S. (1981) The mammalian fetal membranes. *Journal of Reproduction and Fertility* 62, 321–335.

Sachser, N. and Kaiser, S. (1996) Prenatal social stress masculinizes the females' behaviour in guinea pigs. *Physiology and Behavior* 60, 589–594.

Sayers, G., Sayers, M.A., Lewis, H.L. and Long, C.N.H. (1944) Effect of adrenotropic hormone on ascorbic acid and cholesterol content of the adrenal. *Proceedings of the Society for Experimental Biology and Medicine* 55, 238–239.

Sinha, P., Halasz, I., Choi, J.F., McGivern, R.F. and Redei, E. (1997) Maternal adrenalectomy eliminates a surge of plasma dehydroepiandrosterone in the mother and attenuates the prenatal testosterone surge in the male fetus. *Endocrinology* 138, 4792–4797.

Vallée, M., Mayo, W., Dellu, F., Le Moal, M., Simon, H. and Maccari, S. (1997) Prenatal stress induces high anxiety and postnatal handling induces low anxiety in adult offspring: correlation with stress-induced corticosterone secretion. *Journal of Neuroscience* 17, 2626–2636.

Ward, I.L. (1972) Prenatal stress feminizes and demasculinizes the behavior of males. *Science* 175, 82–84.

Ward, I.L. and Weisz, J. (1980) Maternal stress alters plasma testosterone in fetal males. *Science* 207, 328–329.

Weinstock, M. (1997) Does prenatal stress impair coping and regulation of hypothalamic–pituitary–adrenal axis? *Neuroscience and Biobehavioral Reviews* 21, 1–10.

Wiepkema, P.R. (1987) Behavioural aspects of stress. In: Wiepkema, P.R. and van Adrichem, P.W.M. (eds) *Biology of Stress in Farm Animals: An Integrative Approach*. Martinus Nijhoff Publishers, Dordrecht, The Netherlands, pp. 113–133.

Williams, M.T., Hennessy, M.B. and Davis, H.N. (1995) CRF administered to pregnant rats alters offspring behavior and morphology. *Pharmacology, Biochemistry and Behavior* 52, 161–167.

Zarrow, M.X., Philpott, J.E. and Denenberg, V.H. (1970) Passage of ^{14}C-4-corticosterone from the rat mother to the fetus and neonate. *Nature* 226, 1058–1059.

Early Developmental Influences of Experience on Behaviour, Temperament and Stress

13

W.A. Mason

California Regional Primate Research Center, University of California, Davis, California, USA

Introduction

In this chapter I examine the effects of experience early in development on emotional behaviour, temperament and responses to stress in mammals, chiefly rats and primates, from an organismic perspective. A basic tenet of this perspective is that normal development is a process of becoming that begins at conception and continues throughout life. The individual is viewed as an open, highly organized system in active commerce with its environment, whose development is influenced by a host of factors originating internally as well as in the surrounding environment. Complex interactions among these factors constrain and stabilize the system such that in the sequence, rate and form of change, development usually proceeds according to norms characteristic of the species.

Effects of early experience on emotional behaviour and temperament

From an organismic perspective, it is a foregone conclusion that experience will influence behavioural development and that the consequences will depend on the nature of the experience, on when and for how long it is effective, and on the species involved. More than 40 years of animal research has established that the effects of early experience are pervasive and diverse. For example, it affects food selection, mating behaviour, parental behaviour, filial behaviour, aggression, learning and memory,

(eds G.P. Moberg and J.A. Mench)

problem solving, sensorimotor development, perceptual processes, exploratory behaviour and emotionality. A comprehensive early review of the extensive literature resulting from this work can be found in Denenberg (1969b), and more recent overviews are presented by Gandelman (1992) and Michel and Moore (1995).

One important aspect of early experience is its effects on emotional behaviour and temperament. Animal research on this topic has a fairly long history, stimulated initially by the Freudian emphasis on the consequences of events during infancy and childhood for the development of psychological disorders. An early review of research on temperament was restricted to rats and mice, chiefly because they were virtually the only animals for which systematic data were available, and dealt mainly with behaviours relating to the trait of timidity or fearfulness (Hall, 1941). This trait remains a common focus of animal research on early experience and temperament and it will figure prominently in this chapter.

Fearfulness as an abiding trait of temperament is usually distinguished from fear as a transient state. Fearfulness refers to an animal's characteristic mode of responding to novelty, uncertainty, challenge and change (Higley and Suomi, 1989; Clarke and Boinski, 1995). The most widely used method for assessing rearing effects on fearfulness is based on the common observation that many mammals – especially the young and naive – become agitated and distressed if they are left alone in completely novel surroundings. The predictability and strength of this reaction has given rise to novel environment tests, like the open field, as favoured tests for assessing behavioural differences in timidity and fearfulness. These tests are a major source of the data reviewed in this section. Measures of emotional responsiveness depend on species, of course, but frequently include latency to enter the test area, locomotor activity, freezing, defaecation, urination, interactions with test objects and vocalizations. These may be supplemented with data on ethologically derived behaviours indicative of fear, agitation, withdrawal or distress, and (as we will consider in due course) by physiological measures of arousal, particularly heart rate and adrenocortical activity.

The kinds of experiences examined cover a broad spectrum, ranging from discrete forms of stimulation, such as handling, shock, changes in temperature and physical rotation (used chiefly with rats), to continuous variables, such as stable variations in the social and non-social rearing environment (used with many different species). A variety of mammals has been studied, including mice, rats, guinea pigs, monkeys, apes, dogs, cats, rabbits, sheep, goats and humans.

The bulk of research on the effects of early experience on emotionality (i.e. fearfulness) has been carried out with rats. The number of studies showing that emotionality of rats is affected by early experience is huge and the situation in which this has been most often demonstrated is the open field, typically an unfamiliar well-lighted arena devoid of objects and several times more spacious than the animal's familiar living cage. The range of behaviours possible in this situation is restricted and those most often used are locomotion and defaecation. Defaecation appears to have greater validity as a measure of emotional reactivity and, unless otherwise noted, it is the primary measure in the research referred to in this section (Hall, 1941; Denenberg, 1969a; Archer, 1973; Daly, 1973).

The type of early experience investigated most frequently and thoroughly is 'handling', a seemingly innocuous manipulation in which the experimenter removes a rat pup from its mother once a day, placing it alone in a small container for several minutes and then returning it to its home cage. Handling is generally found to reduce emotionality in the open field. The effects of handling are also evident in avoidance conditioning and adrenocortical activity, and they seem to endure indefinitely (Denenberg and Smith, 1963; Levine *et al.*, 1967; Denenberg, 1969b; Daly, 1973; Gandelman, 1992). The persistence and generality of these results provide prima-facie evidence that handling alters the trait of fearfulness or timidity (Boissy, 1995).

The effectiveness of handling varies with age: handling before weaning generally reduces emotionality, whereas handling after weaning has little or no effect (Levine and Otis, 1958; Denenberg, 1966, 1969b; Ader, 1968; Daly, 1973; but see Gertz, 1957; Doty and Doty, 1967). In view of the extensive changes in the behaviour and physiology of the developing rat between the pre- and post-weaning periods, it might be anticipated that age should be influential. It is not apparent, however, how handling causes emotionality to change or why the experience is more effective when it occurs before rather than after the age of weaning.

Several hypotheses have been suggested that might account for the greater effectiveness of early handling. One possibility is that handling is a physical event that acts directly on the pup during a sensitive period in its development, thereby modifying the organization of its emotional reactions. An alternate hypothesis, not necessarily incompatible with the first, is that handling affects the pup indirectly by producing changes in the behaviour of the mother (Richards, 1966; Meier and Schutzman, 1968; Russell, 1971).

Various lines of evidence are consistent with a maternal mediation hypothesis. Maternal behaviour parallels and complements the behaviour and needs of the developing infant, and the parent–offspring relationship is most intimate during the period before weaning, a time when the mother's role in the regulation of vital neonatal functions is pervasive and fundamental (Rosenblatt, 1969; Hofer, 1987). The evidence indicates that unless extreme care is exercised, infant rats are distressed by separation from the mother during this period and respond with increased ultrasonic vocalizations. This activity is likely to continue after the pups are returned to the home cage, especially if they are placed outside the nest. The mother responds with increased pup-directed behaviour (Bell *et al.*, 1974; Brown *et al.*, 1977; Mendoza *et al.*, 1980; Smotherman and Bell, 1980). It is readily conceivable that her actions could bring about abiding changes in the infant's emotional reactivity, perhaps similar to and confounded with the direct effects produced by handling (Smotherman and Bell, 1980). Indeed, offspring that receive high levels of licking, grooming and 'arched-back nursing' from their mothers show substantially lower levels of fearfulness in response to novelty, compared with offspring that receive low frequencies of these behaviours (Caldji *et al.*, 1998).

Maternal influences on the emotional behaviour of her offspring can arise from a variety of sources, of course, including some that are unrelated to maternal behaviour. In a classic study, Thompson demonstrated that prenatal experience had significant effects on emotionality of rats (Thompson, 1957). Non-pregnant females were first trained to expect shock following a buzzer, then made anxious during pregnancy by exposure to the buzzer alone. The offspring of these females, who were raised by foster mothers to control for postpartum maternal influences, were significantly more emotional than the controls when tested at 30–40 days, as indicated by latency to become active and amount of activity in the open field, and latency to leave the home cage and to reach food in a novel area. Prenatal effects have been confirmed in many subsequent rodent studies (Archer and Blackman, 1971).

The effect of handling future mothers during their infancy on responsiveness of their offspring in the open field has also been examined. Contrary to the expectation that early handling of mothers would reduce emotionality of their offspring, defaecation was more frequent among offspring of handled mothers than those of non-handled mothers (Denenberg and Whimbey, 1963). Another study investigating the separate and joint contributions of mothers and littermates to emotional development compared the effects of four rearing conditions during the pre-weaning period (Koch and Arnold, 1972). Pups were raised as single individuals with their mothers, as single individuals alone in an incubator, as members of peer groups in an incubator, or as members of peer groups together with their mothers. Emotionality testing in the open field starting when the subjects were 65 days of age demonstrated that presence of the mother was the critical rearing variable. Rats raised by their mothers, whether singly or with peers, were less emotional, as indicated by latencies to enter the open field, number of squares entered and frequency of urination, than those raised individually or with peers in incubators. In other tests, mother-reared rats gave smaller heart rate responses to auditory stimuli and shock than maternally deprived animals (Koch and Arnold, 1972). The authors of this study point out that the large amount of handling received by the animals raised individually in incubators did not compensate for the lack of social stimulation and conclude that mother–infant interactions are necessary for normal emotional functioning in later life (Koch and Arnold, 1972).

Maternal contributions to emotional development were also investigated in an experiment in which mothers were switched on alternate days between their own and a foster litter during the first 20 days postpartum.

When tested in the open field at 51 days of age, a significantly smaller percentage of offspring from litters raised by alternating mothers defaecated on their first trial, compared with the percentage from control litters. In the same study, some mothers were shocked each day during the first 20 days postpartum; more of their offspring defaecated than the control offspring or offspring whose mothers were rotated between litters. Surprisingly, however, the smallest percentage of animals defaecating was the offspring of mothers who were both shocked and rotated (Denenberg *et al.*, 1962).

As these studies illustrate, many variables are likely to influence behaviour in the open field, although their precise effects are often impossible to anticipate. This is a general feature of novel environment tests. By design, they are ambiguous, unstructured and far beyond the subject's normal experience – qualities that are virtually guaranteed to evoke emotional reactions. Novel environments are good for demonstrating that rearing conditions cause differences in responsiveness and for raising questions about the specific source and significance of these effects. They are poor analytical tools for answering these questions, however.

This caveat also applies to primate research, as became clear to me in the first investigation, conducted with Phillip Green, of the effects of rearing conditions on the emotional reactions of rhesus monkeys to a novel room (Mason and Green, 1962). We compared two groups of adolescent animals, one born in captivity and raised in individual cages for most of their lives and the other captured in the wild. Even though both groups appeared to be distressed by the test situation, their reactions were markedly different. The wild-born monkeys showed much more gross motor activity (e.g. jumping, backward somersaults), urination–defaecation and vocalization than the captive-born monkeys, whereas the latter responded by crouching, clasping themselves, sucking their fingers or toes, engaging in body-rocking and other repetitive stereotyped movements, and remaining immobile for long periods – behaviours that were seldom or never displayed by the wild-born monkeys (Mason and Green, 1962).

The precise source of these differences in reactions to the novel room is not evident, but both maternal variables and levels of general environmental stimulation could have been involved. The wild-born monkeys were raised by their mothers and exposed to levels of environmental complexity and variability that presumably were normal for their species, whereas both factors were drastically reduced in the experience of the captive-born animals inasmuch as they were removed from their mothers within hours after birth and spent the next few months in individual cages in a relatively monotonous laboratory environment. Based on the form and probable function of the responses of the captive-born monkeys, we suggested that their self-directed and stereotyped behaviour patterns in the novel room were manifestations of filial behaviour patterns – actions normally organized around the mother and directed towards her when emotional arousal is high

– which developed in an environment in which the mother had never been available (Mason and Green, 1962; Mason *et al.*, 1968).

In other research, we focused more directly on the developmental consequences of maternal attributes. The effects of maternal mobility on stereotyped rocking were investigated in a study carried out with Gershon Berkson comparing monkeys raised with either mechanically driven mobile mother surrogates or physically identical stationary surrogates. Maternal mobility prevented the development of stereotyped rocking. We also found, however, that mobility affected emotional responsiveness to an unfamiliar environment. Monkeys raised with mobile surrogates entered a novel room more readily, showed more moderate levels of activity and defaecated more frequently than the monkeys raised with stationary surrogates. In other tests they made more frequent and varied contacts with an unfamiliar person, and showed substantially higher levels of visual curiosity when they were tested 10 months after they had been permanently separated from their surrogates (Eastman and Mason, 1975; Mason and Berkson, 1975). As in the original comparison between wild-born monkeys and monkeys raised in captivity, the findings indicated that early experience can produce both quantitative and qualitative differences in reactions to novelty.

The effects of maternal attributes on responsiveness to novel environments were also examined in a study comparing rhesus infants raised with a living mother substitute (a mongrel dog) and an inanimate mother substitute (a plastic animal covered with acrylic fur). Both groups of monkeys were housed outdoors in kennels with their mother substitutes and had frequent opportunities for visual contact with people, dogs, other monkeys and a variety of other ongoing and occasional events. Moreover, from the 3rd through the 15th month of life, every monkey was allowed to roam freely in several different complex outdoor enclosures containing a variety of playthings, puzzles, barriers and climbing devices. The monkeys were observed while they were alone in a novel room at intervals of several months during their first year. Those raised with dogs gave more distress vocalizations and had higher heart rates than monkeys raised with inanimate surrogates. Not surprisingly, responsiveness in both groups diminished with repeated exposures to the room. However, when they were tested much later, as young adults/subadults (mean age = 39 months) in a totally unfamiliar room, the dog-raised monkeys were again more responsive initially as indicated by higher heart rate, locomotion, amount of vocalization and magnitude of their heart rate response to noise (Mason and Capitanio, 1988). Heightened responsiveness was also evident in tests of visual curiosity, carried out during the monkeys' 2nd and 3rd years of life. The monkeys raised with dogs consistently spent more time looking at projected slides, and were more responsive to the novelty, complexity and mode of presentation of the pictures. In keeping with the results of earlier tests, heart rate was also substantially higher in monkeys raised with dogs than in those raised with inanimate surrogates (Wood *et al.*, 1979).

An intriguing pattern emerging from these studies is that the reaction to novelty is more vigorous in monkeys raised with the more responsive maternal figure, whether this is the real mother, a moving inanimate surrogate or a canine mother substitute. Gross motor activity, vocalization, defaecation/urination, and visual and physical contact with novel objects conformed to this pattern in all three studies; heart rate, when recorded, was also higher. Other research has produced similar findings (Green and Gordon, 1964; Green, 1965; Kraemer *et al.*, 1989, 1991; Schneider *et al.*, 1991). The contrasting modes of reactivity may be described in the context of Bowlby's attachment theory as 'protest' and 'despair' (Kraemer *et al.*, 1991) or with reference to the distinct coping patterns of 'flight/fight' and 'conservation/withdrawal' (Engel, 1967; Henry, 1976). Regardless of how they are described, the consistency, generality and persistence of the differences in reactivity between monkeys raised with relatively responsive and unresponsive attachment figures demonstrate that maternal stimulation profoundly influences traits of temperament, as manifested in an individual's characteristic stance towards novelty and change.

The more active mode of responding to change displayed by the monkeys raised with responsive attachment figures may have been influenced by their being repeatedly exposed to *response-contingent feedback* (Lewis and Goldberg, 1969; Mason, 1978). Despite many differences between natural mothers, dogs and mobile surrogates, all present an opportunity for the developing monkey to learn that its behaviour can affect – and in some circumstances, control – the actions of a responsive attachment figure. The cumulative effects of this experience could have led the individual to adopt a distinct mode of perceiving and coping with environmental events (Mason, 1978).

The relevance of response-contingent feedback and control to emotional development and temperament is convincingly demonstrated in another rearing study with young rhesus monkeys. The animals were raised in small peer groups in cages in which one group had control over the delivery of food treats and switching on flashing lights, and another group (yoked controls) experienced precisely the same events but could not control them (Mineka *et al.*, 1986). Like monkeys raised with responsive attachment figures, monkeys that experienced response-contingent feedback and environmental control are less timid than those that had not, as indicated by greater readiness to enter an unfamiliar room and the frequency of contacts with novel objects. It is of interest that these effects are not confined to primates. Rats raised from birth in cages that allow them to control the delivery of food, water and illumination are less emotional in the open field than yoked controls as indicated by activity and defaecation (Joffe *et al.*, 1973).

The scope and diversity of maternal influences on primate emotional development are no doubt greater than these experimental studies might suggest. Nevertheless, in contrast to the mother's immediate effects on her offspring's emotional behaviour, which are well documented, her long-term influences on temperament have been demonstrated in relatively few studies (Clarke and Boinski, 1995). One effect that is firmly established is that offspring are influenced by a female's emotional state during her pregnancy. Rhesus monkeys were repeatedly exposed to novel surroundings and loud noise during post-conception days 90–145 (term is 165 days). To eliminate postnatal maternal effects, infants were separated from their mothers at birth and nursery-reared in a regime that provided ample contact with peers. Prenatally stimulated monkeys, observed in unfamiliar environments at approximately 6 months of age and at 4 years of age, were more distressed than infants of control mothers, as evidenced by higher levels of 'disturbance behaviours' (clinging, self-directed behaviour), and spent more time being inactive and less time locomoting, exploring the environment and engaging in social play (Schneider, 1992; Clarke *et al.*, 1996b). Several studies indicate that the quality of parental behaviour may also influence fearfulness. Adolescent rhesus monkeys raised by punitive mothers display more aggression and less social exploration than offspring of normal mothers (Mitchell *et al.*, 1967; Sackett, 1967), and infant bonnet macaques whose mothers had to search for food under variable-demand conditions showed less 'secure' filial attachments compared with infants whose mothers foraged on a low-demand regimen (Andrews and Rosenblum, 1991). In a study of group-living vervet monkeys, infants and juveniles whose mothers were protective and restrictive were relatively fearful and cautious in novel situations (Fairbanks, 1996).

Based on their experience during development, animals build up a generalized view or 'representation' of their 'normal' or expected environment (Salzen, 1978). Novelty is by definition a departure from the norm, and a source of behavioural and physiological arousal (Berlyne, 1960, 1967). Regardless of species, the most extreme emotional reactions to novelty in mammals occur in individuals whose familiar environment has been radically restricted as the result of early, severe and prolonged limitations on the general level of environmental stimulation. Although these animals usually appear normal if they are observed unobtrusively in their familiar rearing environment, they are likely to respond to novelty with signs of extreme arousal, disorganization and panic. Motor discharge may be exaggerated, as in seizures, tics, whirling fits and frantic running, or it may be minimal, as in persistent crouching, freezing and avoidance of any form of contact with novel objects, surroundings or events. Such effects have been documented for rhesus monkeys, chimpanzees, dogs and cats, and may reflect a general tendency for isolation-reared mammals to overreact to novelty (for reviews, see Mason, 1970; Konrad and Melzack, 1975; Riesen and Zilbert, 1975). With repeated, graduated exposures to novelty some mitigation of the reaction can occur, although residual effects are likely to persist long after the animal has been transferred to a normal environment, perhaps throughout its life.

In summary, this section has reviewed a few of the many studies with rats, primates and other mammals demonstrating that diverse forms of early experience can have profound effects on how individuals respond to novel surroundings, objects and events. Some animals characteristically behave as if they were afraid of anything even slightly beyond their usual experience, whereas in the same circumstances others are more bold or curious than fearful. These tendencies persist over time and across situations and have been demonstrated in a range of different species, suggesting that the effects of early experience on fearfulness are modifying a fundamental mammalian 'system' that 'remembers' the experience.

Early experience and stress

The most obvious candidate for the role of a basic system intimately associated with fear which might be changed by early experience is the *stress system*, a dauntingly complex array of interdependent neuroendocrine structures and functions that collectively maintain an organism's physiology within a normal range (Johnson *et al.*, 1992; Sapolsky, 1992). *Stress* is usually defined in homeostatic terms as a departure from the organism's usual state of equilibrium. *Stressor* refers to any condition or event that causes the departure. The *stress response* refers to a more or less predictable suite of physiological changes that comprise the actual departure from equilibrium and generally act so as to restore the original state. The primary aim of this section is to examine the effects of early experience on the organization of this response.

Contemporary conceptions of the stress response reflect the influence of two major contributors. One is Walter B. Cannon (1939), famous for his research on the sympathetic adrenomedullary system, the source of adrenaline (also known as epinephrine), and for developing the concepts of homeostasis and the flight–fight reaction to emergency situations. The other is Hans Selye (1956, 1973), the most influential promulgator of the stress concept. Selye characterized the reaction to stress as passing through stages which he described as the *general adaptation syndrome*. The first stage in the body's reaction to a stressor is *alarm*, reflected, for example, in activation of the hypothalamic–pituitary–adrenocortical (HPA) system and production of glucocorticoids, such as cortisol and corticosterone. If the stress continues, the organism enters the second stage, resistance or *adaptation*. If homeostasis is not restored, it enters the third and final phase, *exhaustion*, characterized by damage to organs and tissues, suppression of the immune system, emergence of stress-related diseases, and possibly eventuating in death.

Both Cannon and Selye viewed the stress response as relatively nonspecific. For Selye this meant that the magnitude of the response varied with the intensity of the stressor, but was independent of its other qualities. For example, cold, heat, drugs, hormones, sorrow and joy were believed to provoke the same biochemical reaction in the organism, despite great differences in the essential nature of these phenomena (Selye, 1973).

This 'non-specificity' hypothesis, particularly the version propounded by Selye, was challenged by J.W. Mason. Based on his own research and other information that had become available since the classic contributions of Cannon and Selye, Mason suggested that not all stressors were alike. In particular, he emphasized the special potency of psychological variables as activators of the pituitary–adrenocortical system (J.W. Mason, 1968, 1971).

One effect of the more explicit psychological orientation advocated by Mason was to strengthen the conceptual bridge between stress theory and developmental research. The assumption that fear is accompanied by physiological changes has been commonplace for many years, of course. It is also common knowledge that perceptions, expectancies, uncertainty, novelty, lack of control over the environment, and similar psychological processes are capable of eliciting fear and are readily influenced by experience. Only in comparatively recent times, however, has evidence been sought to establish whether specific and enduring changes resulting from such experience might occur in the organization of the stress response.

The first indications that the organization of the stress response was chronically modified by experience were based on the 'early handling' paradigm. These studies established that the pituitary–adrenocortical response to stress was lower in adult rats who were handled during their first 20 days of life than in non-handled controls (Levine, 1967; Levine *et al.*, 1967). Later research confirmed these findings and extended the range of effects to include basal glucocorticoid levels and cellular changes within specific regions of the central nervous system, including the hippocampus and frontal cortex (Meaney *et al.*, 1985, 1988; Bhatnagar and Meaney, 1995). Recent studies have also provided strong presumptive evidence that the mother's behaviour towards her offspring is a principal environmental mediator of these changes (Liu *et al.*, 1997; Caldji *et al.*, 1998).

The search for enduring effects of early experience on the stress response in primates has focused mainly on macaques, and on comparisons between monkeys raised with their biological mothers and with peers only or alone. These conditions are known to produce gross differences in patterns of emotional behaviour, differences that are likely to be associated with changes in some measures of physiological responsiveness like heart rate (Wood *et al.*, 1979; Mason and Capitanio, 1988). Although these associations are intriguing, they obviously do not demonstrate that fundamental regulatory properties of the physiological system have been altered.

The pituitary–adrenocortical system, which has played a major role in stress research from Selye to the present day and is known from studies with rats to be modifiable by early experience, is an obvious place to look for chronic changes in the organization of the stress response in primates. There is no question that this system is highly responsive to stressful situations and events. Cortisol, the chief hormonal product of the adrenal cortex in primates, has repeatedly been shown to be sensitive to novelty, brief social separation and a variety of other stressful psychological variables, not only in monkeys and apes but in a host of other mammalian species (Hennessy, 1997).

Contrary to expectation, a strong and consistent relationship between basal cortisol level and early history has not emerged in primate research. Baseline values are particularly instructive since they purport to reflect the activities of the HPA system in the absence of external perturbations. Although differences between rearing groups have been found, they cover all logical possibilities. Baselines have been reported to be *lower* in mother-reared monkeys compared with animals reared alone or with peers (Champoux *et al.*, 1989), to be *higher* in mother-reared monkeys (Clarke *et al.*, 1998; Shannon *et al.*, 1998), and to show *no effect* of rearing conditions (Meyer and Bowman, 1972; Meyer *et al.*, 1975; Clarke, 1993; Lubach *et al.*, 1995).

Using imposed stressors to investigate the effects of early experience on cortisol responses has also produced mixed results. The stressors used have been diverse, including social isolation and/or novel surroundings (Meyer and Bowman, 1972; Meyer *et al.*, 1975; Champoux *et al.*, 1989; Clarke, 1993; Clarke *et al.*, 1998; Shannon *et al.*, 1998), formation of new social groups (Clarke, 1993), physical restraint (Meyer and Bowman, 1972), and various biochemical challenges (met-enkephalin, dexamethasone, exogenous adrenocorticotropic hormone (ACTH), corticotropin-releasing hormone (CRH)) (Meyer and Bowman, 1972; Champoux *et al.*, 1989; Clarke *et al.*, 1998). Most studies have not found reliable differences in the cortisol response between mother-reared monkeys and animals raised in other conditions.

Only two studies have reported statistically significant rearing effects on cortisol, and both found that values in response to social manipulation were higher in monkeys reared with the mother, compared with other rearing conditions. The outcomes were not simple, however. In one study the effects on cortisol and ACTH were examined over a series of six 1-week social separations and reunions. When the study started, the monkeys were about 10 months old and were living in established peer groups. In this phase of the research, overall levels of cortisol and ACTH were significantly higher in monkeys raised by their mothers than in the monkeys raised with peers. Several years later, however, when their responses to dexamethasone and CRH challenges were tested, no significant rearing effects on cortisol or ACTH were seen (Clarke *et al.*, 1998). (An earlier study with these same subjects also found no rearing effects on cortisol responses to changes in cage location and to the formation of new social groups, although ACTH levels were significantly higher in the mother-reared monkeys (Clarke, 1993).)

In a second study in which significant rearing effects on cortisol responses to stress were found, young monkeys living with their mothers in mixed social groups were compared with two nursery-reared groups, one having experience only with peers (continuous housing from day 37), the other receiving scheduled exposure to peers for 2 h day^{-1} and living continuously with a cloth surrogate (Shannon *et al.*, 1998). Cortisol values were determined after a 30-min period of social isolation when the monkeys were 90, 120 and 150 days of age. Although cortisol concentrations of the mother-reared monkeys were significantly higher compared with those of the peer-and-surrogate animals, their values did not differ significantly from those of monkeys housed continuously with peers, nor did values for the two groups of nursery-reared monkeys differ from each other.

Another approach to exploring the effects of the early social environment on primate stress physiology has used cerebral spinal fluid (CSF) as a means of estimating concentrations in the brain of selected neurotransmitters, chiefly noradrenaline, serotonin and dopamine. As with cortisol and ACTH, the results do not conform to a simple pattern. Mother-reared infants have shown *lower* CSF noradrenaline levels than mother-deprived infants during the first 6 months of life in some research, whereas similar studies covering the same age period report *higher* levels in mother-reared monkeys. Similar variability has been found in differences between maternally-reared and maternally-deprived monkeys in CSF levels of serotonin and dopamine (Higley and Suomi, 1989; Kraemer *et al.*, 1989, 1991; Higley *et al.*, 1990; Clarke *et al.*, 1996a). (See p. 284 for additional information.)

Effects of prenatal stress on concentrations of cortisol and CSF neurotransmitters have also been examined, again with mixed results. As previously noted, repeatedly exposing females in midgestation to noise and novel surroundings increases the emotional behaviour of their nursery-reared offspring (Schneider, 1992; Clarke *et al.*, 1996b). Offspring of such females at 8 months of age responded to separation from companions with significantly higher CSF concentrations of metabolites for noradrenaline (MHPG) and dopamine (DOPAC) compared with controls, although rearing effects on cortisol were not statistically significant. At approximately 15 months of age, however, prenatally stressed monkeys showed significantly higher cortisol responses on three of four samples taken over a 3-month period, and higher levels of ACTH (non-significant). At 18 months, prenatally stressed monkeys showed higher concentrations of both cortisol and ACTH compared with control monkeys, but differences were not statistically significant (Clarke *et al.*, 1994; Schneider *et al.*, 1998).

Neuroanatomical analyses of discrete regions of the rhesus hypothalamus have also found no reliable difference between mother-reared monkeys socialized with peers and socially deprived animals (Ginsberg *et al.*, 1993a, b). Differences have been demonstrated, however, in other regions of the brain, including cortex, striatum, basal ganglia and corpus callosum (Struble and Riesen, 1978; Martin *et al.*, 1990, 1991; Sanchez *et al.*, 1998). These effects appear to be permanent. Although many parts of the brain are involved in emotional behaviour and stress, the specific contributions of these areas to a temperamental trait of fearfulness are not known.

In summary, the search for abiding neuroendocrine changes associated with fearfulness has been carried out mainly with the laboratory rat and the rhesus monkey and has produced contrasting results for the two species. In the rat, studies of the effects of early experience on the HPA system have demonstrated persistent changes at various levels of organization, from concentrations of circulating glucocorticoids in the blood to receptor density in the frontal cortex. This research has also contributed to an improved understanding of the mother's role as the primary mediator of environmental effects and a major regulatory agent in the early development of her offspring. Primate research has also looked for the effects of early experience on the HPA system. A principal comparison has been between monkeys raised with the mother and those raised alone or with early extensive contact with peers. Primate mothers have powerful immediate and long-term effects on the emotional behaviour, physiological responses and temperament of their offspring, of course. However, in contrast to the findings with rats, no consistent pattern has been found indicating enduring experience-produced changes in the organization of the stress response, whether measured by cortisol, ACTH, neurotransmitter substances in CSF, or neuroanatomical changes in the hypothalamus.

Developmental consequences and implications for well-being

A basic premise of this chapter is that fearfulness can be acquired. Individuals differ in their emotional reactions to novelty, challenge and change, some being predictably more timid and fearful than others, even though they have had essentially the same experience. In other cases, however, it is clear that fearfulness can be profoundly and persistently modified by an individual's early experience. The source and nature of these modifications have provided the focus of this chapter. In this section, I examine the developmental consequences of experientially induced changes in fearfulness and the implications of these changes for well-being.

Based on research with a variety of mammalian species, two major experiential sources of fearfulness can be distinguished: the general level of environmental stimulation experienced early in life and the effects of early experience with other members of one's species. Whether these two influences on fearful behaviour involve the same underlying mechanisms is not clear at this point. It seems likely, however, that the ontogenetic processes are overlapping and complementary rather than identical (W.A. Mason, 1971).

The effects on fearfulness of early restrictions on the general level and complexity of environmental stimulation are well documented. Presumably, the effective developmental variable under these conditions is impoverishment of experience during a period when the individual is forming an internal representation of its environment. Although an animal may appear to behave normally within the limits of its rearing environment, novel conditions that exceed these boundaries by seemingly trivial amounts are likely to elicit intense fearful behaviour. Reactions often assume extreme forms, ranging from violent motor discharge to complete withdrawal, behaviours suggesting that the individual is overwhelmed by the situation, in an acute state of hyperarousal or panic. Unusual changes have also been shown in the structure and activities of the central nervous system (Martin *et al.*, 1990, 1991). Although behavioural symptoms diminish with prolonged exposure to more complex environments, manifestations of abnormality do not disappear. Individuals raised in these conditions do not appear to be unusually susceptible to stress-related disease. Nevertheless their well-being is compromised by persistent difficulties in establishing effective relationships with others and accommodating to changing circumstances (Davenport and Rogers, 1970).

A second major influence on the development of fearfulness in mammals is early social experience, particularly with respect to the mother. Maternal influences on mammalian development are pervasive, dynamic and complex, and they vary significantly by species. Experimental studies of maternal effects on fearfulness have been carried out mainly on rats and rhesus monkeys and have emphasized different experimental paradigms for the two species. The majority of studies with rats compare the effects of brief, discrete and recurring perturbations of the infant's ongoing relationship with its mother during a circumscribed period in development, with the effects of an undisturbed relationship. The most common paradigm in primate research focuses on the effects of differences in social experience that are in place continuously over an extended period. Monkeys raised with their biological mothers for the first several months of life (for example, until weaned at 6 months) are compared with animals that are separated from their mothers as neonates and raised in a nursery, either alone or with varying amounts of contact with age-mates. Regardless of the differences between species, the two approaches may be expected to have different developmental consequences and different implications for well-being.

Brief separations of the infant rat or monkey from its mother are usually accompanied by an increase in high-pitched vocalizations and adrenocortical activity. In rats, a principal long-term behavioural effect of repeated separations during the pre-weaning period is a reduction in fearfulness. Squirrel monkey infants experiencing an analogous procedure showed a progressive decline in vocalizations across separations, but continued to display a robust cortisol response. At the end of the series, however, the cortisol responses and relationships with their mothers of separated infants were indistinguishable from those of control infants, suggesting that the separations had no permanent effect on the infants' physiology or behaviour (Coe *et al.*, 1983; Hennessy, 1986). Thus, it does not appear that repeated brief separations from the mother increase fearfulness. Given due concern for age- and species-specific requirements with regard to ambient temperature, food intake, duration of separation and the like, neither does it appear that the experience has long-term detrimental effects on the infant rat or monkey's well-being. On the contrary, the consequences for the rat may be beneficial. It should be noted, however, that potentially harmful effects on immune function have been found in rhesus monkeys exposed to somewhat longer separation intervals (Coe *et al.*, 1989).

Rearing an infant mammal apart from its mother is obviously a more radical procedure than brief separations and would seem to have greater potential for producing significant effects on fearfulness. In reality, the long-term effects of early maternal deprivation on fearfulness are equivocal, at least in primates, the taxon on which most of the data have been collected. The source of ambiguity is that a newborn monkey or ape permanently separated from its mother early in the postnatal period develops a suite of behaviours that appear to be compensatory, modified versions of filial behaviour patterns. These become the animal's characteristic and habitual mode of responding to emotional provocation. Thus, in circumstances in which an infant rhesus monkey raised by its mother seeks her out, clings to her, takes the nipple, and is comforted by these acts, the nursery-reared animal clasps itself, sucks its thumb or big toe, crouches, rocks, and likewise is comforted. The presence of these self-directed behaviours probably explains the paradoxical finding that by traditional criteria immature monkeys raised alone show less emotional arousal than do mother-reared monkeys in mildly stressful situations. Their heart rate is lower, and they vocalize less, defaecate less and engage in fewer gross, non-directed motor activities. At the same time they also show greater behavioural inhibition in their reactions to novel objects or events.

As expected in view of their origins, self-directed versions of filial behaviours are linked to developmental status. They do not appear as habitual responses in monkeys that are separated from their mothers after weaning is completed, and like their mother-directed counterparts, these behaviours become less frequent as development proceeds. Compared with behaviours shown by mother-reared monkeys, however, these behaviours decline more slowly, and under conditions of extremely high arousal, particularly when alternative actions are blocked, they may reappear.

The prevailing view that maternally deprived monkeys are necessarily more fearful than mother-reared animals needs to be re-examined. There is no indication that deprived monkeys are more fearful than mother-reared animals in their familiar rearing environment, nor do they differ in basal cortisol values (Lubach *et al.*, 1995). The behaviours they show most prominently in stressful situations may indeed indicate heightened emotional arousal, but they probably also serve the homeostatic function of reducing arousal, in the manner of the filial responses from which they are derived. A predictable developmental consequence of maternal deprivation is to alter the organization of emotional behaviour. Aspects of stress physiology are also changed, but the precise nature of these changes is not clear.

Few studies have examined the effects of complete maternal deprivation in rats. No doubt the reasons include the formidable difficulties of hand-rearing the newborn rat and the immediate, pervasive, deleterious consequences of eliminating the major regulatory influence on the neonate's physiology and behaviour (Koch and Arnold, 1972; Hofer, 1987). In the long term, early social deprivation increases fearful behaviour as measured by heart rate and behaviour in the open field, but there are no indications so far of modal changes in the organization of emotional behaviour, similar to those occurring in rhesus monkeys (Koch and Arnold, 1972).

In view of the many vital contributions of the mammalian mother to her offspring, it is not surprising that early and prolonged maternal deprivation has deleterious effects on the well-being of both rats and monkeys, as indicated by behavioural and physiological criteria. Changes in some aspects of emotional behaviour and temperament can be included among these effects. In my view, however, the significance of information on the emotional consequences of being raised without a mother relates chiefly to the light it can shed on maternal contributions to individual variations in fearfulness. A few of these effects that have been demonstrated systematically have been considered in this review, but many others have been described in anecdotal accounts of development in natural settings. The biological significance of these effects is clear: in a changing world, presenting fresh opportunities for possible benefit or harm, maternal influences that shape an individual's characteristic stance towards challenge and change are immediately relevant to its well-being and to its future prospects for survival and reproductive success.

Acknowledgements

My thanks to John Capitanio and Sally Mendoza for their comments. Preparation of this chapter was supported by grants MH49033, MH57502 and RR00169 from the US Public Health Service, National Institutes of Health.

Additional information

At the time of writing, I was unaware of research in which the effects of early differences in mother-infant relationships on CSF concentrations of CRH and cortisol of adult bonnet macaques were compared directly. Grown monkeys who were nursed by mothers being exposed to an unpredictable foraging schedule (varying between high and low demand) had significantly higher concentrations of CRH and significantly lower concentrations of cortisol compared with monkeys whose mothers were experiencing an invariant demand schedule (either consistently high or low). (For references see: Coplan, J.D., Andrews, M.W., Rosenblum, L.A., Owens, M.J., Friedman, S., Gorman, J.M. and Nemeroff, C.B. (1996) Persistent elevations of cerebrospinal fluid concentrations of corticotropin-releasing factors in adult nonhuman primates exposed to early-life stressors – Implications for the pathophysiology of mood and anxiety disorders. *Proceedings of the National Academy of Sciences of the United States of America* 93, 1619–1623; Coplan, J.D., Trost, R.C., Owens, M.J., Cooper, T.B., Gorman, J.M., Nemeroff, C.B. and Rosenblum, L.A. (1998) Cerebrospinal fluid concentrations of somatostatin and biogenic amines in grown primates reared by mothers exposed to manipulated foraging conditions. *Archives of Psychiatry* 55, 473–477.)

References

Ader, R. (1968) Effects of early experiences on emotional and physiological reactivity in the rat. *Journal of Comparative and Physiological Psychology* 66, 264–268.

Andrews, M.W. and Rosenblum, L.A. (1991) Attachment in monkey infants raised in variable- and low-demand environments. *Child Development* 62, 686–693.

Archer, J. (1973) Tests for emotionality in rats and mice: a review. *Animal Behaviour* 21, 205–235.

Archer, J.E. and Blackman, D.E. (1971) Prenatal psychological stress and offspring behavior in rats and mice. *Developmental Psychobiology* 4, 193–248.

Bell, R.W., Nitschke, W., Bell, N.J. and Zachman, T.A. (1974) Early experience, ultrasonic vocalizations, and maternal responsiveness in rats. *Developmental Psychobiology* 7, 235–242.

Berlyne, D.E. (1960) *Conflict, Arousal, and Curiosity*. McGraw-Hill, New York.

Berlyne, D.E. (1967) Arousal and reinforcement. In: Levine, D. (ed.) *Nebraska Symposium on Motivation* 15. Lincoln University of Nebraska Press, pp. 1–110.

Bhatnagar, S. and Meaney, M.J. (1995) Hypothalamic–pituitary–adrenal function in chronic intermittently cold-stressed neonatally handled and non handled rats. *Journal of Neuroendocrinology* 7, 97–108.

Boissy, A. (1995) Fear and fearfulness in animals. *Quarterly Review of Biology* 70, 166–191.

Brown, C.P., Smotherman, W.P. and Levine, S. (1977) Interaction-induced reduction in differential maternal responsiveness: an effect of cue-reduction or behavior? *Developmental Psychobiology* 10, 273–280.

Caldji, C., Tannenbaum, B., Sharma, S., Francis, D., Plotsky, P.M. and Meaney, M.J. (1998) Maternal care during infancy regulates the development of neural systems mediating the expression of fearfulness in the rat. *Proceedings of the National Academy of Sciences of the United States of America* 95, 5335–5340.

Cannon, W.B. (1939) *The Wisdom of the Body*. W.W. Norton & Company, New York.

Champoux, M., Coe, C.L., Schanberg, S.M., Kuhn, C.M. and Suomi, S.J. (1989) Hormonal effects of early rearing conditions in the infant rhesus monkey. *American Journal of Primatology* 19, 111–117.

Clarke, A.S. (1993) Social rearing effects on HPA axis activity over early development and in response to stress in rhesus monkeys. *Developmental Psychobiology* 26, 433–446.

Clarke, A.S. and Boinski, S. (1995) Temperament in nonhuman primates. *American Journal of Primatology* 37, 103–125.

Clarke, A.S., Wittwer, D.J., Abbott, D.H. and Schneider, M.L. (1994) Long-term effects of prenatal stress on HPA axis activity in juvenile rhesus monkeys. *Developmental Psychobiology* 27, 257–269.

Clarke, A.S., Hedeker, D.R., Ebert, M.H., Schmidt, D.E., McKinney, W.T. and Kraemer, G.W. (1996a) Rearing experience and biogenic amine activity in infant rhesus monkeys. *Biological Psychiatry* 40, 338–352.

Clarke, A.S., Soto, A., Bergholz, T. and Schneider, M.L. (1996b) Maternal gestational stress alters adaptive and social behavior in adolescent rhesus monkey offspring. *Infant Behavior and Development* 19, 451–461.

Clarke, A.S., Kraemer, G.W. and Kupfer, D.J. (1998) Effects of rearing condition on HPA axis response to fluoxetine and desipramine treatment over repeated social separations in young rhesus monkeys. *Psychiatry Research* 79, 91–104.

Coe, C.L., Glass, J.C., Wiener, S.G. and Levine, S. (1983) Behavioral, but not psysiological, adaptation to repeated separation in mother and infant primates. *Psychoneuroendocrinology* 8, 401–409.

Coe, C.L., Lubach, G.R., Ershler, W.B. and Klopp, R.G. (1989) Influence of early rearing on lymphocyte proliferation responses in juvenile rhesus monkeys. *Brain, Behavior, and Immunity* 3, 47–60.

Daly, M. (1973) Early stimulation of rodents: a critical review of present interpretations. *British Journal of Psychology* 64, 435–460.

Davenport, R.K. and Rogers, C.M. (1970) Differential rearing of the chimpanzee: a project survey. In: Bourne, G.H. (ed.) *The Chimpanzee,* 3. Karger Publishing, New York, pp. 337–360.

Denenberg, V.H. (1966) Animal studies on developmental determinants of behavioral adaptability. In: Harvey, O.J. (ed.) *Experience, Structure, and Adaptability.* Springer, New York, pp. 123–147.

Denenberg, V.H. (1969a) Open-field behavior in the rat: what does it mean? *Annals of the New York Academy of Sciences* 159, 852–859.

Denenberg, V.H. (1969b) The effects of early experience. In: Hafez, E.S.E. (ed.) *The Behaviour of Domestic Animals.* Bailliere, Tindall & Cassell, London, pp. 96–130.

Denenberg, V.H. and Smith, S.A. (1963) Effects of infantile stimulation and age upon behavior. *Journal of Comparative and Physiological Psychology* 56, 307–312.

Denenberg, V.H. and Whimbey, A.E. (1963) Behavior of adult rats is modified by the experiences their mothers had as infants. *Science* 142, 1192–1193.

Denenberg, V.H., Ottinger, D.R. and Stephens, M.W. (1962) Effects of maternal factors upon growth and behavior of the rat. *Child Development* 33, 65–71.

Doty, B.A. and Doty, L.A. (1967) Effects of handling at various ages on later open-field behaviour. *Canadian Journal of Psychology/Review of Canadian Psychology* 21, 463–470.

Eastman, R.F. and Mason, W.A. (1975) Looking behavior in monkeys raised with mobile and stationary artificial mothers. *Developmental Psychobiology* 8, 213–221.

Engel, G.L. (1967) A psychological setting of somatic disease: the 'giving up–given up complex. *Proceedings of the Royal Society of Medicine* 60, 553–555.
Fairbanks, L.A. (1996) Individual differences in maternal style – causes and consequences for mothers and offspring. *Advances in the Study of Behavior* 25, 579–611.
Gandelman, R. (1992) *Psychobiology of Behavioral Development.* Oxford University Press, New York.
Gertz, B. (1957) The effect of handling at various age levels on emotional behavior of adult rats. *Journal of Comparative and Physiological Psychology* 50, 613–616.
Ginsberg, S.D., Hof, P.R., McKinney, W.T. and Morrison, J.H. (1993a) The noradrenergic innervation density of the monkey paraventricular nucleus is not altered by early social deprivation. *Neuroscience Letters* 158, 130–134.
Ginsberg, S.D., Hof, P.R., McKinney, W.T. and Morrison, J.H. (1993b) Quantitative analysis of tuberoinfundibular tyrosine hydroxylase- and corticotropin-releasing factor-immunoreactive neurons in monkeys raised with differential rearing conditions. *Experimental Neurology* 120, 95–105.
Green, P.C. (1965) Influence of early experience and age on expression of affect in monkeys. *Genetic Psychology* 106, 157–171.
Green, P.C. and Gordon, M. (1964) Maternal deprivation: its influence on visual exploration in infant monkeys. *Science* 145, 292–294.
Hall, C.S. (1941) Temperament: a survey of animal studies. *Psychological Bulletin* 28, 909–943.
Hennessy, M.B. (1986) Multiple, brief maternal separations in the squirrel monkey: changes in hormonal and behavioral responsiveness. *Physiology and Behavior* 36, 245–250.
Hennessy, M.B. (1997) Hypothalamic–pituitary–adrenal responses to brief social separation. *Neuroscience and Biobehavioral Reviews* 21, 11–29.
Henry, J.P. (1976) Mechanisms of psychosomatic disease in animals. *Advances in Veterinary Sciences and Comparative Medicine* 20, 115–145.
Higley, J.D. and Suomi, S.J. (1989) Temperamental reactivity in non-human primates. In: Kohnstamm, G.A., Bates, J.E. and Rothbart, M.K. (eds) *Temperament in Childhood.* John Wiley & Sons, New York, pp. 153–167.
Higley, J.D., Suomi, S.J. and Linnoila, M. (1990) Developmental influences on the serotonergic system and timidity in the nonhuman primate. In: Coccaro, E.F. and Murphy, D.L. (eds) *Serotonin in Major Psychiatric Disorders: Progress in Psychiatry Series, 21.* American Psychiatric Press, Washington, DC, pp. 29–46.
Hofer, M.A. (1987) Early social relationships: a psychobiologist's view. *Child Development* 58, 633–647.
Joffe, J.M., Rawson, R.A. and Mulick, J.A. (1973) Control of their environment reduces emotionality in rats. *Science* 180, 1383–1384.
Johnson, E.O., Kamilaris, T.C., Chrousos, G.P. and Gold, P.W. (1992) Mechanisms of stress: a dynamic overview of hormonal and behavioral homeostatis. *Neuroscience and Biobehavioral Reviews* 16, 115–130.
Koch, M.D. and Arnold, W.J. (1972) Effects of early social deprivation on emotionality in rats. *Journal of Comparative and Physiological Psychology* 78, 391–399.
Konrad, K. and Melzack, R. (1975) Novelty-enhancement effects associated with early sensory-social isolation. In: Riesen, A.H. (ed.) *The Developmental Neuropsychology of Sensory Deprivation.* Academic Press, New York, pp. 253–276.

Kraemer, G.W., Ebert, M.H., Schmidt, D.E. and McKinney, W.T. (1989) A longitudinal study of the effect of different social rearing conditions on cerebrospinal fluid norepinephrine and biogenic amine metabolites in rhesus monkeys. *Neuropsychopharmacology* 2, 175–189.

Kraemer, G.W., Ebert, M.H., Schmidt, D.E. and McKinney, W.T. (1991) Strangers in a strange land: a psychobiological study of infant monkeys before and after separation from real or inanimate mothers. *Child Development* 62, 548–566.

Levine, S. (1967) Maternal and environmental influences on the adrenocortical response to stress in weanling rats. *Science* 156, 258–260.

Levine, S. and Otis, L.S. (1958) The effects of handling before and after weaning on the resistance of albino rats to later deprivation. *Canadian Journal of Psychology* 12(2), 103–108.

Levine, S., Haltmeyer, G.C., Karas, G.G. and Denenberg, V.H. (1967) Physiological and behavioral effects of infantile stimulation. *Physiology and Behavior* 2, 55–59.

Lewis, M. and Goldberg, S. (1969) Perceptual-cognitive development in infancy: a generalized expectancy model as a function of the mother–infant interaction. *Merrill-Palmer Quarterly of Behavior and Development* 15, 81–100.

Liu, D., Diorio, J., Tannenbaum, B., Caldji, C., Fancis, D., Freedman, A., Sharma, S., Pearson, D., Plotsky, P.M. and Meaney, M.J. (1997) Maternal care, hippocampal glucocorticoid receptors, and hypothalamic–pituitary–adrenal responses to stress. *Science* 277, 1659–1661.

Lubach, G.R., Coe, C.L. and Ershler, W.B. (1995) Effects of early rearing environment on immune responses of infant rhesus monkeys. *Brain, Behavior, and Immunity* 9, 31–46.

Martin, L., Lewis, M., Gluck, J. and Cork L. (1990) Aberrant compartmental organization of the striatum in isolation reared monkeys. *Journal of Neuropathology and Experimental Neurology* 49, 284.

Martin, L.J., Spicer, D.M., Lewis, M.H., Gluck, J.P. and Cork, L.C. (1991) Social deprivation of infant rhesus monkeys alters the chemoarchitecture of the brain: I. Subcortical regions. *Journal of Neuroscience* 11, 3344–3349.

Mason, J.W. (1968) A review of psychoendocrine research on the pituitary-adrenal cortical system. *Psychosomatic Medicine* 30, 576–607.

Mason, J.W. (1971) A re-evaluation of the concept of non-specificity in stress theory. *Journal of Psychiatric Research* 8, 323–333.

Mason, W.A. (1970) Information processing and experiential deprivation: a biologic perspective. In: Young, F.A. and Lindsley, D.B. (eds) *Early Experience and Visual Information Processing in Perceptual and Reading Disorder.* National Academy of Sciences, Washington, pp. 302–323.

Mason, W.A. (1971) Motivational factors in psychosocial development. In: Arnold, W.J. and Page, M.M. (eds) *Nebraska Symposium on Motivation, 8.* University of Nebraska Press, Lincoln, Nebraska, pp. 35–67.

Mason, W.A. (1978) Social experience and primate cognitive development. In: Burghardt, G.M. and Bekoff, M. (eds) *The Development of Behavior: Comparative and Evolutionary Aspects.* Garland Press, New York, pp. 233–251.

Mason, W.A. and Berkson, G. (1975) Effects of maternal mobility on the development of rocking and other behaviors in rhesus monkeys: a study with artificial mothers. *Developmental Psychobiology* 8, 197–211.

Mason, W.A. and Capitanio, J.P. (1988) Formation and expression of filial attachment in rhesus monkeys raised with living and inanimate mother substitutes. *Developmental Psychobiology* 21, 401–430.

Mason, W.A. and Green, P.C. (1962) The effects of social restriction on the behavior of rhesus monkeys: IV: responses to a novel environment and to an alien species. *Journal of Comparative and Physiological Psychology* 55, 363–368.

Mason, W.A., Davenport, R.K., Jr and Menzel, E.W., Jr (1968) Early experience and the social development of rhesus monkeys and chimpanzees. In: Newton, G. and Levine, S. (eds) *Early Experience and Behavior.* Charles C. Thomas, Springfield, Illinois, pp. 440–480.

Meaney, M.J., Aitken, D.H., Bodnoff, S.R., Iny, L.J., Tatarewicz, J.E. and Sapolsky, R.M. (1985) Early postnatal handling alters glucocorticoid receptor concentrations in selected brain regions. *Behavioral Neuroscience* 99, 765–770.

Meaney, M.J., Aitken, D.H., van Berkel, C., Bhatnagar, S. and Sapolsky, R.M. (1988) Effect of neonatal handling on age-related impairments associated with the hippocampus. *Science* 239, 766–768.

Meier, G.W. and Schutzman, L.H. (1968) Mother–infant interactions and experimental manipulation: confounding or misidentification? *Developmental Psychobiology* 1, 141–145.

Mendoza, S.P., Coe, C.L., Smotherman, W.P., Kaplan, J. and Levine, S. (1980) Functional consequences of attachment: a comparison of two species. In: Bell, R.W. and Smotherman, W.P. (eds) *Maternal Influences and Early Behavior.* Spectrum Publications, New York, pp. 235–252.

Meyer, J.S. and Bowman, R.E. (1972) Rearing experience, stress and adrenocorticosteroids in the rhesus monkey. *Physiology and Behavior* 8, 339–343.

Meyer, J.S., Novak, M.A., Bowman, R.E. and Harlow, H.F. (1975) Behavioral and hormonal effects of attachment object separation in surrogate-peer-reared and mother-reared infant rhesus monkeys. *Developmental Psychobiology* 8, 425–435.

Michel, G.F. and Moore, C.L. (1995) *Developmental Psychobiology: An Interdisciplinary Science.* The MIT Press, Cambridge, Massachusetts.

Mineka, S., Gunnar, M. and Champoux, M. (1986) Control and early socioemotional development: infant rhesus monkeys reared in controllable versus uncontrollable environments. *Child Development* 57, 1241–1256.

Mitchell, G.D., Arling G.L. and Moller, G.W. (1967) Long-term effects of maternal punishment on the behavior of monkeys. *Psychonomic Science* 8, 209–210.

Richards, M.P.M. (1966) Infantile handling in rodents: a reassessment in the light of recent studies of maternal behaviour. *Animal Behaviour* 14(4), 582.

Riesen, A.H. and Zilbert, D.E. (1975) Behavioral consequences of variations in early sensory environments. In: Riesen, A.H. (ed.) *The Developmental Neuropsychology of Sensory Deprivation.* Academic Press, New York, pp. 211–252.

Rosenblatt, J.S. (1969) The development of maternal responsiveness in the rat. *American Journal of Orthopsychiatry* 39, 36–56.

Russell, P.A. (1971) Infantile stimulation in rodents: a consideration of possible mechanisms. *Psychological Bulletin* 75, 192–202.

Sackett, G.P. (1967) Some persistent effects of different rearing conditions on preadult social behavior of monkeys. *Journal of Comparative and Physiological Psychology* 64, 363–365.

Salzen, E.A. (1978) Social attachment and a sense of security – a review. *Social Science Information* 17, 555–627.

Sanchez, M.M., Hearn, E.F., Do, D., Rilling, J.K. and Herndon, J.G. (1998) Differential rearing affects corpus callosum size and cognitive function of rhesus monkeys. *Brain Research* 812, 38–49.

Sapolsky, R.M. (1992) Neuroendocrinology of the stress-response. In: Becker, J.B., Breedlove, S.M. and Crews, D. (eds) *Behavioral Endocrinology*. The MIT Press, Cambridge, Massachusetts, pp. 287–324.

Schneider, M.L. (1992) Prenatal stress exposure alters postnatal behavioral expression under conditions of novelty challenge in rhesus monkey infants. *Developmental Psychobiology* 25, 529–540.

Schneider, M.L., Kraemer, G.W. and Suomi, S.J. (1991) The effects of vestibular-proprioceptive stimulation on motor maturation and response to challenge in rhesus monkey infants. *Occupational Therapy Journal of Research* 11, 135–152.

Schneider, M.L., Clarke, A.S., Kraemer, G.W., Roughton, E.C., Lubach, G.R., Rimm-Kaufman, S., Schmidt, D. and Ebert, M. (1998) Parental stress alters brain biogenic amine levels in primates. *Development and Psychopathology* 10, 427–440.

Selye, H. (1956) *The Stress of Life*. McGraw-Hill, New York.

Selye, H. (1973) The evolution of the stress concept. *American Scientist* 61, 692–699.

Shannon, C., Champoux, M. and Suomi, S.J. (1998) Rearing condition and plasma cortisol in rhesus monkey infants. *American Journal of Primatology* 46, 311–321.

Smotherman, W.P. and Bell, R.W. (1980) Maternal mediation of early experience. In: Smotherman, W.P. and Bell, R.W. (eds) *Maternal Influences and Early Behavior*. Spectrum Publications, New York, pp. 201–210.

Struble, R.G. and Riesen, A.H. (1978) Changes in cortical dendritic branching subsequent to partial social isolation in stumptailed monkeys. *Developmental Psychobiology* 11, 479–486.

Thompson, W.R. (1957) Influence of prenatal maternal anxiety on emotionality in young rats. *Science* 125, 698–699.

Wood, B.S., Mason, W.A. and Kenney, M.D. (1979) Contrasts in visual responsiveness and emotional arousal between rhesus monkeys raised with living and those raised with inanimate substitute mothers. *Journal of Comparative and Physiological Psychology* 93, 368–377.

14

Genetic Selection to Reduce Stress in Animals

T.G. Pottinger

NERC Institute of Freshwater Ecology, Windermere Laboratory, Far Sawrey, Ambleside, Cumbria, UK

Introduction

Concerns regarding the welfare of animals now permeate all levels of society and are directed towards a diverse range of activities. These range from debate regarding the conditions under which domestic animals are maintained on farms (in particular intensively reared species), to the zoo environment and captive animals in general, through the use of animals in toxicological testing (Smith and Boyd, 1991) and the exploitation of animals for sport (Bateson and Bradshaw, 1998). They even extend to the assessment of the welfare of free-living wild animals (Kirkwood *et al.*, 1994) and consideration of the well-being of fish, within both intensive aquaculture systems (Farm Animal Welfare Council, 1996) and natural fisheries (Pottinger, 1995).

It is widely agreed that there are difficulties inherent in defining precisely what is meant by welfare or well-being in different animal groups, and how it should be measured (Curtis, 1985; Dawkins, 1985; Sanford, 1992; Rushen and de Passillé, 1992) and there is a possibility that defining the subjective state of 'suffering' in physiological or behavioural terms is an unrealistic goal (Barnard and Hurst, 1996). However, there is some degree of consensus that the presence of stress may be associated with a deterioration in the well-being of animals. Although, as discussed by Moberg (1996), the presence of stress *per se* need not be equated directly with suffering, the exposure of animals to stimuli or conditions which lead to frequent or chronic activation of the stress response is likely to lead to adverse effects on growth, behaviour, reproduction and disease resistance (Johnson *et al.*, 1992).

Why reduce stress responsiveness?

For most groups of animals in environments that are controlled by, or subject to the influence of, human activities, the most appropriate strategy to reduce the degree of stress experienced by the animals is to modify the animals' environment to reduce the frequency or severity of the stressor(s). However, in some cases this is not an option. For example, in intensive rearing environments changes in management practices to reduce the degree of stress might not be feasible because of significant economic and practical constraints.

In addition to the nature of the holding environment, the degree to which the animal itself is pre-adapted to the environment is important. Although for most species of economic significance to agriculture the process of domestication has been taking place for thousands of years, resulting in a greater tolerance of environments and stimuli which would be stressful to wild stock, newly exploited species (e.g. many of those important to aquaculture) are not far removed from the wild state and cannot be considered domesticated. In such species, the stress of common husbandry procedures and environmental factors is exacerbated. Reducing the stress responsiveness of such animals to intensive rearing conditions may be considered a strategy to enhance or accelerate the process of domestication.

Where no further improvements in the rearing environment can be achieved, or where novel species are being subjected to intensive rearing practices, consideration may be given to altering the sensitivity of the animal itself to potentially stressful stimuli (e.g. Hester *et al.*, 1996). Given that overstimulation of the hypothalamic–pituitary–adrenal (HPA) axis is associated with deleterious effects on growth, reproductive performance and disease resistance (Johnson *et al.*, 1992), reduction of stress responsiveness may reduce 'inappropriate' effects of frequent or continuous exposure to stressful stimuli. The benefits of reducing sensitivity to stressors might therefore include:

- increased production (better growth, higher feed conversion);
- improved reproductive performance (optimization of productivity);
- reduced incidence of disease (improvement in welfare, reduction in therapeutic costs); and
- an overall enhancement of the quality of life of captive animals (improvement in welfare).

How might a reduction in responsiveness be achieved? Domestication represents the outcome of a long-term ongoing selective breeding programme, to some extent both deliberate and inadvertent. It has been proposed that selective breeding may offer a route by which the incidence of behaviours and responses associated with welfare problems (and by association, performance) might be reduced (Brown, 1959; Mench, 1992).

Selective breeding as a strategy for the reduction of stress responsiveness

Although various pharmacological methods are available that allow elements of the neuroendocrine stress response to be blocked or modulated, these are not appropriate approaches for application on a large scale. Instead, considerable effort has been directed at evaluating the prospects for altering the inherent responsiveness of animals to stress by selective breeding. A number of points must be borne in mind when considering the practicalities of modifying the response of an animal to stress. These same considerations apply to any attempt to enhance a specific trait by means of a directed breeding programme.

Is there a broad spectrum of stress responsiveness within the population?

The range of inter-individual variability must offer scope for selecting individuals that differ markedly from the population mean. Even in what are considered to be thoroughly domesticated species, such as domestic fowl, considerable genetic diversity still exists both between and within populations (Jones, 1996). Numerous studies have established that within populations considerable variation is seen in the magnitude of the stress response of individuals to identical stressful stimuli (Berger *et al.*, 1987; Pottinger *et al.*, 1992; von Borell and Ladewig, 1992; Cummins and Gevirtz, 1993; Marsland *et al.*, 1995). Some of this variation may be attributed to phenotype, arising from the influence of external factors (see below), while a significant proportion is undoubtedly a reflection of genotype.

Is stress responsiveness an individual characteristic?

To enable the identification of individuals which display favourable traits, the observed inter-individual variability must represent differences between individuals that are consistent and stable with time rather than simply reflecting random variation in the responsiveness of the individual to stressful stimuli (or imprecision in the measurement). As a number of studies demonstrate, the magnitude of responsiveness of an individual to a stressor is evidently a characteristic of that individual during their lifetime. In human infants, consistent individual differences in adrenocortical function can be detected within 6 months of birth (Lewis and Ramsay, 1995), and in rats the plasma catecholamine stress response is characteristic of the individual for at least 12 months (Taylor *et al.*, 1989). In pigs, possession of an 'active' or 'passive' coping strategy to stress is an individual characteristic (Schouten and Wiegant, 1997) and, in fish, consistent inter-individual differences in interrenal (the adrenal homologue) responsiveness to stress are apparent for up to 2 years (Pottinger *et al.*, 1992).

What intrinsic and extrinsic factors modify responsiveness?

Assessment of individual variation in responsiveness must also take into account factors that may be reflected in the phenotypic response. For example, one of the most significant modifiers of phenotypic stress responsiveness in vertebrates is sexual maturity. In mammals, there is a distinct sexual dimorphism in the response of the HPA axis to stress (Vamvakopoulos and Chrousos, 1993). Suppression of responsiveness in males is linked to elevated androgen levels (Boissy and Bouissou, 1994; Handa *et al.*, 1994) while hyperresponsiveness in females is related to elevated oestrogen levels (Leśniewska *et al.*, 1990; Spinedi *et al.*, 1994). A similar phenomenon occurs in other animals, such as fish (Pottinger *et al.*, 1995, 1996), leading to the possibility that underlying differences in responsiveness may be obscured by phenotypic modification.

Social status may also modify the apparent responsiveness of an individual to a stressor (e.g. see Sapolsky, 1988). In addition, the early experience of the animal may be a potent modifier of stress responsiveness. In rats subjected to postnatal handling (e.g. Meaney *et al.*, 1993a), life-long responsiveness to a wide variety of stressors is reduced as a consequence of alterations in the sensitivity of the hypothalamus to corticosteroid feedback (Meaney *et al.*, 1993b). However, recent data showing differential effects of prenatal stress in two genetically distinct strains of rats suggest that the effects of early experience may also be dependent on genotype (Stöhr *et al.*, 1998).

What element of the stress response should be modified?

The physiological stress response is an immensely complex neuroendocrine response (Matteri *et al.*, Chapter 3, this volume), the ultimate control of which lies within the higher centres of the nervous system. The questions of how responsiveness should be assessed and which element of the response is most closely linked to the adverse effects of stress are critically important. Individual differences in stress responsiveness may be complex and may represent dissimilarities in different elements of the response dynamic. Lewis (1992) suggested that differences between individuals may occur in threshold (the amount of stimulation required to produce a response), in dampening (the capacity to terminate the response to a particular stimulus), and in reactivation (the ability to respond to an additional subsequent stimulus). Walker *et al.* (1992) emphasized that individual differences in the response to a stressor can occur at multiple levels within the 'sensory, cognitive, neurochemical and endocrine' cascade. This level of complexity has implications for the choice of trait upon which selection pressure is to be exerted. Both physiological and behavioural indices of stress have been employed as selection traits in selective breeding programmes, as will be discussed below.

Does stress responsiveness have a genetic, or heritable, component?
Is genetic manipulation of stress responsiveness actually feasible? That is, is the stress responsiveness of an individual animal genetically determined wholly or in part? The evidence for a significant genetic contribution to stress responsiveness in vertebrates is overwhelming and will be considered below.

Does possession of a high or low sensitivity to stressors benefit the animal?
The stress response is essentially an adaptive response, conserved throughout vertebrate phylogeny, which facilitates the animal's ability to cope with challenging circumstances. Perhaps unsurprisingly, given the complexity of the issues, there is no firm consensus as to whether modification of stress responsiveness can benefit an animal within an intensive rearing environment.

Selective breeding to modify stress responsiveness – animal models

The occurrence of individual differences in stress responsiveness that remain consistent with time may represent stable modulation of stress responsiveness at the phenotypic level, for example as a consequence of early experience (see above) or sex of the animal. However, evidence that stress responsiveness is at least in part controlled by the genotype is provided by strains or lines of animals that display levels of responsiveness that differ significantly from each other. Representative studies in rodents, domestic animals, poultry and fish will be considered. The reasons for the development of strains and lines of rodents with differing stress responsiveness differs from the practical imperatives that have driven such undertakings in economically significant animal groups. Nevertheless, consideration of these models is appropriate given the insight into the genetics of the stress response that they have provided.

Rodent models

The need for model systems to study the neurobiological processes that underly inter-individual differences in responses to stressful situations led to the development of a number of strains/lines of rodents divergent for a particular measure of stress responsiveness. In many cases, selection was on behavioural traits and based on perceived differences in emotionality, although, as pointed out by Ramos and Mormède (1998), consideration of emotionality in non-primate mammals is often restricted to situations inducing fear and/or anxiety. In addition, there are various strains of rodents which, although not necessarily selected on the basis of a stress-dependent trait, have subsequently been shown to display disparate physiological responsiveness to stressors (e.g. Shanks *et al.*, 1990; Gómez *et al.*, 1996). Some of the better known rodent models will be considered briefly below.

Roman high avoidance (RHA)/Roman low avoidance (RLA) rats

A description of the attributes of these strains is provided by Ramos and Mormède (1998), Walker *et al.* (1992) and Sutanto and de Kloet (1994). These strains were originally selected for divergent responses to active avoidance conditioning in a shuttle box. RHA rats rapidly acquire a conditioned avoidance response whereas RLA rats fail to acquire the response. The lines demonstrate differing behavioural responses ('coping mechanisms') to a stressor; RHA rats display a flight response to threat while RLA rats display a 'freezing' or immobility response. Differences are also apparent in endocrine aspects of the response to stress. RLA rats exhibit lower resting levels of adrenocorticotropic hormone (ACTH) but higher corticosterone secretion following exposure to stressors. Overall, RLA animals are more stress responsive than RHA. Another pair of lines originally selected on a similar basis to the RHA/RLA strains, the Syracuse high avoidance (SHA) and Syracuse low avoidance (SLA) rats, also show physiological correlates of the selection trait; SLA rats show a greater hyperglycaemic response to a novel environment than SHA rats (Flaherty and Rowan, 1989).

Maudsley reactive (MR) and non-reactive (MNR) rats

These lines were originally selected on the basis of their behaviour in an open-field test and are considered to provide contrasting examples of emotional reactivity (Broadhurst, 1975). By introducing the rat to an unfamiliar environment, which the open field provides, a behavioural stress response is elicited. Freezing coupled with defaecation are linked to high (anxious) emotionality, and exploratory behaviour coupled with low levels of defaecation (confident) are considered to indicate low emotionality (Denenberg, 1969). These lines display a similar behavioural divergence to the RLA/RHA rats. MR rats are more susceptible to restraint-induced ulcers than the non-reactive strain. MNR rats have higher noradrenaline in several tissues including plasma, but no differences are evident in the corticosteroid responses of these strains to stressors (Walker *et al.*, 1992; Ramos and Mormède, 1998).

Other rodent models

Other well-described rodent lines include the Wistar-Kyoto (WKY) stress-ulcer prone rats. These rats are hyperresponsive to stress in terms of high susceptibility to stress-induced gastric ulceration and show higher plasma

ACTH levels in response to restraint than other strains. The strain is characterized by immobility in the forced swim test and scores low on exploratory behaviour (see Redei *et al.*, 1994; Ramos and Mormède, 1998).

Rodent models are perhaps the best characterized, in terms of behaviour, neurochemistry and endocrinology, of the animal models which have been genetically selected for stress responsiveness. However, they represent model systems and are employed primarily in basic research into the mechanistic basis of behaviour and emotionality. For more practical applications of selective breeding for modified stress responsiveness, domestic animals must be considered.

Domestic mammals

In contrast to the rodent models in which behavioural responses to stressful stimuli are given prominence, work on the genetic control of the stress response in farm animals has concentrated on the practical performance-related benefits of modifying the stress response. Pigs are widely reared in intensive systems, and stress susceptibility is perceived as a major problem (Pfeiffer and von Lengerken, 1991). Individual pigs display a wide range of adrenocortical responsiveness to stress, the magnitude of which tends to be an individual characteristic (Hennessy *et al.*, 1988). Pigs displaying responses within the lower range display better growth and feed conversion than animals within the higher range of responsiveness (Hennessy and Jackson, 1987). It is possible to identify pigs with different 'coping' strategies, active or passive, using the behavioural response to mild restraint (Schouten and Wiegant, 1997), and these types are characterized by related differences in their physiological response to stress. However, selective breeding for low stress responsiveness has not been pursued actively for pigs as it has, for example, with poultry. Rather, selective breeding has been directed at attempts to eliminate a gene associated with high susceptibility to stress (porcine stress syndrome, PSS), the so-called halothane stress gene (von Lengerken and Pfeiffer, 1991). Stress-susceptible individuals are evidently less able to control excessive lactic acid during stress, leading to mortality arising from acidosis, vasoconstriction, hyperkalaemia, hypotension and reduced cardiac output (see Bäkström and Kauffman, 1995 for detailed discussion).

Poultry models

Unlike selective breeding efforts in pigs, in poultry efforts have been directed at modification of overall sensitivity to stressors.

Selection for endocrine traits in the domestic turkey

As long ago as 1959 the possible implications of environmental stress for poultry production were considered, and selective breeding for stress-resistant fowl was suggested as a possible means of ameliorating stress-related problems (Brown, 1959). Since then, there has been considerable progress made in understanding the physiology and pathology of stress in domestic fowl.

Freeman (1976) reviewed the early work on selective breeding for stress responsiveness in poultry, emphasizing that there is sufficient individual variability in the HPA axis response to stress within any one strain of poultry to allow for divergent selection. The pivotal role of the adrenal in the adverse effects of stress, and the relationship between high blood corticosterone and predisposition to viral and mycoplasmal diseases (Gross and Siegel, 1985), led to the adoption of a strategy to breed turkeys selectively (*Meleagris gallopavo*) for alterations in adrenal corticosterone production (Brown and Nestor, 1974). The plasma corticosterone elevation elicited by application of a low temperature stressor was employed as the selection trait and a divergent selection response was sought. Within nine generations a very marked divergence in the corticosterone response of the high (HL) and low (LL) lines was achieved (Brown and Nestor, 1974), with post-stress plasma corticosterone levels of 195 and 72 ng ml^{-1} respectively. Within nine generations the LL birds were significantly heavier, laid more eggs of higher fertility and hatchability, and displayed lower mortality than HL birds (Brown and Nestor, 1973). The realized heritabilities (h^2) for the plasma corticosterone response to stress for the HL and LL lines were reported to be 0.25 and 0.14 respectively, indicating that approximately 25% and 14% of the response of the progeny can be explained by genetic factors that influence the trait. Blood corticosterone levels in turkeys from the HL line were significantly more responsive to ACTH injections than in the LL birds, suggesting that adrenal sensitivity to ACTH accounted for at least a component of the overall difference in responsiveness between the two lines (Brown and Nestor, 1973).

Similar conclusions were reached by Carsia and Weber (1986) who measured the blood corticosterone response to acute stress in several strains (not selected for stress responsiveness) of chickens (*Gallus domesticus*). These authors found that there were significant between-strain differences in corticosterone levels following stress and that these were apparently linked to the adrenal weight/body weight ratio, greater steroid biosynthetic capacity, and greater sensitivity of the interrenal cells to ACTH in the high-responding strain. Differences in adrenal sensitivity to ACTH were also found to contribute to inter-strain differences in the HPA axis response to stress in five genetically distinct rat strains (Gómez *et al.*, 1996).

Selection for endocrine traits in the domestic chicken

Similar studies have been carried out using chickens. Gross and Colmano (1971) describe the divergent selection of two lines of chickens based on the plasma corticosterone response of the birds to social stress. Although no evidence is presented in this particular paper that the selection procedure was successful, with little or no divergence apparent in the F1 birds, subsequent papers confirmed the eventual effectiveness of the selection strategy within six generations (e.g. Gross *et al.*, 1984; Gross and Siegel, 1985). Interestingly, it was concluded that these lines had been selected on the basis of differences in the birds' perception of environmental situations (in terms of severity of the stressor) in that the two lines showed similar, rather than divergent, responses to non-social stressors (Siegel, 1989). This conclusion draws attention to the potential problems associated with the selection of a stimulus-specific response and the importance of selecting on a trait that integrates the animal's response to a wide range of stressors.

Selection for endocrine and behavioural traits in Japanese quail

The Japanese quail (*Coturnix coturnix japonica*) has been utilized in a number of studies, which have ultimately demonstrated strong links between genetically defined physiological and behavioural indices of stress responsiveness. Divergent selection for high or low blood corticosterone responses to immobilization stress for 12 generations resulted in high stress (HS) and low stress (LS) lines that differed significantly in responsiveness from the unselected control line (Satterlee and Johnson, 1988). As in chickens (Carsia and Weber, 1986), differences in stress reactivity between the selected quail lines appeared to reside at least in part within the adrenal tissue. These were manifested as an increase in adrenal mass and in the responsiveness of adrenocortical cells to ACTH in the HS line relative to the LS line (Carsia *et al.*, 1988).

Evaluation of the performance of quail from the HS and LS lines in an open-field test, which evokes a fear response, revealed a strong correlation between adrenocortical reactivity to stress and behavioural responses. Birds of the HS line displayed greater levels of fearfulness (as evidenced by freezing and locomotory delay) than birds of the LS line (Jones *et al.*, 1992a). A similar conclusion was reached in a related study in which tonic immobility (an index of fear) was quantified in birds of the HS and LS lines subject to a prolonged multifactorial stressor. HS birds were more susceptible to induction of tonic immobility and remained immobile for longer than LS birds (Jones *et al.*, 1992b). These data support earlier conclusions that there is a positive association (though not necessarily a causal one) between adrenocortical responsiveness to stressors and fearfulness (Faure, 1980) and suggest that selection on either stress responsiveness or fearfulness would provide a broad response, as opposed to a narrow stimulus-specific response of limited practical use. Subsequent studies reinforced this conclusion and are considered below.

Fear is considered to be a potent stressor with adverse effects on production, particularly in an intensive rearing environment. Fear is often considered to be a stimulus associated with human presence (Barnett *et al.*, 1992). The complexities of defining the term 'fear', the numerous stimuli that may represent sources of fear in the intensive rearing environment, and the impact of fear on performance are discussed by Jones (1996) in the context of poultry production. Possible strategies for reducing fear in the intensive rearing environment include environmental enrichment, habituation to humans and genetic selection (Jones, 1996). In the quail lines selected for high and low adrenocortical responsiveness to immobilization discussed above, it has also been shown that, in addition to a correlation between adrenocortical responsiveness and performance in the open-field test (Jones *et al.*, 1992a), the two lines diverge significantly in their response to human contact. HS birds perceive human contact as a more aversive stimulus than do LS birds and this is reflected as a positive correlation between the adrenocortical response to capture stress and the avoidance score (Jones *et al.*, 1994). Subsequent studies demonstrated that selection for reduced adrenocortical responsiveness and, concomitantly, reduced fearfulness do not influence the ease of capture of the birds. The lack of a difference between HS and LS lines in the difficulty of capture and handling was interpreted by the authors to indicate that there had been no enhancement of the active flight response in the LS line and to have implications for the management of birds within intensive rearing systems (Satterlee and Jones, 1997). By comparing the behavioural responses of the two lines to restraint, it was also shown that LS birds employed an active, and HS birds a passive, coping strategy (Jones and Satterlee, 1996).

The robust nature of the linkage between adrenocortical responsiveness and fearfulness in Japanese quail is demonstrated by the fact that a second independently selected pair of quail lines, divergent for tonic immobility (Mills and Faure, 1991), show similar responses to a range of behavioural and physiological assessments of fearfulness and stress to those of the HS and LS lines (Jones, 1996).

Fish models

The intensive cultivation of fish as a worldwide industry is a relatively new branch of food production and one that features a unique set of challenges. Fish are poikilothermic animals that inhabit an aquatic environment, and as such are sensitive to alterations in their immediate environment to a greater extent than terrestrial homeotherms. As with other intensively reared species, environmental stressors can have profoundly adverse effects on the growth (Pickering, 1993), reproductive performance (Campbell *et al.*, 1992) and disease resistance (Pickering and Pottinger, 1989) of aquacultured fish. Selective breeding has been widely applied within aquaculture in attempts to improve production characteristics (Tave, 1993). It was in the context of selection for improvements in disease resistance that the stress response in fish was first considered as a trait upon which selection pressure should be directed. Given the causal links between activation of the HPI (hypothalamic–pituitary–interrenal) axis and immunosuppression, it was proposed that the stress responsiveness of an individual fish should be used as an indirect selection criterion for the disease susceptibility of that individual (Refstie, 1982, 1986; Fevolden *et al.*, 1991).

Selection for endocrine traits in salmonid fish

In common with other vertebrates, there is evidence that the responsiveness of fish to stress is, at least in part, genetically influenced. Significant differences in the magnitude of the cortisol response were observed among five strains of hatchery-reared rainbow trout (*Oncorhynchus mykiss*) maintained under identical conditions and exposed to an episodic confinement stressor (Pickering and Pottinger, 1989). In a study in which cortisol levels were not measured but a variety of physiological indices were quantified following exposure to a stressor, significant differences in response to challenge were noted among six genetically distinct strains of coho salmon (*Oncorhynchus kisutch*; McGeer *et al.*, 1991).

For individual fish, the magnitude of the blood cortisol response to confinement was found to be stable with time in at least a proportion of a population of rainbow trout tested over a 28 month period (Pottinger *et al.*, 1992). This response permitted the identification of high-responding (HR) and low-responding (LR) individuals within the population. A considerable degree of divergence in the magnitude of mean post-stress cortisol levels of HR and LR fish was observed, with post-stress cortisol levels in the HR group being, on average, twice those of the LR group. Because response differentials remained consistent even under varied rearing conditions, the authors concluded that the responsiveness of the selected fish was not environmentally determined. Subsequently, a series of pooled gamete matings was carried out within each response group (Pottinger *et al.*, 1994) and the stress responsiveness of the two progeny groups was assessed using the confinement stress paradigm. The progeny of HR parents displayed a significantly greater plasma cortisol response to confinement than the progeny of LR parents. Following confinement, HR progeny displayed a more sustained reduction in lymphocytes than LR fish, accompanied by a post-stress increase in circulating neutrophils significantly greater in LR than in HR fish. Unexpectedly, highly significant differences in the stress-induced elevation of blood somatolactin (SL) were observed in the HR and LR progeny, with HR fish displaying a significantly greater elevation of SL in response to stress than the LR group (Rand-Weaver *et al.*, 1993). There is as yet no evidence functionally to link the SL response to stress with activity of the HPI axis, suggesting that the selection process employed by these authors operated more broadly than originally anticipated.

Although these results provide evidence that stress responsiveness is heritable in fish, and that selective breeding for low stress lines is feasible, this study (Pottinger *et al.*, 1994) employed a simplistic approach to the assessment of heritability. In subsequent, ongoing studies, individual rainbow trout, selected as HR or LR, have been employed to generate families whose relative performance under aquaculture conditions is being monitored. In these more recent studies a highly significant and robust divergence in stress responsiveness (as indicated by blood cortisol elevation) has been achieved in the F1 generation, with a parent–offspring heritability (h^2) of 0.41 (T.G. Pottinger and T.R. Carrick, unpublished). Preliminary data indicate that LR fish grow better under aquaculture conditions than do HR fish.

The only comparable studies on the genetic selection of stress responsiveness in fish have been carried out in Norway. In these studies the original aim was to utilize stress responsiveness as an indirect measure of disease resistance in fish, and thus an indirect selection trait for improved resistance to common pathogens. Within fish populations, mortality due to specific pathogens displays significant genetic variation (Fjalestad *et al.*, 1993). However, direct selection for fish that are resistant to more than one specific disease would be a complex and probably impractical strategy. Therefore, selection criteria other than survival following a single pathogen challenge are required (Fevolden *et al.*, 1992). One such criterion might be the stress responsiveness of the individual, which is functionally related to disease susceptibility. This approach requires that a genetic link between responsiveness to stress and disease resistance is demonstrated (Fevolden *et al.*, 1991). In both rainbow trout and Atlantic salmon (*Salmo salar*), highly significant sire effects on both plasma cortisol and plasma glucose levels are apparent. Estimates of h^2 for the cortisol and glucose responses to stress were determined in rainbow trout and Atlantic salmon populations used to establish high and low response lines (Fevolden *et al.*, 1991, 1993a). Estimates of heritability across the year classes for both cortisol ($h^2 = 0.05$) and glucose ($h^2 = 0.03$) were not significantly different from zero in the salmon, whereas in the rainbow trout, heritability estimates for cortisol were of low to medium magnitude ($h^2 = 0.27$ across year classes) while heritability for glucose was low ($h^2 = 0.07$). These values are lower than have been achieved in current studies using rainbow trout and lower than those reported for corticosteroid stress responsiveness in other species. However, Fevolden *et al.* (1993b) suggested that these values may be improved upon by optimizing the design of the breeding programme, by employing a standardized and reproducible stressor, and by using individuals with demonstrably consistent responsiveness to stress in order to generate progeny.

The assumption underlying efforts to select fish for low responsiveness to stress is that such fish will show an enhanced performance under aquacultural conditions. As yet, there is little evidence available with which to test this assumption. However, data from the Norwegian studies on selected rainbow trout and Atlantic salmon lines suggested that selection for stress responsiveness also influences specific disease resistance, offering the eventual prospect of reducing the prevalence of disease outbreaks (Fevolden *et al.*, 1991, 1993b).

Conclusions

In summary, the evidence provided by the literature is strongly suggestive, if not confirmatory, that a significant proportion of the stress responsiveness of an individual animal is inherited. This therefore allows for the possibility of modifying the stress responsiveness of an individual by selective breeding. That this approach is feasible is clearly demonstrated by the existence of lines of animals bred for divergence in a number of behavioural or physiological traits associated with the stress response. While there are undoubtedly ethical questions posed by the manipulation of an animal's genome, these are not peculiar to attempts to modify stress responsiveness but apply to all selective breeding programmes.

In theory, reducing the impact of stress on an individual by reducing the responsiveness of that individual to stressors should provide a broad range of benefits. If the levels of stress normally encountered within the rearing environment are sufficient to impact adversely on growth, reproductive performance and disease resistance, then improvements across all these performance characteristics should be apparent. Some of the work on selection of poultry for stress responsiveness suggests that this is indeed the case. However, it is possible that selection for reduced stress responsiveness may confound one or more economically important traits such as growth or fecundity. In that case, the practical application of selective breeding for stress tolerance may require the application of selection indices to allow co-selection of a number of desirable traits simultaneously.

Acknowledgements

The author gratefully acknowledges the financial support of the Natural Environment Research Council of the UK, and the European Community (Contract FAIR-CT95-0152).

References

Bäkström, L. and Kauffman, R. (1995) The porcine stress syndrome: a review of genetics, environmental factors, and animal well-being implications. *Agri-Practice* 16, 24–30.

Barnard, C.J. and Hurst, J.L. (1996) Welfare by design: the natural selection of welfare criteria. *Animal Welfare* 5, 405–433.

Barnett, J.L., Hemsworth, P.H. and Newman, E.A. (1992) Fear of humans and its relationships with productivity in laying hens at commercial farms. *British Poultry Science* 33, 699–710.

Bateson, P. and Bradshaw, E.L. (1998) Physiological effects of hunting red deer (*Cervus elaphus*). *Proceedings of the Royal Society of London, Series B – Biological Sciences* 264, 1707–1714.

Berger, M., Bossert, S., Krieg, J.-C., Dirlich, G., Ettmeier, W., Schreiber, W. and von Zerssen, D. (1987) Interindividual differences in the susceptibility of the cortisol system: an important factor for the degree of hypercortisolism in stress situations. *Biological Psychiatry* 22, 1327–1339.

Boissy, A. and Bouissou, M.F. (1994) Effects of androgen treatment on behavioural and physiological responses of heifers to fear-eliciting situations. *Hormones and Behavior* 28, 66–83.

Broadhurst, P.L. (1975) The Maudsley reactive and nonreactive strains of rats: a survey. *Behavior Genetics* 5, 299–319.

Brown, K.I. (1959) Stress and its implications in poultry production. *World's Poultry Science Journal* 15, 255–263.

Brown, K.I. and Nestor, K.E. (1973) Some physiological responses of turkeys selected for high and low adrenal response to cold stress. *Poultry Science* 52, 1948–1954.

Brown, K.I. and Nestor, K.E. (1974) Implications of selection for high and low adrenal response to stress. *Poultry Science* 53, 1297–1306.

Campbell, P.M., Pottinger, T.G. and Sumpter, J.P. (1992) Stress reduces the quality of gametes produced by rainbow trout. *Biology of Reproduction* 47, 1140–1150.

Carsia, R.V. and Weber, H. (1986) Genetic-dependent alterations in adrenal stress response and adrenocortical cell function of the domestic fowl (*Gallus domesticus*). *Proceedings of the Society for Experimental Biology and Medicine* 183, 99–105.

Carsia, R.V., Weber, H. and Satterlee, D.G. (1988) Steroidogenic properties of isolated adrenocortical cells from Japanese quail selected for high serum corticosterone response to immobilization. *Domestic Animal Endocrinology* 5, 231–240.

Cummins, S.E. and Gevirtz, R.N. (1993) The relationship between daily stress and urinary cortisol in a normal population: an emphasis on individual differences. *Behavioural Medicine* 19, 129–134.

Curtis, S.E. (1985) What constitutes animal well-being? In: Moberg, G.P. (ed.) *Animal Stress*. American Physiological Society, Bethesda, Maryland, pp. 1–14.

Dawkins, M.S. (1985) The scientific basis for assessing suffering in animals. In: Singer, P. (ed.) *In Defence of Animals*. Basil Blackwell, Oxford, pp. 27–40.

Denenberg, V.H. (1969) Open-field behaviour in the rat: what does it mean? *Annals of the New York Academy of Sciences* 159, 852–859.

Farm Animal Welfare Council (1996) *Report on the Welfare of Farmed Fish*. Ministry of Agriculture Fisheries and Food, London.

Faure, J.M. (1980) To adapt the environment to the bird or the bird to the environment? In: Moss, R. (ed.) *The Laying Hen and its Environment*. EEC, Brussels, pp. 19–42.

Fevolden, S.E., Refstie, T. and Røed, K.H. (1991) Selection for high and low cortisol stress response in Atlantic salmon (*Salmo salar*) and rainbow trout (*Oncorhynchus mykiss*). *Aquaculture* 95, 53–65.

Fevolden, S.E., Refstie, T. and Røed, K.H. (1992) Disease resistance in rainbow trout (*Oncorhynchus mykiss*) selected for stress response. *Aquaculture* 104, 19–29.

Fevolden, S.E., Refstie, T. and Gjerde, B. (1993a) Genetic and phenotypic parameters for cortisol and glucose stress response in Atlantic salmon and rainbow trout. *Aquaculture* 118, 205–216.

Fevolden, S.E., Nordmo, R., Refstie, T. and Røed, K.H. (1993b) Disease resistance in Atlantic salmon (*Salmo salar*) selected for high or low responses to stress. *Aquaculture* 109, 215–224.

Fjalestad, K.T., Gjedrem, T. and Gjerde, B. (1993) Genetic improvement of disease resistance in fish: an overview. *Aquaculture* 111, 65–74.

Flaherty, C.F. and Rowan, G.A. (1989) Rats (*Rattus norvegicus*) selectively bred to differ in avoidance behaviour also differ in response to novelty stress, in glycemic conditioning, and in reward contrast. *Behavioral and Neural Biology* 51, 145–164.

Freeman, B.M. (1976) Stress and the domestic fowl: a physiological re-appraisal. *World's Poultry Science Journal* 32, 249–256.

Gómez, F., Lahmam, A., de Kloet, E.R. and Armario, A. (1996) Hypothalamic–pituitary–adrenal response to chronic stress in five inbred rat strains: differential responses are mainly located at the adrenocortical level. *Neuroendocrinology* 63, 327–337.

Gross, W.B. and Colmano, G. (1971) Effect of infectious agents on chickens selected for plasma corticosterone response to social stress. *Poultry Science* 50, 1213–1217.

Gross, W.B. and Siegel, P.B. (1985) Selective breeding of chickens for corticosterone response to social stress. *Poultry Science* 64, 2230–2233.

Gross, W.B., Dunnington, E.A. and Siegel, P.B. (1984) Environmental effects on the well-being of chickens from lines selected for responses to social strife. *Archiv für Geflugelkunde* 48, 3–7.

Handa, R.J., Nunley, K.M., Lorens, S.A., Louie, J.P., McGivern, R.F. and Bollnow, M.R. (1994) Androgen regulation of adrenocorticotropin and corticosterone secretion in the male rat following novelty and foot shock stressors. *Physiology and Behavior* 55, 117–124.

Hennessy, D.P. and Jackson, P.N. (1987) Relationship between adrenal responsiveness and growth rate. In: APSA Committee (eds) *Manipulating Pig Production.* Australasian Pig Science Association, Werribee, Victoria, Australia, p. 23.

Hennessy, D.P., Stelmasiak, T., Johnston, N.E., Jackson, P.N. and Outch, K.H. (1988) Consistent capacity for adrenocortical response to ACTH administration in pigs. *American Journal of Veterinary Research* 49, 1276–1283.

Hester, P.Y., Muir, W.M. and Craig, J.V. (1996) Group selection for adaptation to multiple-hen cages: hematology and adrenal function. *Poultry Science* 75, 1295–1307.

Johnson, E.O., Kamilaris, T.C., Chrousos, G.P. and Gold, P.W. (1992) Mechanisms of stress: a dynamic overview of hormonal and behavioural homeostasis. *Neuroscience and Biobehavioral Reviews* 16, 115–130.

Jones, R.B. (1996) Fear and adaptability in poultry: insights, implications and imperatives. *World's Poultry Science Journal* 52, 131–174.

Jones, R.B. and Satterlee, D.G. (1996) Threat-induced behavioural inhibition in Japanese quail genetically selected for contrasting adrenocortical response to mechanical restraint. *British Poultry Science* 37, 465–470.

Jones, R.B., Satterlee, D.G. and Ryder, F.H. (1992a) Open-field behaviour of Japanese quail chicks genetically selected for low or high plasma corticosterone response to immobilization stress. *Poultry Science* 71, 1403–1407.

Jones, R.B., Satterlee, D.G. and Ryder, F.H. (1992b) Fear and distress in Japanese quail chicks of two lines genetically selected for low or high adrenocortical response to immobilization stress. *Hormones and Behavior* 26, 385–393.

Jones, R.B., Satterlee, D.G. and Ryder, F.H. (1994) Fear of humans in Japanese quail selected for low or high adrenocortical response. *Physiology and Behavior* 56, 379–383.

Kirkwood, J.K., Sainsbury, A.W. and Bennett, P.M. (1994) The welfare of free-living wild animals: methods of assessment. *Animal Welfare* 3, 257–273.

Leśniewska, B., Miśkowiak, B., Nowak, M. and Malendowicz, L.K. (1990) Sex differences in adrenocortical structure and function. XXVII. The effect of ether stress on ACTH and corticosterone in intact, gonadectomized, and testosterone- or estradiol-replaced rats. *Research in Experimental Medicine* 190, 95–103.

Lewis, M. (1992) Individual differences in response to stress. *Pediatrics* 90, 487–490.

Lewis, M. and Ramsay, D.S. (1995) Stability and change in cortisol and behavioural response to stress during the first 18 months of life. *Developmental Psychobiology* 28, 419–428.

Marsland, A.L., Manuck, S.B., Fazzari, T.V., Stewart, C.J. and Rabin, B.S. (1995) Stability of individual differences in cellular immune responses to acute psychological stress. *Psychosomatic Medicine* 57, 295–298.

McGeer, J.C., Baranyi, L. and Iwama, G.K. (1991) Physiological responses to challenge tests in six stocks of coho salmon (*Oncorhynchus kisutch*). *Canadian Journal of Fisheries and Aquatic Sciences* 48, 1761–1771.

Meaney, M.J., Bhatnagar, S., Larocque, S., McCormick, C., Shanks, N., Sharma, S., Smythe, J., Viau, V. and Plotsky, P.M. (1993a) Individual differences in the hypothalamic–pituitary–adrenal stress response and the hypothalamic CRF system. *Annals of the New York Academy of Sciences* 697, 70–85.

Meaney, M.J., Bhatnagar, S., Diorio, J., Larocque, S., Francis, D., O'Donnell, D., Shanks, N., Sharma, S., Smythe, J. and Viau, V. (1993b) Molecular basis for the development of individual differences in the hypothalamic–pituitary–adrenal stress response. *Cellular and Molecular Neurobiology* 13, 321–347.

Mench, J.A. (1992) The welfare of poultry in modern production systems. *Poultry Science Reviews* 4, 107–128.

Mills, A.D. and Faure, J.M. (1991) Divergent selection for duration of tonic immobility and social reinstatement behaviour in Japanese quail chicks. *Journal of Comparative Psychology* 105, 25–38.

Moberg, G.P. (1996) Suffering from stress: an approach for evaluating the welfare of an animal. *Acta Agriculturae Scandinavica, Section A, Animal Science Supplement* 27, 46–49.

Pfeiffer, H. and von Lengerken, G. (1991) Reduction of stress and stress susceptibility in pigs as prerequisite for high results. 1. Stress reduction serves effective exploitation of the genetic merit in pigs. *Archiv für Tierzucht* 2, S141–150.

Pickering, A.D. (1993) Growth and stress in fish production. *Aquaculture* 111, 51–63.

Pickering, A.D. and Pottinger, T.G. (1989) Stress responses and disease resistance in salmonid fish: effects of chronic elevation of plasma cortisol. *Fish Physiology and Biochemistry* 7, 253–258.

Pottinger, T.G. (1995) *Fish Welfare Review.* Angling Governing Bodies Liaison Group and the British Field Sports Society, Windermere, UK.

Pottinger, T.G., Pickering, A.D. and Hurley, M.A. (1992) Consistency in the stress response of individuals of two strains of rainbow trout, *Oncorhynchus mykiss. Aquaculture* 103, 275–289.

Pottinger, T.G., Moran, T.A. and Morgan, J.A.W. (1994) Primary and secondary indices of stress in the progeny of rainbow trout (*Oncorhynchus mykiss*) selected for high and low responsiveness to stress. *Journal of Fish Biology* 44, 149–163.

Pottinger, T.G., Balm, P.H.M. and Pickering, A.D. (1995) Sexual maturity modifies the responsiveness of the pituitary-interrenal axis to stress in male rainbow trout. *General and Comparative Endocrinology* 98, 311–320.

Pottinger, T.G., Carrick, T.R., Hughes, S.E. and Balm, P.H.M. (1996) Testosterone, 11-ketotestosterone and estradiol-17β modify baseline and stress-induced interrenal and corticotropic activity in trout. *General and Comparative Endocrinology* 104, 284–295.

Ramos, A. and Mormède, P. (1998) Stress and emotionality: a multidimensional and genetic approach. *Neuroscience and Biobehavioral Reviews* 22, 33–57.

Rand-Weaver, M., Pottinger, T.G. and Sumpter, J.P. (1993) Plasma somatolactin concentrations in salmonid fish are elevated by stress. *Journal of Endocrinology* 138, 509–515.

Redei, E., Pare, W.P., Aird, F. and Kluczynski, J. (1994) Strain differences in hypothalamic–pituitary–adrenal activity and stress ulcer. *American Journal of Physiology* 266, R353–R360.

Refstie, T. (1982) Preliminary results: differences between rainbow trout families in resistance against vibriosis and stress. *Developmental and Comparative Immunology* 2 (Suppl.), 205–209.

Refstie, T. (1986) Genetic differences in stress response in Atlantic salmon and rainbow trout. *Aquaculture* 57, 374.

Rushen, J. and de Passillé, A.M.B. (1992) The scientific assessment of the impact of housing on animal-welfare – a critical review. *Canadian Journal of Animal Science* 72, 721–743.

Sanford, J. (1992) Guidelines for detection and assessment of pain and distress in experimental animals: initiatives and experience in the United Kingdom. In: Short, C.E. and Van Poznak, A. (eds) *Animal Pain.* Churchill Livingstone, New York, pp. 515–524.

Sapolsky, R.M. (1988) Individual differences and the stress response: studies of a wild primate. *Advances in Experimental Medicine and Biology* 245, 399–411.

Satterlee, D.G. and Johnson, W.A. (1988) Selection of Japanese quail for contrasting blood corticosterone response to immobilization. *Poultry Science* 67, 25–32.

Satterlee, D.G. and Jones, R.B. (1997) Ease of capture in Japanese quail of two lines divergently selected for adrenocortical response to immobilization. *Poultry Science* 76, 469–471.

Schouten, W.G.P. and Wiegant, V.M. (1997) Individual responses to acute and chronic stress in pigs. *Acta Physiologica Scandinavica* 161 (Suppl. 640), 88–91.

Shanks, N., Griffiths, J., Zalcman, S., Zacharko, R.M. and Anisman, H. (1990) Mouse strain differences in plasma corticosterone following uncontrollable footshock. *Pharmacology, Biochemistry and Behaviour* 36, 515–519.

Siegel, P.B. (1989) The genetic-behaviour interface and well-being of poultry. *British Poultry Science* 30, 3–13.

Smith, J.A. and Boyd, K.M. (1991) *Lives in the Balance. The Ethics of Using Animals in Biomedical Research.* Oxford University Press, Oxford.

Spinedi, E., Salas, M., Chisari, A., Perone, M., Carino, M. and Gaillard, R.C. (1994) Sex differences in the hypothalamo-pituitary-adrenal axis response to inflammatory and neuroendocrine stressors. *Neuroendocrinology* 60, 609–617.

Stöhr, T., Wermeling, D.S., Szuran, T., Pliska, V., Domeney, A., Welzl, H., Weiner, I. and Feldon, J. (1998) Differential effects of prenatal stress in two inbred strains of rats. *Pharmacology, Biochemistry and Behaviour* 59, 799–805.

Sutanto, W. and de Kloet, E.R. (1994) The use of various animal models in the study of stress and stress-related phenomena. *Laboratory Animals* 28, 293–306.

Tave, D. (1993) *Genetics for Fish Hatchery Managers.* AVI Publishing Company, Westport, Connecticut.

Taylor, J., Weyers, P., Harris, N. and Vogel, W.H. (1989) The plasma catecholamine stress response is characteristic for a given animal over a one-year period. *Physiology and Behavior* 46, 853–856.

Vamvakopoulos, N.C. and Chrousos, G.P. (1993) Evidence of direct estrogenic regulation of human corticotropin-releasing hormone gene expression. Potential implications for the sexual dimorphism of the stress response and immune/inflammatory reaction. *Journal of Clinical Investigation* 92, 1896–1902.

von Borell, E. and Ladewig, J. (1992) Relationship between behaviour and adrenocortical response pattern in domestic pigs. *Applied Animal Behaviour Science* 34, 195–206.

von Lengerken, G. and Pfeiffer, H. (1991) Reduction of stress susceptibility in pigs as prerequisite for quality and quantitatively high results. II. Possibilities to reduce stress susceptibility and meat quality deficiencies in pigs. *Archiv für Tierzucht* 34, 241–248.

Walker, C.-D., Aubert, M.L., Meaney, M.J. and Driscoll, P. (1992) Individual differences in the activity of the hypothalamus-pituitary-adrenocortical system after stressors: use of psychogenetically selected rat lines as a model. In: Driscoll, P. (ed.) *Genetically Defined Animal Models of Neurobehavioural Dysfunctions.* Birkhauser, Boston, pp. 276–296.

Human–Animal Interactions and Animal Stress

15

P.H. Hemsworth[1,2] and J.L. Barnett[2]

[1]Animal Welfare Center, University of Melbourne and [2]Agriculture Victoria, Victorian Institute of Animal Science, Werribee, Victoria, Australia

Introduction

Humans interact with animals in many walks of life. In situations in which these interactions are close and frequent, the quality of these interactions may have considerable consequences for either partner. For example, keeping pets is common in most households. The potential benefits for humans include promoting the development of social competency and responsibility in children (Edney, 1992) and providing companionship, love and affection for children and adults (Leslie *et al.*, 1994). While domestication of pets has generally provided these animals with obvious benefits such as food, good health, protection and shelter, little is known of the effects of human–animal interactions on pets themselves. Perhaps surprisingly, more is known of the effects of human–animal interactions on farm animals.

Human–animal interactions are a common feature of modern livestock production. Research has shown that the quality of the relationship that is developed between stockpeople and their animals can have surprising effects on the animals. This chapter will utilize this research on human–farm animal interactions to consider the effects of human–animal interactions on animal stress. This discussion obviously has implications for animals in other walks of life in which close interaction occurs between humans and animals.

Human–animal interactions in intensive farming systems

Humans and animals are in regular and at times close contact in modern intensive farming systems. While there may be considerable automation in these production systems, stockpeople are required to monitor animals and their conditions regularly and impose routine husbandry procedures. Consequently the amount of human contact that these animals receive is considerable. For example, a stockperson in modern meat chicken units may manage tens of thousands of birds at a time and, although the stockperson may not physically interact with his or her animals, they will often be in close visual contact with most of the birds as often as six times daily during routine inspection of birds and their conditions. Lactating dairy cows are in close contact with humans during lactation, and a stockperson may handle several hundred cows twice daily at the time of milking. In intensive pig units, a stockperson may closely interact with several hundred breeding pigs, often handling many animals several times a day at critical stages of reproduction.

There are several levels of interaction between stockpeople and their farm animals. Many of these interactions are associated with regular observation of the animals and their conditions. Thus this type of interaction often involves only visual contact between the stockperson and the animals, perhaps without the stockperson entering the animals' pen. Visual and auditory interactions may also be used to move animals. In some industries, such as the pig and dairy industries, tactile interactions are often used by stockpeople to move animals for routine husbandry procedures during breeding and lactation. Tactile interactions also occur in situations in which animals must be restrained and subjected to management or health procedures.

Research in the livestock industries has provided evidence of relationships between these human–animal interactions and animal stress and productivity. For example, observations in the Australian pig industry have revealed sequential relationships between the attitudes of stockpeople towards interacting with their pigs, the behaviour of the stockpeople towards their pigs, the behavioural response of breeding pigs to humans and the reproductive performance of pigs (Hemsworth *et al.*, 1989). There are also reciprocal relationships operating within this pathway: it is not unidirectional but bi-directional, with the behavioural response of the animal to humans (fear), for example, feeding back on the attitudes and thus behaviour of the stockperson (Hemsworth and Coleman, 1998). A model of these human–animal relationships in intensive livestock industries is presented in Fig. 15.1.

A particularly important result of this research on human–animal interactions in the livestock industries is the consistent finding of a negative correlation between fear of humans, assessed on the basis of the behavioural response of animals to humans, and animal productivity. The negative relationship found in broiler (meat) chicken farms between fear and productivity (Fig. 15.2) is typical of the fear–productivity relationship observed in other industries such as the pig and dairy industries. Because of its apparent implication for both animal productivity and welfare, this consistent finding of a negative fear–productivity relationship in a number of livestock industries has stimulated considerable research over the last two decades, particularly on pigs. As will be discussed later in this chapter, the causal basis of this negative fear–productivity relationship has been demonstrated in a number of experimental studies, predominantly on pigs and poultry, in which handling treatments resulting in high levels of fear of humans depress growth

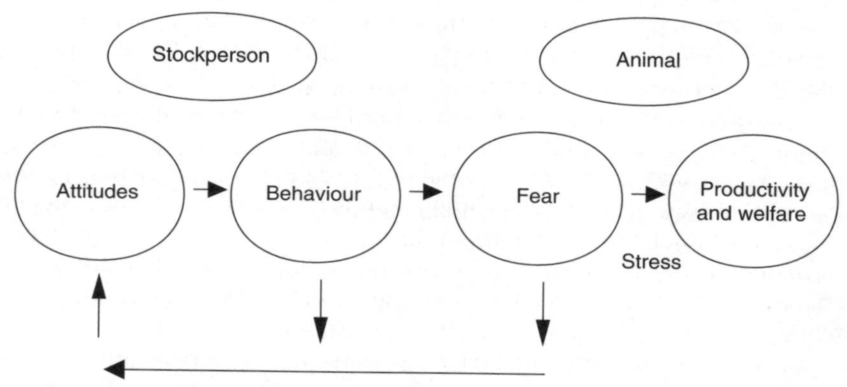

Fig. 15.1. A model of human–animal interactions in the livestock industries.

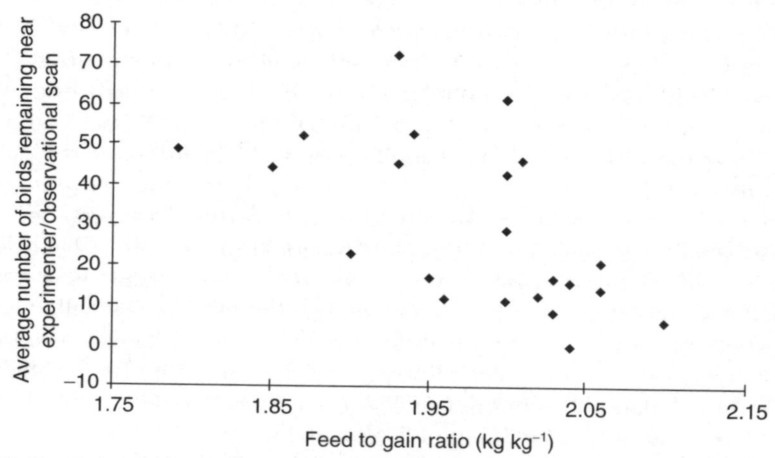

Fig. 15.2. Relationship (based on farm averages) between fear of humans and feed conversion at 22 commercial broiler chicken farms (compiled from Hemsworth *et al.*, 1994a: $r = -0.57$, df = 20, $P < 0.01$).

and reproductive performance. These handling studies have also implicated both acute and chronic stress in the mechanism whereby high fear of humans may limit animal productivity. The objective of this chapter is to review these results and, in doing so, demonstrate the effects of human–animal interactions on animal stress.

Defining and measuring fear of humans

There is some controversy about the definition and measurement of fear (see Hinde, 1970; Murphy, 1978; Hemsworth and Coleman, 1998) and thus it is necessary in this chapter to define its use and describe its measurement. The term 'fear' is a useful and convenient term to describe the behavioural responses of animals to humans. These responses mainly include escape–avoidance responses. A common view of fear (Jones, 1987; Hemsworth and Coleman, 1998) is that it can be considered as an intervening variable, linked to both a range of stimuli which may pose some risk or danger to the welfare of the animal and to a series of responses, both behavioural and physiological, by the animal that enable it to respond appropriately to this source of danger. Gray (1987) defined fear as a form of emotional reaction to a stimulus that the animal works to terminate, escape from, or avoid. McFarland (1981) considered fear as a motivational state that is aroused by certain specific stimuli and that normally gives rise to defensive behaviour or escape.

Observations on farm animals in the presence of humans indicate that these animals commonly display behavioural patterns that can be labelled fear responses, such as withdrawal from or avoidance of humans as well as immobility responses such as freezing or crouching (Hemsworth and Barnett, 1987; Jones, 1987; Mills and Faure, 1990). What is surprising when studying the behavioural response of farm animals to humans is the magnitude of these responses, given that these animals have been domesticated over many generations and that there is generally substantial contact between humans and farm animals in modern intensive livestock systems.

In both experimental and commercial situations, fear of humans can be assessed on the basis of the behavioural response of the animal to an experimenter in a standard test, i.e. either the avoidance response to an approaching experimenter or, conversely, the approach behaviour to a stationary experimenter. For example, in studies with pigs and cattle, the approach behaviour of the animal to a stationary experimenter in a standard arena has been used to assess the animal's fear of humans (Hemsworth *et al.*, 1981a, 1996c). In these tests, although the degree of novelty of the test arena is reduced because of the similarity of the arena to the animals' home pen, animals introduced into this new environment will be motivated to explore and familiarize themselves with the environment once the initial fear responses have waned. Therefore, since the animals may be motivated to both avoid and explore the arena and the human stimulus, the animals' fear of humans will have a major influence on the animals' approach to the human stimulus. The imposition of handling treatments designed to produce different levels of fear of humans generally produce the expected variations in the approach behaviour of pigs and cattle to a stationary experimenter in a standard test (Hemsworth *et al.*, 1981a, 1986, 1987, 1994b; Gonyou *et al.*, 1986; Hemsworth and Barnett, 1991). Measurement of the flight distance from humans, which generally is defined and measured as the distance at which an animal withdraws or escapes as a human approaches (Hediger, 1964), has also been used to assess fear of humans in a number of farm animals such as cattle and sheep (Hutson, 1982; Breuer *et al.*, 1998).

Since poultry initially show little locomotion in a novel arena, levels of fear of humans have often been assessed on the avoidance responses of birds to an approaching human when tested in their home pen. Orientation away and withdrawal from the approaching human have been equated with high fear levels. Handling treatments intuitively expected to reduce the fear of humans by poultry reduce the incidence of orientation away from the experimenter and reduce the withdrawal from the experimenter in several standard behavioural tests (Jones and Faure, 1981; Barnett *et al.*, 1992; Jones and Waddington, 1992; Hemsworth *et al.*, 1994a, 1996b).

Factors regulating the behavioural response of farm animals to humans

There are marked between-species and within-species differences in the behavioural responses of animals to humans, for example the flight distance to humans. Murphey *et al.* (1981) reported marked differences in the flight distance of *Bos indicus* and *Bos taurus* breeds of cattle to humans. Hearnshaw *et al.* (1979) found marked differences in the behaviour of cross-bred Brahman cattle and British breeds to the combination of restraint and handling. Indeed, the latter authors reported that the behavioural response to restraint in a squeeze shute (or stall) in the close presence of humans, often referred to as temperament, is moderately heritable in *Bos indicus* cattle. Furthermore, the flight distance of extensively grazed farm animals is generally reported to be greater than that of intensively managed farm animals, perhaps because there is less contact with humans in the extensive situations. For example, there are reports of flight distances of 6–11 m for extensively grazed or rangeland sheep, and 31 m for extensively grazed beef cattle in comparison with 2–8 m for feedlot beef cattle and 0–7 m for dairy cattle (Grandin, 1980, 1993; Hutson, 1982; Purcell *et al.*, 1988). A number of studies on farm animals have been conducted to examine the factors affecting their behavioural responses to humans, and the results of these studies will be considered now.

Fear of unfamiliarity

The differences in the avoidance responses of animals to humans may, in part, reflect inherent differences among species in their fear of unfamiliar stimuli (neophobia). Selection for neophobia will probably affect the general fearfulness of the naive animal rather than influence specific responses to specific novel stimuli. Therefore, differences in neophobia are more likely to affect the initial responses of inexperienced animals to humans since humans would be perceived as novel stimuli. However, over time, experience with humans should modify these responses to the extent that these responses to humans become stimulus specific. Murphy and Duncan (1977, 1978) studied two strains of chickens, termed 'flighty' and 'docile' on the basis of their behavioural responses to humans, and found that early handling resulted in different behavioural responses to humans in the two strains of birds, with the docile birds showing a more rapid reduction in their withdrawal responses to humans with regular exposure to humans than the flighty birds. These strain differences may be stimulus-specific, since observations indicated that the docile birds did not necessarily show less withdrawal than the flighty birds to novel stimuli like a mechanical scraper and an inflating balloon (Murphy, 1976).

Further evidence that the handling effects on the behavioural response of animals to humans may be specific to humans and not generalized to a range of fear-provoking stimuli is provided by a series of studies by Jones and colleagues (Jones, 1991; Jones and Waddington, 1992). These authors examined the effects of regular handling on the behavioural responses of young quail and domestic chickens to both novel stimuli, such as a blue light, and humans. They found that handling predominantly affected the avoidance response of birds to humans rather than to the novel stimuli. These data indicate that experience with humans results in stimulus-specific effects rather than effects on general fearfulness.

Fear of humans

A major component of the experienced animal's response to humans is learned. For example, handling studies on pigs and cattle have shown that negative or aversive tactile interactions imposed briefly but regularly by humans will quickly produce high levels of fear of humans. Negative interactions imposed daily for as little as 15–30 s consistently result in pigs showing marked avoidance of humans when subsequently tested with a stationary experimenter (Hemsworth *et al.*, 1981a, 1986, 1987, 1996a; Gonyou *et al.*, 1986; Hemsworth and Barnett, 1991). In contrast, brief positive handling results in low fear levels. These results indicate that, through conditioning, animals may learn to avoid stockpeople from whom they have received aversive stimulation and conversely may show greater approach or less avoidance of stockpeople from whom they have received positive interactions. The process of habituation of fear responses may also be operating in the latter case.

Cows will quickly learn to avoid humans who use mainly negative interactions. Brief exposure to a handling treatment involving slapping or brief shocks with a battery-operated prodder resulted in cows rapidly showing avoidance responses to humans (Munksgaard *et al.*, 1995: de Passillé *et al.*, 1996). Moderate or forceful slaps given whenever heifers failed to avoid the approach of a handler subsequently resulted in reductions in the approach of the heifers to a stationary experimenter and an increase in the flight distance of the heifers from an approaching experimenter (Breuer *et al.*, 1998). Regular interactions with feedlot cattle in which humans slowly approached and squatted to encourage approach by the animals resulted, presumably through habituation, in reductions in the animals' fear of humans (Hemsworth *et al.*, 1996c). Similarly, a number of handling studies involving the imposition of positive tactile interactions such as pats, strokes and fondling, have shown that handled cattle display less avoidance of humans in a range of testing situations (Boissy and Bouissou, 1988; Boivin *et al.*, 1992).

Chickens and laying hens are particularly sensitive to visual contact with humans. Regular treatments involving the experimenter placing his/her hand either on or in the chicken's cage and allowing birds to observe other birds being handled result in reductions in the subsequent avoidance of humans by young chickens (Jones, 1993). Interestingly, visual contact without tactile contact is more effective in reducing fear than picking up and stroking the bird, suggesting that tactile handling by humans may contain aversive elements for birds such as close approach and restraint. A handling study on laying hens by Barnett *et al.* (1994) also clearly demonstrates the influential effects of visual contact with humans on fear responses of birds to humans. Regular visual contact with humans, involving positive elements such as slow and deliberate movements by the experimenter, markedly reduced the subsequent avoidance behaviour of mature laying hens to humans in comparison with minimal human contact that at times contained elements of sudden and unexpected human contact.

These studies indicate that learning processes such as conditioning and habituation are influential in affecting the animals' fear of humans. They are supported by observations in the animal industries in which the behaviour of stockpeople has been shown to be predictive of the average level of fear of humans shown by animals at the farm. For example, observations in the pig and dairy industries indicate that negative interactions by pig and dairy stockpeople are positively correlated with the level of fear of humans, as assessed by the amount of approach by pigs and cows to a stationary experimenter (Table 15.1). Negative tactile interactions include moderate to forceful slaps, hits, kicks and pushes, while positive tactile interactions include pats, strokes and a hand resting on the animal's back. Observations on stockpeople at broiler chicken farms reveal that the visual cues from

Table 15.1. Stockperson behaviour–animal behaviour correlations in the dairy, pig and poultry industries.

	Negative stockperson behaviour and fear
Pig industry	
Hemsworth *et al.* (1989)	0.45[a]
Hemsworth *et al.* (1994b)	0.01
Coleman *et al.* (unpublished data)	0.40[a]
Dairy industry	
Hemsworth *et al.* (1995)	0.31
Hemsworth *et al.* (unpublished data)	0.30[a]
Poultry industry	
Cransberg (1996)	0.43[a]
Hemsworth *et al.* (1996b)	0.32

[a]Indicates significant correlation ($P < 0.05$) between the two variables.
Negative stockperson behaviour in the pig and dairy industries was assessed by the use of negative tactile interactions while in the poultry industry it was speed of movement. Fear in pigs and cows was assessed by the time spent near a stationary experimenter while fear of humans in poultry was assessed by the avoidance of an approaching experimenter.

the stockperson may influence the fear responses of commercial birds to humans. For example, speed of movement of the stockperson is positively correlated with the level of fear of humans by birds (Table 15.1).

Therefore, the studies reviewed in this section indicate that conditioned approach–avoidance responses develop as a consequence of associations between the stockperson and aversive and rewarding elements of the handling bouts. For pigs and cattle, the main aversive properties of humans, which will increase the animal's fear of humans, include hits, slaps and kicks by the stockperson, while the rewarding properties, which will decrease the animal's fear of humans, include pats, strokes and the hand of the stockperson resting on the animal. For poultry, the main aversive properties of humans, which will increase the animal's fear of humans, appear to include fast speed of movement and unexpected movement or appearance of the stockperson. Habituation may also occur over time as the animal's fear of humans is gradually reduced by repeated exposure to humans in a neutral context; that is, the human's presence has neither rewarding nor punishing elements.

Fear and animal productivity

Handling studies and observations on commercial animals indicate that high levels of fear of humans may limit the productivity of farm animals.

Studies at our laboratory have consistently shown that handling treatments inducing high levels of fear of humans will reduce the growth and reproductive performance of pigs. Some of these results are summarized in Table 15.2.

Seabrook and Bartle (1992) also reported depressions in the growth of pigs following aversive handling. In contrast, Paterson and Pearce (1989, 1992) and Pearce *et al.* (1989) found no effects of regular aversive handling on the growth performance and stress physiology of young pigs. There is no obvious explanation for this lack of effects in the studies by Paterson and colleagues. It is possible that differences between studies in the nature, amount and imposition of handling treatments may be responsible for these apparently contradictory results. For example, as is discussed in the next section, a behavioural response of animals to an apparently aversive stimulus (e.g. withdrawal from aversive handling by humans) in some situations may be an effective strategy to enable the animals to cope with this stimulus

Table 15.2. The effects of handling treatments on productivity and stress physiology of pigs in six studies.

Experiment and parameters	Positive treatment	Minimal treatment[1]	Negative treatment
Hemsworth *et al.* (1981a)			
Growth rate (11–22 weeks in g day^{-1})	709^{b}	—	669^{a}
Cortisol concentrations (ng ml^{-1})[2]	2.1^{x}	—	3.1^{y}
Gonyou *et al.* (1986)			
Growth rate (8–18 weeks, g day^{-1})	897^{b}	881^{ab}	837^{a}
Adrenal cortex (mm^2)	23.2^{a}	24.9^{ab}	33.1^{b}
Hemsworth *et al.* (1986)			
Pregnancy rate of gilts (%)	88^{b}	57^{ab}	33^{a}
Cortisol concentrations (ng ml^{-1})[2]	1.7^{a}	1.8^{ab}	2.4^{b}
Hemsworth *et al.* (1987)			
Growth rate (7–13 weeks, g day^{-1})	455^{b}	458^{b}	404^{b}
Cortisol concentrations (ng ml^{-1})[2]	1.6^{x}	1.7^{x}	2.5^{y}
Hemsworth and Barnett (1991)			
Growth rate (from 15 kg for 10 weeks in g day^{-1})	656	—	641
Cortisol concentrations (ng ml^{-1})[2]	1.5	—	1.1
Hemsworth *et al.* (1996a)			
Growth rate (from 63 kg for 4 weeks in kg day^{-1})	0.97^{ab}	1.05^{b}	0.94^{b}
Adrenal weights (g)	3.82^{x}	4.03^{x}	4.81^{y}

^{a,b} and ^{x,y} denote significant differences at $P < 0.05$ and $P < 0.01$, respectively.
[1]Treatment involving minimal human contact.
[2]Average of blood samples remotely collected at hourly intervals from 0800 to 1700 h.

without having to resort to any long-term physiological adjustment. It is also possible that there may be genetic differences between pigs in their ability to cope with chronic stressors; however there is little evidence of this in the literature. Barnett *et al.* (1988) found that, although there were differences between two genotypes of pigs in their basal cortisol concentrations, both genotypes exhibited similar pituitary–adrenal axis responses (in terms of percentage increases in cortisol in the long term) to being restrained by tethers.

Based on farm averages, negative correlations between fear of humans and reproductive performance of pigs have been found in two of three studies in the pig industry (Table 15.3). Fear was assessed in these studies on the basis of the behavioural response of recently mated female pigs to a stationary experimenter. These observations indicate that the productivity of pigs was lowest at the farms on which pigs were most fearful of humans. Based on a regression analysis using farm means, Hemsworth *et al.* (1989) found that variation in fear of humans accounted for about 20% of the variation in reproductive performance across the study farms.

Handling of a negative nature has generally been shown to reduce the growth performance of chickens. Gross and Siegel (1979, 1980, 1982) found that young chickens that received frequent human contact, apparently

Table 15.3. Animal behaviour–animal productivity correlations in the dairy, pig and poultry industries.

	Fear and animal productivity
Pig industry	
Hemsworth *et al.* (1981b)	–0.51[a]
Hemsworth *et al.* (1989)	–0.55[a]
Hemsworth *et al.* (1994b)	–0.01
Dairy industry	
Hemsworth *et al.* (1995)	–0.46[a]
Hemsworth *et al.* (unpublished data)	–0.25
Broiler industry	
Hemsworth *et al.* (1994a)	–0.57[b]
Cransberg (1996)	–0.10
Hemsworth *et al.* (1996b)	–0.39
Egg industry	
Barnett *et al.* (1992)	–0.58[b]

[a] and [b] indicate significant correlations ([a] = $P < 0.05$ and [b] = $P < 0.01$) between the two variables.
Fear of humans by pigs and cows was assessed by the time spent near a stationary experimenter while fear of humans by poultry was assessed by the avoidance of an approaching experimenter. The productivity variables were reproduction (piglets born per sow year^{-1}) in pigs, milk yield in cows, feed conversion in chickens and egg production in hens.

of a positive nature, had improved growth rates and feed efficiency and were more resistant to infection than birds that received minimal human contact or birds that had been deliberately scared, by a person shouting and banging on the birds' cages. In a study with adult laying hens, Barnett *et al.* (1994) found that regular visual contact, involving positive elements such as slow and deliberate movements, reduced the subsequent avoidance behaviour of mature laying hens and resulted in higher egg production than a treatment that involved minimal and at times negative human contact. Field observations on poultry have demonstrated significant negative relationships, based on farm averages, between the level of fear of humans and the productivity of commercial broiler chickens and laying hens (Table 15.3, also see Fig. 15.2). The avoidance behaviour of birds to an experimenter was used to assess fear in these studies. These correlations indicate that both egg production of laying hens and the efficiency of feed conversion of broiler chickens are negatively correlated with the level of fear of humans by birds. Bredbacka (1988) also reported that egg mass production was lower in hens that showed increased avoidance of a human.

Recent observations on commercial dairy cows also indicate the existence of a significant negative fear–productivity relationship (Table 15.3). The amount of approach of dairy cows to a stationary experimenter is positively correlated with the milk yield of the farm, indicating that milk yield is lowest at farms on which cows are highly fearful of humans. Rushen *et al.* (1997) reported that cows suffered a 10% reduction in their milk yield when, during milking, they were in the close presence of a handler who had previously aversively handled them. While fear was not quantified, Seabrook (1972) has also suggested that milk yield may be reduced in situations where cows are unwilling to approach the stockperson. Therefore, handling studies and observations in the animal industries on fear–productivity relationships generally show that high levels of fear of humans may limit the productivity of farm animals.

Fear and animal stress

The impetus to examine fear and stress more closely has been the consistent negative relationships between fear and productivity in the livestock industries. This section considers the role of stress in the effects of fear of humans on animal productivity.

Biological responses to stress

The close presence of humans is likely to result in fearful animals utilizing biological responses, both behavioural and physiological, to attempt to cope with the challenge of this stressor. A startle response may initially occur, followed by an orientation response and, if necessary, avoidance or defensive responses such as freezing, running away or preparation for fight (e.g. growling, threatening stance, etc.). In concert with these behavioural responses there is a series of physiological responses, with the full elicitation of this series dependent on the time of exposure to the stressor and the success of the biological responses in coping with the challenge.

These physiological responses are reviewed in detail in Matteri *et al.* (Chapter 3, this volume) and Moberg (Chapter 1, this volume), but it is useful briefly to consider some general features and effects of these responses before discussing the consequences of human–animal interactions on animal productivity and welfare. The physiological responses of animals to a stressor involve a series of three events. The first is an immediate response and is an adrenaline-dependent mechanism that includes the production of glucose from liver glycogen (glycogenolysis) for an immediate energy supply. This immediate or 'emergency' response is the 'fight or flight' response proposed by Walter Cannon (Cannon, 1914) and is the principal regulatory mechanism that allows the animal to meet physical or emotional challenges by its effects on metabolic rate, cardiac function, blood pressure, peripheral circulation, respiration, visual acuity, and energy availability and use (Frieden and Lipner, 1971; Harper, 1973). This initial reaction lasts for only a short period of time and if the stressor is not removed a second series of events occurs. This second response, called the acute stress response, is part of Hans Selye's 'general adaptation syndrome' (Selye, 1946, 1976) and is a corticosteroid-dependent mechanism. A major function of this second series of events is to provide glucose from food or muscle protein (gluconeogenesis). Therefore, during this stage a steady state is achieved in which the increased demand for energy is met by increased metabolic performance. This physiological state of stress disappears on removal of the stressor with no ill effects other than a depletion of energy reserves. This is an effective mechanism whereby the animal can adapt to changes in its environment. If the stressor continues, the response proceeds to the third series of events which is the chronic stress response. Again, this series of events is a corticosteroid-dependent mechanism but, while in the acute phase the effects are potentially beneficial, in this chronic phase there are detrimental effects on growth, reproduction and health (see Moberg, 1985; Toates, 1995).

The hormones of choice to demonstrate acute and chronic stress responses are glucocorticoids. This is because there is a large body of literature (see Selye, 1976) indicating the central role of the hypothalamic–pituitary–adrenal (HPA) axis in adaptation (e.g. effects of catecholamines and glucocorticoids on energy mobilization) and conversely in failure to adapt (e.g. adverse effects of glucocorticoids on growth, reproduction and health) to environmental change (Selye, 1976; Broom and Johnson, 1993; Hemsworth *et al.*, 1996a).

Although some key elements of Selye's general adaptation syndrome (GAS) have been reviewed and amended, his contribution has been critical to the development of our current understanding of the animal's response to stressors. For example, responses to stressors can include an increase in adrenal gland weight (as can occur in sheep (Panaretto and Ferguson, 1969a, b) and pigs (Hemsworth *et al.*, 1996a)), hypertrophy of the adrenal cortex (Selye and Stone, 1950) and an increase in free glucocorticoid concentrations (Westphal, 1971). The last point is particularly important in demonstrating a chronic stress response. Glucocorticoids circulate in three fractions, one tightly bound to a globulin protein (corticosteroid-binding globulin or transcortin), one loosely bound to an albumin protein, and a free fraction. While total cortisol concentrations are a reliable indicator of biological activity in acute stress responses, it is the free fraction that reflects biological activity in a chronic stress response (Westphal, 1971; McDonald, 1979). Thus, in the following section, more reliance can be placed on those studies in which free cortisol concentrations have been measured to indicate a chronic stress response. Unfortunately, most studies still rely on total hormone concentrations.

In many of our handling studies conducted on pigs, dairy cows and poultry, cortisol concentrations in the presence of humans and 'at rest' have been measured as evidence of acute and chronic stress responses. In turn, these stress responses have been interpreted in terms of risks to animal productivity and indeed animal welfare. Many of these handling studies have used indwelling venous catheters with extensions to enable remote blood sampling to avoid confounding effects of the presence of humans on cortisol profiles. These handling studies, and particularly those on pigs and dairy cattle, have shown that animals that are highly fearful of humans may experience both acute stress responses in the presence of humans and chronic stress responses in the absence of humans. These results will be considered in the following sections.

Animal handling and acute stress responses

There is good evidence in a number of farm animal species that animals that are fearful of humans show a greater acute stress response in the presence of humans than animals that are less fearful. For example, as shown in Fig. 15.3, pigs that had been negatively handled for 2 min three times per week from 11 to 22 weeks of age had a greater increase in cortisol concentrations 10 min after brief exposure to a human than pigs that had been positively handled (Hemsworth *et al.*, 1981a). As expected, these aversively handled pigs showed marked avoidance of humans when subsequently tested with a stationary experimenter. Several other handling studies at our laboratory have shown similar acute stress responses in fearful pigs in the presence of humans (Hemsworth *et al.*, 1986, 1987).

Several handling studies on young dairy cows have also been conducted at our laboratory. Breuer *et al.* (1998) examined the effects of positive

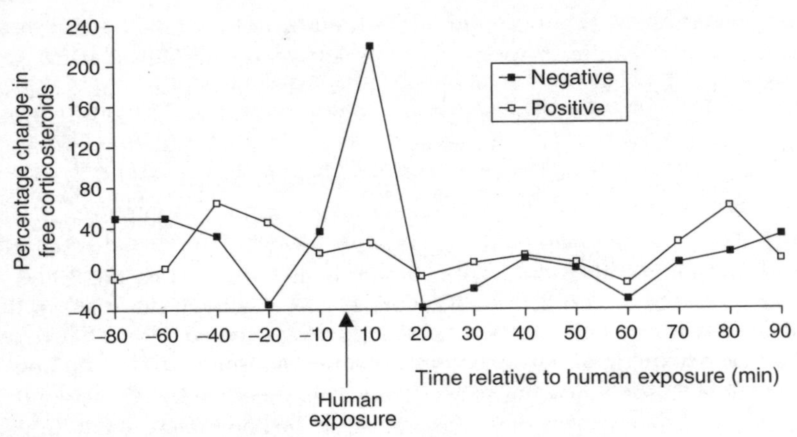

Fig. 15.3. Percentage change in free plasma corticosteroid concentrations in negatively (■) and positively (□) handled gilts before and after a 2-min exposure to a human (from Hemsworth *et al.*, 1981a).

and negative handling twice daily over a 5-week period on dairy heifers. Heifers that had been negatively handled showed increased avoidance of both a stationary and an approaching experimenter. In this study, blood samples were remotely collected immediately prior to and then 5 min after 30 s of human contact in the animal's home pen in order to examine the short-term cortisol response to human presence. The more fearful, negatively handled heifers showed increased cortisol concentrations after 5 min of brief human exposure compared with their positively handled counterparts.

It is relevant that data from a recent study of 66 Australian dairy farms have shown significant positive correlations, based on farm averages, between the level of fear of humans by cows and milk cortisol concentrations ($r = 0.33$, $n = 66$, $P < 0.01$; Hemsworth *et al.*, unpublished data). Milk samples were regularly collected from each farm during lactation and fear of humans was assessed on the basis of the average avoidance of a stationary experimenter shown by cows. Verkerk *et al.* (1998) reported that milk cortisol concentrations are a sensitive measure of plasma concentrations immediately prior to milking, since free cortisol moves readily between compartments by simple diffusion. Thus, activation of the HPA axis in fearful cows in response to humans around the time of milking is likely to be reflected in milk cortisol concentrations. It is also of interest that in the above on-farm observations negative interactions by stockpeople were significantly correlated with milk cortisol concentrations ($r = 0.31$, $n = 66$, $P < 0.05$).

Goats reared by humans show less avoidance of humans in a range of behavioural tests designed to assess fear of humans than goats reared by their dams (Lyons *et al.*, 1988). These more fearful goats also have higher corticosteroid concentrations in response to considerable human exposure than human-reared goats. Single blood samples were collected using jugular venepuncture in the goats to measure corticosteroid concentration both at rest (basal) and after about 16 min of human contact in two successive behavioural tests. These samples were collected within 3 min of entry to the animal's home pen to avoid the effects of handling associated with sampling on corticosteroid concentrations.

The effects of human contact on fear and stress physiology of poultry have been examined in several studies. Hemsworth *et al.* (1994a) examined the effects of differing degrees of previous exposure to humans on the behavioural and adrenocortical responses of broiler chickens to approach and restraint by an experimenter. A higher proportion of birds that had received minimal human contact during rearing withdrew as an experimenter approached in a standard test than birds that had received regular human contact. Furthermore, those birds that had received minimal human contact had higher plasma corticosterone concentrations after 12 min of handling than birds in the latter treatment. The effects of harvesting procedures on the acute stress responses of broiler chickens have been studied by Duncan *et al.* (1986). Birds were caught either by hand or by a specially designed machine. While the maximum heart rates of birds were similar, heart rates remained high for longer in manually caught birds than in the mechanically harvested birds. The greater acute stress responses in the manually caught birds may be because either the procedure was more aversive or the previous human contact during rearing was aversive. In a study on adult laying hens, Barnett *et al.* (1994) found that increased human contact, predominantly involving visual contact, reduced the subsequent avoidance responses of adult laying hens to humans and their corticosterone response to human presence.

Animal handling and chronic stress responses

While farm animals that are fearful of humans show a substantial acute stress response in the presence of humans, there is also good evidence of a chronic stress response in these fearful animals on the basis of a sustained elevation in free cortisol concentrations. A number of studies have shown that pigs highly fearful as a consequence of regular aversive handling have a sustained elevation in free cortisol concentrations. The typical elevation seen in fearful pigs is depicted in Fig. 15.4 (Hemsworth *et al.*, 1981a). These handling treatments were imposed twice daily for 11 weeks. Pigs that had been aversively handled showed marked avoidance of humans when subsequently tested with a stationary experimenter. Similar effects on the stress physiology of pigs have been seen in a number of other handling studies at our laboratory, and some of these results are summarized in

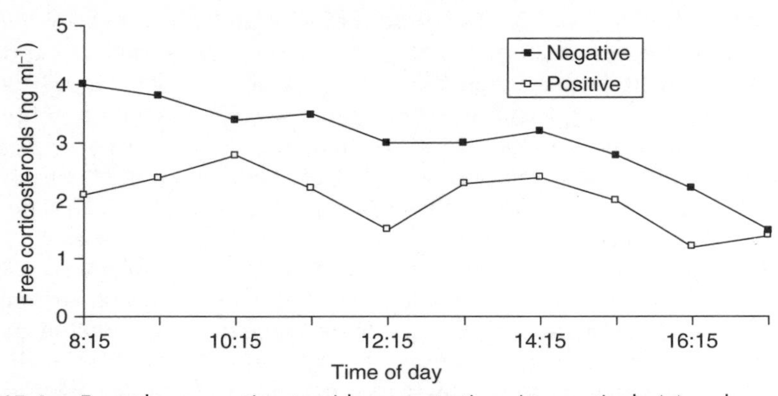

Fig. 15.4. Free plasma corticosteroid concentrations in negatively (■) and positively (□) handled gilts between 8:15 and 17:15 h (from Hemsworth *et al.*, 1981a).

Table 15.2. The effects of high fear levels on elevated basal corticosteroid concentrations are also supported by changes in adrenal gland weight and morphology in several of these studies. It is important to note that indwelling venous catheters were used to obtain blood samples in the absence of humans in these studies.

There is similar evidence in poultry that high fear levels may reduce their productivity (Bredbacka, 1988; Barnett *et al.*, 1992; Hemsworth *et al.*, 1994a; Jones, 1996). Although the mechanism(s) responsible is unclear, as seen in fearful pigs, a chronic stress response or even a series of acute stress responses in the presence of humans may be responsible for the depressed productivity in fearful poultry. Support for this suggestion is provided by the known effects of corticosteroids on nitrogen balance, protein catabolism, and energy retention or excretion in chickens (Siegel and van Kampen, 1984). Further, exogenous elevations of circulating corticosterone concentrations adversely affect growth rate and efficiency in chickens (Bellamy and Leonard, 1965; Adams, 1968; Bartov *et al.*, 1980; Siegel and van Kampen, 1984; Saadoun *et al.*, 1987). In a study on laying hens (Barnett *et al.*, 1994) in which handled birds had a lower corticosterone response to the close presence of humans (see previous section), indirect evidence of a chronic stress response in the fearful birds was provided by reduced reproduction (egg production) and reduced immunological responsiveness.

The effect of regular handling on basal cortisol concentrations has recently been studied in dairy heifers. Breuer *et al.* (1998) found that twice-daily negative handling over a 5-week period resulted in heifers showing a sustained ($P < 0.05$) increase in free cortisol concentrations in the afternoon in the absence of humans, relative to less fearful heifers which were positively handled (Fig. 15.5). Interestingly, total cortisol concentrations did not significantly differ between the two groups of heifers. These animals

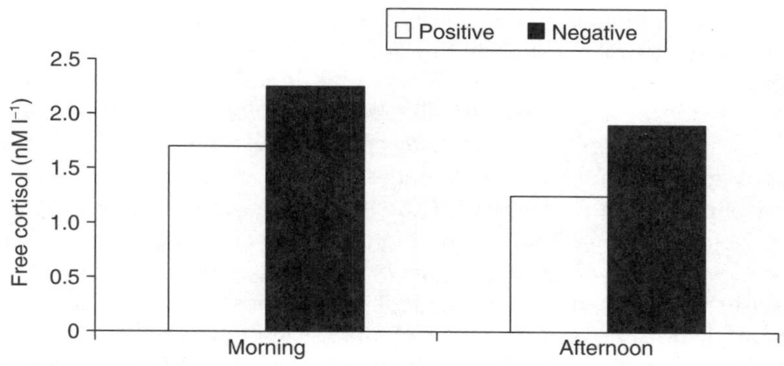

Fig. 15.5. Free cortisol concentrations in negatively (■) and positively (□) handled heifers in the morning and afternoon (from Breuer *et al.*, 1998).

were remotely blood sampled using indwelling jugular catheters. This elevation of basal free cortisol concentrations following 5 weeks of negative handling suggests that these fearful heifers were experiencing a chronic stress response.

Therefore, based on the pig as a model, there is evidence that farm animals that are highly fearful of humans and are also in close contact with humans may initially display short-term physiological responses such as an acute stress response and behavioural responses such as avoidance. However, these responses may be ineffective in commercial situations in which farm animals are in close contact with humans and thus may be unable to avoid or alleviate the challenge of the close presence of the human. In such situations, the animal may have to resort to a long-term physiological response, that is a chronic stress response. This response is likely to come at a physiological cost to the animal: prolonged activation of the HPA axis may result in decreased metabolic efficiency and thus decreased growth performance, impaired immunity and reduced reproductive performance. There is considerable evidence in the literature that stress hormones, and in particular a sustained elevation in corticosteroids, may adversely affect growth and reproductive performance by disrupting protein metabolism and key reproductive endocrine events (Klasing, 1985; Moberg, 1985; Clarke *et al.*, 1992). Similarly, immunosuppression may occur as a consequence of the long-term activation of the HPA axis (Selye, 1976; Toates, 1995).

Fear and animal welfare

In addition to the obvious effects of stress on animal productivity, the concept of stress has implications for animal welfare. While definitions of welfare vary, a common definition of welfare within scientific circles is: 'The welfare of an individual is its state as regards its attempts to cope with its environment' (Broom, 1986). In this definition, the 'state as regards attempts to cope' refers to both how much has to be done by the animal in order to cope with the environment and the extent to which the animal's coping attempts are succeeding. Attempts to cope include the functioning of body repair systems, immunological defences, physiological stress response and a variety of behavioural responses. Therefore, using such a definition, the risks to the welfare of an animal by an environmental challenge can be assessed at two levels: firstly the magnitude of the behavioural and physiological responses and secondly the biological cost of these responses (Broom and Johnson, 1993; Hemsworth and Coleman, 1998). These behavioural and physiological responses include the stress response, while the biological cost includes adverse effects on the animal's ability to grow, reproduce and remain healthy.

Fear of humans may also limit the welfare of farm animals in a number of ways. The most obvious concern is that the avoidance responses of fearful animals may lead to injuries sustained in trying to avoid humans during routine inspections and handling. Gregory and Wilkins (1989) found that about one-third of laying hens culled at the end of their laying period had broken bones when sampled immediately prior to slaughter. While handling during capture and prior slaughter processes may result in broken bones, prior handling and fear may contribute to previous breakages or responses at capture and thus the likelihood of injury at this time (Reed *et al.*, 1993). Breuer *et al.* (1998) found that 44% of negatively handled heifers were lame within the first 8 weeks of lactation compared with 11% of positively handled heifers. High fear of humans, resulting in marked avoidance responses in the presence of humans, may have contributed to this high incidence of lameness in negatively handled animals.

Chronic stress, as a consequence of high fear levels (see previous section), may adversely affect the health and thus welfare of animals through its effects on immunosuppression (Kelley, 1985). Thus, a further opportunity whereby human–animal interactions may adversely affect animal welfare is through animal stress. An indirect effect on animal welfare that is not always recognized relates to the stockperson's attitude and behaviour towards the animal. In situations in which the attitude and behaviour of the stockperson towards the animals are negative, the welfare of the animals may be at risk because the stockperson's commitment to the surveillance of and the attendance to welfare issues may be less than optimal.

Improving human–animal interactions in agriculture

This model of human–animal interactions in agriculture as depicted in Fig. 15.1 indicates the potential to improve animal productivity and welfare by targeting the stockperson's key attitudes and behaviour for improvement. Techniques that may be useful in this regard include staff selection and training procedures addressing these important human attributes. These two approaches will be briefly discussed in this section because they offer the animal industries considerable opportunities to reduce animal fear and thus remove any limitations imposed by animal fear and stress on animal productivity.

Selecting stockpeople to improve their interactions with their animals

The opportunities to select stockpeople to improve human–animal interactions have been reviewed and discussed recently by Hemsworth and Coleman (1998). Only limited research has been specifically conducted to identify those characteristics that make a person well suited to work with animals in the livestock industries. There is a wide variety of tests that focus on work characteristics such as personality, vocational preference and work motivation. The specific needs for selecting stockpersons, however, mean that many of these tests are not appropriate for the subject population involved (Hemsworth and Coleman, 1998).

As discussed earlier in this chapter, research on human–animal interactions in the livestock industries has shown that there are aspects of stockperson attitudes and behaviour that affect productivity in intensive farming industries. A battery of tests that targets those specific attitudes and behaviours as well as those generic characteristics that are predictive of work performance may well be appropriate. At present, there have been no research studies investigating the range of possible factors that may be relevant to stockperson performance. Where people have had previous experience working in a particular industry, research on human–animal interactions has shown that the person's attitude towards working with animals in that industry is a good predictor of their behaviour and, ultimately, animal productivity. While some studies have suggested that personality may be relevant (Ravel *et al.*, 1996) and that empathy may be important (Coleman *et al.*, 1998), these variables need to be studied further. As yet no research has focused on other variables such as attitudes to work, work motivation and work preference.

An increasing recognition of the need to employ stockpeople in the livestock industries who are adaptable, motivated and who will treat animals well is likely to result in a demand to develop a selection procedure that can be widely used. No such procedures presently exist in the agricultural industries.

Training stockpeople to improve their interactions with their animals

An understanding of stockperson behaviour is necessary if we wish to influence or change this behaviour. Hemsworth and Coleman (1998) have discussed these opportunities, and readers are referred to this book for a more comprehensive description of the rationale underlying behavioural change.

To change human behaviour ultimately requires targeting both the beliefs that underlie the behaviour and the behaviour in question and then maintaining these changed beliefs and behaviour. Such an approach involves exposing stockpeople, with timely, ongoing reinforcement, to the information that will produce changes in both their beliefs and behaviour. This type of training is called cognitive–behavioural therapy, and one intervention study in the livestock industries will be briefly reviewed in order to review the potential to improve the attitudes and behaviour of stockpeople to improve animal productivity. This training programme was specifically designed to target those attitudes and behaviours of the stockperson that have a direct effect on animal fear and productivity. As considered earlier, research in the pig and dairy industries has identified some of the key human characteristics that affect the animal's fear of humans, and the intervention programme targeted these key attitudinal and behavioural profiles in stockpeople in order to improve human–animal interactions.

Hemsworth *et al.* (1994b) examined the usefulness of a cognitive–behavioural intervention programme targeting the important attitudinal and behavioural profiles of stockpeople to improve the productivity of commercial pigs. Thirty-five farms were initially studied and two treatments were then imposed: an intervention treatment, consisting of a cognitive–behavioural intervention procedure designed to improve the attitude and behaviour of stockpeople towards pigs; and a control treatment, where no intervention was attempted. Farms were randomly allocated to treatment within fear strata. The effectiveness of the intervention programme was assessed by monitoring the changes in the attitudinal and behavioural profiles of stockpeople and the behaviour and productivity of pigs at the two groups of farms.

The intervention treatment involved a 1-h individual session with each stockperson in which the cognitive–behavioural intervention procedure was used. This procedure was specifically designed to target those attitudes and behaviours of the stockperson that had a direct effect on pig behaviour and productivity. In order to reinforce the information presented in this session, stockpeople were provided with posters to place in working areas of the piggeries. Also regular newsletters were used to summarize the important points of the cognitive–behavioural intervention programme and to prompt an assessment by the stockperson as to whether or not changes in his or her behaviour and the behaviour of his or her pigs were being achieved. The control treatment involved two researchers, the same two who imposed the cognitive–behavioural intervention procedure, visiting the farm for a 1-h session in which only general developments in the pig industry were discussed.

Relative to the control treatment, the intervention treatment resulted in a greater increase in positive attitudes towards patting and talking to pigs. Corresponding with this relative improvement in the attitude of stockpeople at the intervention farms was a significant reduction in the percentage of negative physical interactions displayed by the stockperson. These relative improvements in the attitude and behaviour of stockpeople at the intervention farms corresponded to a significant relative reduction in the level of fear of humans by pigs at these farms. There were increases in the time that pigs spent within 0.5 m of the experimenter and in the number of interactions with the experimenter in the standard test for pigs at the intervention farms relative to pigs at the control farms. Furthermore, there was a marked tendency for an increase in the number of pigs born per sow per year at the intervention farms relative to the control farms (7% increase in reproductive performance at the former group of farms).

Research currently being conducted by the authors and colleagues indicates similar opportunities in the dairy industry to improve animal productivity by targeting the key human–animal interactions for improvement. Presumably these training effects occur as a consequence of reduced stress removing production limitations caused by high fear of humans.

Therefore, the results of these intervention studies in the pig and dairy industries, together with numerous studies on the effects of handling and the observed relationships between these human and animal variables in the livestock industries, indicate that cognitive–behavioural interventions that successfully target the key attitudes and behaviour of stockpeople that regulate the animal's fear of humans, offer the livestock industries good opportunities to improve the productivity of their animals. While such effects on animal performance may also occur through improvements in job-related characteristics of the stockperson, such as job satisfaction, work motivation and technical knowledge and skills (Hemsworth and Coleman, 1998), reductions in level of fear of humans and stress in the animal are likely to be influential pathways in achieving such changes in animal performance. Indeed, the short time frame following intervention in which changes in animal behaviour and productivity occur indicates that this latter mechanism is clearly important.

Conclusion

Human–animal interactions are a common feature of modern intensive farming systems. Research in the livestock industries has provided evidence of sequential relationships between stockperson attitudes, stockperson behaviour, animal fear, and animal productivity and welfare. A particularly important result of this research is the consistent finding of a negative correlation between fear of humans and animal productivity. The causal basis of a negative fear–productivity relationship in the livestock industries has been demonstrated in a number of experimental studies, predominantly on pigs and poultry, in which handling treatments resulting in high fear levels depress growth and reproductive performance. These handling studies also implicate the role of both acute and chronic stress in the negative fear–productivity relationship. Thus, these studies on both experimental and commercial farm animals provide evidence of the role of stress in mediating the effects of human–animal interactions on the productivity and welfare of commercial farm animals.

While welfare has not been considered in detail in this review, it is clear that human–animal interactions have considerable implications for animal welfare. For example, fear of humans may lead to injury and chronic stress, both of which may affect morbidity and mortality. The attitude and behaviour of the stockpeople towards farm animals may also affect animal welfare by influencing the capacity of stockpeople to care properly for their animals. The importance of the human–animal relationship to both animal productivity and welfare cannot be over-emphasized. For example, a recent study at our laboratory (Pedersen *et al.*, 1997) provided limited evidence that positive handling by stockpeople may ameliorate the chronic stress response associated with an aversive housing system.

References

Adams, B.M. (1968) Effect of cortisol on growth and uric acid excretion in the chick. *Journal of Endocrinology* 40, 145–151.

Barnett, J.L., Hemsworth, P.H., Cronin, G.M., Winfield, C.G., McCallum, T.H. and Newman, E.H. (1988) The effects of genotype on physiological and behavioural responses related to the welfare of pregnant pigs. *Applied Animal Behaviour Science* 20, 287–296.

Barnett, J.L., Hemsworth, P.H. and Newman, E.A. (1992) Fear of humans and its relationships with productivity in laying hens at commercial farms. *British Poultry Science* 33, 699–710.

Barnett, J.L., Hemsworth, P.H., Hennessy, D.P., McCallum, T.M. and Newman, E.A. (1994) The effects of modifying the amount of human contact on the behavioural, physiological and production responses of laying hens. *Applied Animal Behaviour Science* 41, 87–100.

Bartov, I., Jensen, L.S. and Veltman, J.R. (1980) Effect of corticosterone and prolactin on fattening in broiler chicks. *Poultry Science* 59, 1328–1334.

Bellamy, D. and Leonard, R.A. (1965) Effect of cortisol on the growth of chicks. *General Comparative Endocrinology* 5, 402–410.

Boissy, A. and Bouissou, M.F. (1988) Effects of early handling on heifers' subsequent reactivity to humans and to unfamiliar situations. *Applied Animal Behaviour Science* 20, 259–273.

Boivin, X., LeNeindre, P. and Chupin, J.M. (1992) Establishment of cattle–human relationships. *Applied Animal Behaviour Science* 32, 325–335.

Bredbacka, P. (1988) Relationships between fear, welfare and productive traits in caged white leghorn hens. *Proceedings of the International Congress on Applied Ethology in Farm Animals*, pp. 74–79.

Breuer, K., Coleman, G.J. and Hemsworth, P.H. (1998) The effect of handling on the stress physiology and behaviour of nonlactating heifers. *Proceedings of the Australian Society for the Study of Animal Behaviour, 29th Annual Conference Palmerston North, New Zealand.* Institute of Natural Resources, Massey University, New Zealand, pp. 8–9.

Broom, D.M. (1986) Indicators of poor welfare. *British Veterinary Journal* 142, 524–526.

Broom, D.M. and Johnson, K.G. (1993) *Stress and Animal Welfare.* Chapman and Hall, London.

Cannon, W.B. (1914) The emergency function of the adrenal medulla in pain and the major emotions. *American Journal of Physiology* 33, 356–372.

Clarke, I.J., Hemsworth, P.H., Barnett, J.L. and Tilbrook, A.J. (1992) Stress and reproduction in farm animals. In: Sheppard, K.E., Boublik, J.H. and Funder, J.W. (eds) *Stress and Reproduction*, Serono Symposium Publications, Vol. 86. Raven Press, New York, pp. 239–251.

Coleman, G.C., Hemsworth, P.H., Hay, M. and Cox, M. (1998) Predicting stockperson behaviour towards pigs from attitudinal and job-related variables and empathy. *Applied Animal Behaviour Science* 58, 63–75.

Cransberg, P.H. (1996) The relationship between human factors and the productivity and welfare of commercial broiler chickens. MSc thesis, La Trobe University, Victoria, Australia.

Duncan, I.J.H., Slee, G.S., Kettlewell, P., Berry, P. and Carlisle, A.J. (1986) Comparison of the stressfulness of harvesting broiler chickens by machine and by hand. *British Poultry Science* 27, 109–114.

Edney, A.T.B. (1992) Companion animals and human health. *Veterinary Record* 130(14), 285–288.

Frieden, E. and Lipner, H. (1971) *Biochemical Endocrinology of the Vertebrates.* Prentice-Hall, Englewood Cliffs, New Jersey.

Gonyou, H.W., Hemsworth, P.H. and Barnett, J.L. (1986) Effects of frequent interactions with humans on growing pigs. *Applied Animal Behaviour Science* 16, 269–278.

Grandin, T. (1980) Livestock behavior as related to handling facilities design. *International Journal for the Study of Animal Problems* 1, 33–52.

Grandin, T. (1993) Behavioural principles of cattle handling under extensive conditions. In: Grandin, T. (ed.) *Livestock Handling and Transport.* CAB International, Wallingford, UK, pp. 43–57.

Gray, J.A. (1987) *The Psychology of Fear and Stress*, 2nd edn. Cambridge University Press, Cambridge.

Gregory, N.G. and Wilkins, L.J. (1989) Broken bones in domestic fowl: handling and processing damage in end-of-lay battery hens. *British Poultry Science* 30, 555–562.

Gross, W.B. and Siegel, P.B. (1979) Adaptations of chickens to their handler, and experimental results. *Avian Diseases* 23, 708–714.

Gross, W.B. and Siegel, P.B. (1980) Effects of early environmental stresses on chicken body weight, antibody response to RBC antigens, feed efficiency and response to fasting. *Avian Diseases* 24, 549–579.

Gross, W.B. and Siegel, P.B. (1982) Influences of sequences of environmental factors on the response of chickens to fasting and to *Staphylococcus aureus* infection. *American Journal of Veterinary Research* 43, 137–139.

Harper, H.A. (1973) *Review of Physiological Chemistry,* 14th edn. Large Medicine Publications, California.

Hearnshaw, H., Barlow, R. and Want, G. (1979) Development of a 'temperament' or 'handling difficulty' score for cattle. *Proceedings of the Inaugural Conference of Australian Animal Breed Genetics* 1, 164–166.

Hediger, H. (1964) The animals relationship with man. In: *Wild Animals in Captivity.* Dover Publications, NewYork.

Hemsworth, P.H. and Barnett, J.L. (1987) Human–animal interactions. *Veterinary Clinics of North America, Food Animal Practice* 3, 339–356.

Hemsworth, P.H. and Barnett, J.L. (1991) The effects of aversively handling pigs, either individually or in groups, on their behaviour, growth and corticosteroids. *Applied Animal Behaviour Science* 30, 61–72.

Hemsworth, P.H. and Coleman, G.J. (1998) *Human–Livestock Interactions: The Stockperson and the Productivity and Welfare of Intensively-farmed Animals.* CAB International, Wallingford, UK.

Hemsworth, P.H., Barnett, J.L. and Hansen, C. (1981a) The influence of handling by humans on the behaviour, growth and corticosteroids in the juvenile female pig. *Hormones and Behavior* 15, 396–403.

Hemsworth, P.H., Brand, A. and Willems, P.J. (1981b) The behavioural response of sows to the presence of human beings and their productivity. *Livestock Production Science* 8, 67–74.

Hemsworth, P.H., Barnett, J.L. and Hansen, C. (1986) The influence of handling by humans on the behaviour, reproduction and corticosteroids of male and female pigs. *Applied Animal Behaviour Science* 15, 303–314.

Hemsworth, P.H., Barnett, J.L. and Hansen, C. (1987) The influence of inconsistent handling on the behaviour, growth and corticosteroids of young pigs. *Applied Animal Behaviour Science* 17, 245–252.

Hemsworth, P.H., Barnett, J.L., Coleman, G.J. and Hansen, C. (1989) A study of the relationships between the attitudinal and behavioural profiles of stockpeople and the level of fear of humans and the reproductive performance of commercial pigs. *Applied Animal Behaviour Science* 23, 301–314.

Hemsworth, P.H., Coleman, G.C., Barnett, J.L. and Jones, R.B. (1994a) Behavioural responses of humans and the productivity of commercial broiler chickens. *Applied Animal Behaviour Science* 41, 101–114.

Hemsworth, P.H., Coleman, G.J. and Barnett, J.L. (1994b) Improving the attitude and behaviour of stockpeople towards pigs and the consequences on the behaviour and reproductive performance of commercial pigs. *Applied Animal Behaviour Science* 39, 349–362.

Hemsworth, P.H., Barnett, J.L., Breuer, K., Coleman, G.C. and Matthews, L.R. (1995) An investigation of the relationships between handling and human contact and the milking behaviour, productivity and welfare of commercial dairy cows. *Research Report on Dairy Research and Development Council Project,* Attwood, Australia.

Hemsworth, P.H., Barnett, J.L. and Campbell, R.G. (1996a) A study of the relative aversiveness of a new daily injection procedure for pigs. *Applied Animal Behaviour Science* 49, 389–401.

Hemsworth, P.H., Coleman, G.C., Cransberg, P.H. and Barnett, J.L. (1996b) Human factors and the productivity and welfare of commercial broiler chickens. *Research Report on Chicken Meat Research and Development Council Project*, Attwood, Australia.

Hemsworth, P.H., Price, E.O. and Bogward, R. (1996c) Behavioural responses of domestic pigs and cattle to humans and novel stimuli. *Applied Animal Behavioural Science* 50, 43–55.

Hinde, R.A. (1970) *Behaviour: A Synthesis of Ethology and Comparative Psychology*. McGraw-Hill, Kogakusha, Japan.

Hutson, G.D. (1982) 'Flight distance' in Merino Sheep. *Animal Production* 35, 231–235.

Jones, R.B. (1987) Fear and fear responses: a hypothetical consideration. *Medical Science Research* 15, 1287–1290.

Jones, R.B. (1991) Taming and the effects of variations in the nature of chick–human interactions. *Poultry Science* 70, 61 (Abstract).

Jones, R.B. (1993) Reduction of the domestic chick's fear of humans by regular handling and related treatments. *Animal Behaviour* 46, 991–998.

Jones, R.B. (1996) Fear and adaptability in poultry: insights, implications and imperatives. *World's Poultry Science Journal* 52, 131–174.

Jones, R.B. and Faure, J.M. (1981) The effects of regular handling on fear responses in the domestic chick. *Behavioural Processes* 6, 135–143.

Jones, R.B. and Waddington, D. (1992) Modification of fear in domestic chicks, *Gallus gallus domesticus* via regular handling and early environmental enrichment. *Animal Behaviour* 43, 1021–1033.

Kelley, K.W. (1985) Immunological consequences of changing environmental stimuli. In: Moberg, G.P. (ed.) *Animal Stress*. American Physiological Society, Baltimore, Maryland, pp. 193–223.

Klasing, K.C. (1985) Influence of stress on protein metabolism. In: Moberg, G.P. (ed.) *Animal Stress*, American Physiological Society, Bethesda, Maryland, pp. 269–280.

Leslie, B.E., Meek, A.H., Kawash, G.F. and McKeown, D.B. (1994) An epidemiological investigation of pet ownership in Ontario. *Canadian Veterinary Journal* 35, 218–222.

Lyons, D.M., Price, E.D. and Moberg, G.P. (1988) Individual differences in temperament of domestic dairy goats: constancy and change. *Animal Behaviour* 36, 1323–1333.

McDonald, I.R. (1979) The concept of stress as a physiological state. *Deer Farming in Victoria, Proceedings of a Symposium*. Agricultural Note Series. Department of Agriculture, Melbourne 15, 37–49.

McFarland, D. (1981) *The Oxford Companion to Animal Behaviour*. Oxford University Press, Oxford.

Mills, A.D. and Faure, J.M. (1990) Panic and hysteria in domestic fowl: a review. In: Zayan, R. and Dantzer, R. (eds) *Social Stress in Domestic Animals*. Kluwer Academic Publishers, Dordrecht, The Netherlands, pp. 248–272.

Moberg, G.P. (1985) Influence of stress on reproduction: measure of well-being. In: Moberg, G.P. (ed.) *Animal Stress.* American Physiological Society, Bethesda, Maryland, pp. 245–267.

Munksgaard, L., de Passillé, A.M., Rushen, J., Thodberg, K. and Jensen, M.B. (1995) The ability of dairy cows to distinguish between people. *Proceedings of the 29th International Congress of the International Society of Applied Ethology, Exeter, 3–5 August,* University Federation for Animal Welfare, Great Britain, pp. 19–20 (Abstract).

Murphey, R.M., Moura Duarte, F.A. and Torres Penendo, M.C. (1981) Responses of cattle to humans in open spaces: breed comparisons and approach–avoidance relationships. *Behavior Genetics* 2, 37–47.

Murphy, L.B. (1976) A study of the behavioural expression of fear and exploration in two stocks of domestic fowl. PhD dissertation, Edinburgh University, UK.

Murphy, L.B. (1978) The practical problems of recognizing and measuring fear and exploration behaviour in the domestic fowl. *Animal Behaviour* 26, 422–431.

Murphy, L.B. and Duncan, L.J.H. (1977) Attempts to modify the responses of domestic fowl towards human beings. 1. The association of human contact with a food reward. *Applied Animal Ethology* 3, 321–334.

Murphy, L.B. and Duncan, L.J.H. (1978) Attempts to modify the responses of domestic fowl towards human beings. II. The effect of early experience. *Applied Animal Ethology* 4, 5–12.

Panaretto, B.A. and Ferguson, K.A. (1969a) Pituitary adrenal interactions in shorn sheep exposed to cold, wet conditions. *Australian Journal of Agricultural Research* 20, 99–113.

Panaretto, B.A. and Ferguson, K.A. (1969b) Comparison of the effects of several stressing agents on the adrenal glands of normal and hypophysectomized sheep. *Australian Journal of Agricultural Research* 20, 115–124.

de Passillé, A.M., Rushen, J., Ladewig, J. and Petherick C. (1996) Dairy calves' discrimination of people based on previous handling. *Journal of Animal Science* 74, 969–974.

Paterson, A.M. and Pearce, G.P. (1989) Boar-induced puberty in gilts handled pleasantly or unpleasantly during rearing. *Applied Animal Behaviour Science* 22, 225–233.

Paterson, A.M. and Pearce, G.P. (1992) Growth, response to humans and corticosteroids in male pigs housed individually and subjected to pleasant, unpleasant or minimal handling during rearing. *Applied Animal Behaviour Science* 34, 315–328.

Pearce, G.P., Paterson, A.M. and Pearce, A.N. (1989) The influence of pleasant and unpleasant handling and the provision of toys on the growth and behaviour of male pigs. *Applied Animal Behaviour Science* 23, 27–37.

Pedersen, V., Barnett, J.L., Hemsworth, P.H., Newman, E.A. and Schirmer, B. (1997) The effects of handling on behavioural and physiological responses to housing in tether-stalls in pregnant pigs. *Animal Welfare* 7, 137–150.

Purcell, D., Arave, C.W. and Walters, J.L. (1988) Relationship of three measures of behaviour to milk production. *Applied Animal Behaviour Science* 21, 307–313.

Ravel, A., D'Allaire, S.D., Bigras-Poulin, M. and Ward, R. (1996) Personality traits of stockpeople working in farrowing units on two types of farms in Quebec. *Proceedings of 14th Congress of the International Pig Veterinary Society, 7–10 July,*

Bologna, Italy, Faculty of Veterinary Medicine, University of Bologna, pp. 514 (Abstract).

Reed, H.J., Wilkins, S.D., Austin, S.D. and Gregory, N.G. (1993) The effect of environmental enrichment on fear reactions and depopoulation trauma in adult caged hens. *Applied Animal Behaviour Science* 36, 39–46.

Rushen, J., de Passillé, A.M. and Munksgaard, L. (1997) Dairy cows' fear of people reduces milk yield and affects behaviour at milking. *Proceedings of the 31st International Congress of the International Society for Applied Ethology, Prague, Czech Republic,* Research Institute of Animal Production, Prague and Institute of Animal Biochemistry and Genetics, Slovak Academy of Sciences, Slovakia, pp. 219 (Abstract).

Saadoun, A., Simon, J.R. and Leclercq, B. (1987) Effect of exogenous corticosterone in genetically fat and lean chickens. *British Poultry Science* 28, 519–528.

Seabrook, M.F. (1972) A study to determine the influence of the herdsman's personality on milk yield. *Journal of Agriculture Labour Science* 1, 45–59.

Seabrook, M.F. and Bartle, N.C. (1992) The practical implications of animals' responses to man. *British Society of Animal Production, Winter Meeting, 23–25 March, 1992, Scarborough, UK,* paper number 34.

Selye, H. (1946) The general adaptation syndrome and the diseases of adaptation. *Journal of Clinical Endocrinology* 6, 117–230.

Selye, H. (1976) *Stress in Health and Disease.* Butterworths, Boston.

Selye, H. and Stone, H. (1950) *On the Experimental Morphology of the Adrenal Cortex.* Charles C. Thomas, Springfield, Illinois.

Siegel, H.S. and van Kampen, M. (1984) Energy relationships in growing chickens given daily injections of corticosterone. *British Poultry Science* 25, 477–485.

Toates, F. (1995) *Stress. Conceptual and Biological Aspects.* John Wiley & Sons, Chichester, UK.

Verkerk, G.A., Phipps, A.M., Carragher, J.F., Matthews, L.R. and Stelwagen, K. (1998) Characterization of milk cortisol concentrations as a measure of short-term stress responses in lactating dairy cows. *Animal Behaviour* 7, 77–86.

Westphal, U. (1971) *Steroid–Protein Interactions.* Springer-Verlag, Berlin.

Alleviating Stress in Zoo Animals with Environmental Enrichment

16

K. Carlstead[1] and D. Shepherdson[2]

[1]National Zoological Park, Smithsonian Institution, Washington DC, USA; [2]Oregon Zoo, Portland, Oregon, USA

Introduction

Perceptions of the psychological needs of wild animals living in zoos have changed greatly in the last few decades. The use of environmental enrichment to meet these perceived needs has grown accordingly. The documented benefits of enrichment for zoo animals range from changes in behaviour, such as decreasing aggression, increasing activity and decreasing abnormal behaviour, to improved reproduction and health and increased survival of captive-bred animals released into the wild (Markowitz *et al.*, 1978; Chamove *et al.*, 1982; Wilson, 1982; Tripp, 1985; Miller-Schroeder and Paterson, 1989; Carlstead and Shepherdson, 1994; Vargas and Anderson, 1999). Environmental enrichment involves the practice of increasing the physical, social and temporal complexity of captive environments. Physical complexity refers to the variety of structural, visual, auditory, olfactory and gustatory stimuli in an animal's environment, whereas temporal complexity refers to the degree to which this stimulation changes, introducing novelty and variability into fixed surroundings. Social complexity is an aspect of enrichment that for many species is of paramount importance (Shapiro *et al.*, 1996, 1998) and requires careful consideration of appropriate social group size and composition. Mixing species in an enclosure is also a means of increasing social complexity. However, changes in conspecific and species composition can also be the cause of stressful competition or dominance relationships (Bloomsmith *et al.*, 1988). Because zoos have long recognized the importance of keeping animals in harmonious, species-appropriate social groupings, and make every attempt to do so, we will limit our discussion of enrichment in this chapter to the attempts of zoos to increase environmental complexity by providing *inanimate enrichment.*

It is well known that stress can suppress reproductive function, impair immune function, cause aberrant behaviour and influence the development of animals (Engel, 1967; Epple, 1978; Moberg, 1985; Suomi, 1987). It is also known that not all stress is deleterious and that some level of acute stress may actually benefit animals by facilitating reproductive activation (Antelman and Caggiula, 1980), enhancing learning, alertness and exploration, or improving immune responses (Weiss *et al.*, 1989; Konarska *et al.*, 1989; Moodie and Chamove, 1990). The ill-defined concept 'the stress of captivity' has long been named as the cause of health, welfare and propagation problems in zoo animals (e.g. Hediger, 1964). In captivity, the deleterious effects of stress may occur if an animal is unable, in the short term, to cope with environmental threats and challenges, or if it is asked to cope repeatedly. Increasing opportunities in captive environments for animals to perform behavioural coping responses that allow control over or escape from threatening stimulation conceivably helps alleviate the 'stress of captivity'. The question we ask in this chapter is: what is the evidence that increasing inanimate environmental complexity actually modulates stress in zoo animals?

Unlike domesticated farm and laboratory animals, zoo animals have not been intentionally selected for adaptation to captive conditions. Indeed, strenuous efforts are made specifically to avoid such genetic changes in captive populations of wild animals. Zoos aim to raise animals that behave 'normally' according to criteria based on the behaviour of their wild-living conspecifics. This follows from one of the central missions of modern zoos, namely to conserve endangered species through captive propagation and display (conservation education). Through 'Species Survival Plans' (SSPs), zoos cooperatively and collectively coordinate husbandry and propagation of endangered species. The primary goal of these programmes is to maintain genetic diversity within captive populations (Ballou and Lacy, 1995). Also, in recognition of the fact that some of these captive populations might at some point in the future be tapped as a source for reintroductions into the wild, maintaining behavioural diversity is recognized as an important goal (Beck, 1991; May, 1991; Markowitz, 1997). Thus, successful captive propagation of endangered species requires that most individuals live productive lives in which a full range and diversity of 'natural' reproductive and parental behaviours occur, and that captive-born offspring develop into normally breeding adults.

Increasingly, zoos are looking to the practice of enriching environments to ensure that behavioural diversity is maintained in captivity. This is, perhaps, a lofty expectation. Although in many respects there are fewer constraints on what is considered acceptable environmental enrichment practice in zoos as compared with farms and laboratories, a higher standard is expected because animals are on public display. As the practice of enrichment in zoos grows, the mechanisms through which increased environmental complexity benefits animals in zoos continue to be empirically investigated. This will lead to the development of many types of enrichment that are specific not only to species, but to age groups within a species, and even to individuals.

Although the field of environmental enrichment has done much to promote the welfare of zoo animals, this does not guarantee that deleterious effects of captivity-induced stress do not persist. Captive conditions are often inadvertently stressful, especially for many endangered species if little is known from field studies about their behavioural needs. Identifying and rectifying sources of stress in captive environments which interfere with the behaviour, reproduction and health of endangered species is a continual challenge to zoo animal managers. Although much empirical and anecdotal research has been conducted in zoos on the environmental needs of animals and their performance in captivity (e.g. growth rates, nutrition, maternal behaviour), research on stress *per se* has not been a high priority. The reasons for this are twofold. First, because one of the central missions of modern zoos is to conserve endangered species through captive propagation, zoos have made major commitments towards research designed to benefit populations such as genetics, demographic analysis, reproductive physiology and assisted reproductive techniques. Second, zoos have become proficient in research that assesses the behavioural improvements of various enrichment techniques, perhaps because the display of interesting animal behaviour is one of the exhibition goals of zoos. However, the focus has been on behavioural benefits. In many cases physiological, immunological and reproductive benefits have only been assumed, since these types of benefits are more difficult, intrusive and expensive to assess.

This chapter reviews the evidence that environmental enrichment alleviates stress in zoo animals. Our discussion will be divided into three sections concerning: (i) potential stressors in zoo environments; (ii) behavioural benefits of environmental enrichment techniques that are used in zoos; and (iii) physiological evidence that environmental enrichment reduces stress in captive animals.

Stressors in the zoo environment

Given the great range of species, exhibit types and husbandry protocols existing in today's zoos it is perhaps not surprising that up to now few systematic studies have been conducted on potential stressors and their effects on zoo animals. To identify stress in zoo animals, zoo biologists still rely heavily on extrapolation from results of more controlled studies in laboratories, as well as anecdotal observations of zoo animal managers, keepers, veterinarians and pathologists. Further muddying the picture are the diverse backgrounds of individual zoo animals. For example, in one group of chimpanzees some individuals may be wild caught, others may be captive born, some may have been reared by their natural parents whereas others were hand reared, and yet others may have been reared by foster parents. All of these types of early experience might be expected to exert their influence on an individual's responsiveness to stress (Huck and Price, 1975). Environments that are stressful for one species or one individual may not be stressful for another species, or even another individual of the same species. In some respects these are characteristics that zoo environments share more with the wild than with other captive husbandry systems.

In spite of these problems, a number of environmental stimuli common to zoo environments have been empirically identified as potential stressors. One of the most obvious causes of chronic stress in confined wild animals is the inability to respond to fearful situations with active avoidance and/or escape responses. Because most zoo animals have restricted freedom of movement compared with their wild counterparts, they are often unable to withdraw effectively from a source of aversive stimulation. Hediger (1964) describes situations in which captive animals injured themselves or failed to breed because of the inability to escape from caretakers or visitors.

Olfactory and auditory cues (which we may not perceive) from predators, prey, neighbouring species or competing conspecifics could be fear-inducing to animals that readily attend to such cues. For example, Buchanan-Smith *et al.* (1993) found that both wild and captive-born cotton-top tamarins (*Saguinus oedipus*) respond to faecal scents of predators with behaviour indicative of high anxiety. Another example is provided by Carlstead *et al.* (1993). Leopard cats (*Felis bengalensis*) housed in a barren cage with only an animal transport carrier as furniture showed sustained elevations in urinary cortisol when housed in a holding facility for lions and tigers, which they could hear and smell, but not see. Cortisol was not chronically elevated when the same individuals were housed with the same cage furnishings at a research laboratory and in a quarantine building. It was hypothesized that the presence of the larger cats, which in the wild often prey on the smaller species, was a chronic stressor for the leopard cats.

Machine sounds may be another inadvertent source of aversion to many species. Rappaport and King (1987) document a case of zoo kinkajou (*Potos flavus*) stereotypy that was correlated with noise. Another study describes temporal variation in stereotypic pacing patterns of two male fennec foxes (*Fennecus zerda*) housed in small zoo enclosures. Pacing would increase during and after machine noises such as a vacuum cleaner in the building or a lawn mower outside (Carlstead, 1991). The fennec fox exhibits were vacuumed daily with the animals still in them, eliciting at least an hour of pacing in one of the males after the cleaning event. Thus the response to machine noises in general may have been a conditioned fear response as a result of the vacuuming. Various aspects of persistent acoustic stimuli such as intensity, frequency distribution, infrasound and content may be common stressors for captive animals (Stoskopf, 1983). Siberian tigers (*Panthera tigris altaica*) at the Bucharest zoo have been reported to be prone to developing gastroenteritis due to a failure to adapt to unfamiliar quarters. However, a persistent high noise level lasting several months caused by repairs in an adjacent courtyard was also sufficient to induce gastroenteritis in some of the tigers (Cociu *et al.*, 1974). Construction noise has also been identified as a potential source of stress although the few studies conducted to date have failed to demonstrate anything other than short-term responses to specific auditory events such as blasting noise (Mellen and Shepherdson, Oregon, 1997, personal communication).

Zoo visitors are, almost by definition, a prominent feature of most zoo environments, and a number of studies have documented negative responses to zoo visitors. For example, Carlstead *et al.* (1999) found a correlation in black rhinoceros (*Diceros bicornis*) between the degree of public access around the rhino's enclosure and the animal's score for fearful behaviour (timid, shy, anxious, sleeps a lot). Zoos with enclosures that have greater public access also have higher mortality rates among their black rhinos. In other studies, the presence of zoo visitors has been associated with a number of behavioural changes that suggest that this is a potential stressor. These include increased vigilance, closer social spacing, reductions in affiliative behaviour and increases in aggression (in primates: Glatston *et al.*, 1984; Hosey and Druck, 1987; Chamove *et al.*, 1988; in ungulates: Thompson, 1989). Zoo animals frequently react strongly to their human caretakers (keepers, veterinarians, etc.) in both positive and negative ways, although to date there has been little empirical study of this in zoo animals (Hemsworth and Barnett, Chapter 15, this volume). Persistent uncertainty about the actions of caretakers or veterinarians may contribute to the stressfulness of some captive situations if the animals have no reliable predictive cues. This has been demonstrated in pigs in an intensive farming system. When signals preceding feeding time were made unreliable, increased aggression occurred in response to random sounds on the farm. Pigs receiving reliably signalled meals did not become behaviourally aroused in the presence of the same sounds (Carlstead, 1986).

Other potential sources of stress that are yet to be quantitatively evaluated include preventing access to stimuli within the perceptual field of an animal, such as food items, potential mates, oestrus females or prey species. Stereotypic behaviour is often described as increasing under such circumstances (Meyer-Holzapfel, 1968; Boorer, 1972; Mason, 1993), but it is not known how frustrating or stressful this is to animals. Unpredictable invasive or traumatic handling by vets, researchers, trainers or keepers, such as blood sampling, medical examinations and direct contact training such as is done with elephants (see below), could also be stressors. New studies in which behavioural observations are combined with recently developed non-invasive physiological measures (such as faecal steroid assays) show great promise in revealing some of the complex relationships between stress and environmental variables in the zoo environment.

Behavioural benefits of environmental enrichment techniques used in zoos

Environmental enrichment techniques in zoos generally follow one or more of the following guiding principles: (i) increasing control or contingency between animal action and environmental reaction; (ii) presenting cognitive challenges such as learning what a trainer is requesting or solving a problem; (iii) meeting specific behavioural needs such as a need for shelter/hiding or foraging; (iv) providing an environment in which exploration is stimulated and rewarded with new and useful information; and (v) stimulating social interaction (Shepherdson, 1998). These categories are not exclusive; an environment that provides more information for an animal to gather through exploration may also be an environment that is more likely to provide appropriate control or contingency between action (behaviour) and reaction (consequence). Indeed, the inevitable interdependence of many of these factors is a significant problem in understanding their relative importance. In any case, a heavy emphasis is placed on the importance of providing enrichment that is appropriate to the specific biology (to the extent to which it is known) of the species under consideration.

Often enrichment is essentially therapeutic in that relatively minor changes are made to existing exhibits in order to improve environments thought to be inadequate by current standards. Increasingly, however, new exhibits are constructed with the principles of enrichment in mind from the planning stage onwards. It has to be said that much of the activity that constitutes enrichment in zoos is currently based on informed intuition rather than empirical evaluation. No apology needs to be made for this; the inadequacies of current scientific models of animals' needs and welfare should not be used as an excuse to do nothing when intuitive understanding suggests a need.

Attempts to enhance zoo environments using *inanimate* enrichment, following from the principles described above, have resulted in a wide diversity of imaginative and creative applications. Most of these, however, fall into one or more of the following categories: (i) increasing physical and/or temporal complexity; (ii) providing cognitive challenges; and (iii) satisfying motivation to perform specific appetitive behaviours. In most of the cases that have been evaluated, and for which the results have been published, success is measured in terms of behavioural improvements. The following examples give an idea of the range of enrichment activities commonly employed in zoos, the behavioural changes that result, and the rationale for believing that these may indicate reductions in stress.

Increasing environmental complexity

Increasing physical and temporal complexity may add biologically relevant information to an animal's enclosure, resulting in increased opportunities for exploration. It can provide hitherto absent behavioural opportunities through increased diversity of substrates. It can also provide greater contingency, or control, by providing a context within which an animal can learn to increase its chance of achieving a desired goal through the performance of appropriate behaviour. This latter factor, control, may be particularly important (Sambrook and Buchanan-Smith, 1997). Laboratory studies certainly have shown that lack of control over noxious stimulation (e.g. electric shock) is stress-inducing (Weiss, 1968). Laboratory studies have also demonstrated that lack of early experience in controlling environmental events can produce an animal that later in life is less able to adapt to stressful events and less likely actively to investigate and learn about novel situations (e.g. Joffe *et al.*, 1973; Widman *et al.*, 1992). Early experience with a socially or physically enriched environment has also been demonstrated to promote normal species-specific behaviour, increase activity and reduce emotionality (Denenberg and Morton, 1962, 1964; Denenberg, 1964; Pfaffenberger *et al.*, 1976; Uphouse, 1980; Chamove *et al.*, 1982; Mineka and Henderson, 1985). Fortunately there is also some empirical evidence that the emotional, cognitive and anatomical deficits associated with prolonged early experience with impoverished environments, while long lasting, are subject to at least partial improvement when 'therapy' in an enriched environment is provided (Inglis, 1975; Warren *et al.*, 1982; Stein *et al.*, 1983).

Biologically appropriate complexity can be increased in many ways, for example, by adding substrates such as dirt, litter, mulch, vegetation or trees. These substrates elicit foraging and investigatory behaviour by concealing food, smells, naturally occurring insects or other wildlife, etc. In so doing, this method of enrichment may increase the information content of an animal's habitat. Barriers and landscaping can provide privacy, promote territorial behaviour, provide escape routes, and thus improve social interactions. Climbing structures allow more efficient use of space and provide shade and temperature gradients for choice of microclimate. They can also provide hiding places from conspecifics, public and keepers.

Zoo studies that demonstrate the importance of this kind of enrichment include a multi-zoo gorilla (*Gorilla gorilla*) study by Miller-Schroeder and Paterson (1989). They found that the larger and more complex the enclosure was, especially with respect to sight barriers, privacy refuges and dens that allowed escape from conspecifics and the public, the better were the female's chances of rearing her infants normally. In a similar multi-institutional study, Mellen *et al.* (1998) found a negative correlation between stereotypic pacing in small felids and enclosure complexity. Wood (1998) found that a variety of enrichment objects, including manipulable objects and scatter feeding, reduced unwanted behaviours like aggression and abnormal behaviour in zoo chimpanzees (*Pan troglodytes*). Tripp (1985) came to a similar conclusion for zoo-housed orang-utans (*Pongo pygmaeus*), as did Wilson (1982) and Bowen (1980) for gorillas. Chamove *et al.* (1982) provided a more complex environment for several species of zoo-housed primates by providing a straw substrate in which food was scattered, and found that abnormal behaviours were reduced in exchange for normal rates of foraging behaviour. Studies in which food was hidden in log and/or brush piles for hunting dogs (Ings *et al.*, 1997), leopard cats (*Felis bengalensis*) (Shepherdson *et al.*, 1993) and several species of bear in zoos (Carlstead *et al.*, 1991; Forthman *et al.*, 1992) demonstrate the effectiveness of this strategy in motivating these carnivores to forage and in reducing abnormal behaviours such as pacing.

Cognitive challenge

It is assumed that captive environments provide less stimulation and opportunities for animals to exercise their cognitive abilities to solve problems than do wild environments. Many have argued that this is important for the well-being of captive animals and cite studies on the deleterious effects of understimulating environments as evidence (e.g. Butler, 1957; Gendreau *et al.*, 1968). These effects include decreased stimulation-seeking behaviour, lethargy and drowsiness, and an inability to cope with environmental change. Cognitive challenges, by definition, put captive animals in a position in which they can learn actively to control and explore some aspect of their environment.

Markowitz (e.g. 1982) published a number of studies in which zoo animals were given the opportunity to learn to control some part of their environment through use of a mechanical apparatus that produces a food reward. For example, a mandrill (*Papio sphinx*) exhibiting excessive aggression towards cage mates was given a chance to learn to play a computer game with visitors, which, if he won, gave him a food reward. This opportunity resulted in reduced rates of stereotypic behaviour and aggressive interaction (Yanofski and Markowitz, 1978). Puzzle devices that require animals to learn how to remove food items have also been associated with significant effects on rates of abnormal behaviour. Bloomsmith *et al.* (1988) found a reduction in stereotypic behaviour, coprophagy and aggression in a group of chimpanzees consequent upon the introduction of puzzle feeders, as did Gilloux *et al.* (1992), and Maki *et al.* (1989).

More recently, the potential of training, not just as a management tool but as cognitive enrichment for captive animals, has begun to be realized (Laule and Desmond, 1998). Earlier studies demonstrated some behavioural benefits of training. Kastelein and Wiepkema (1988) found that 60 min of training per day reduced the amount of time a Stellar sea lion (*Eumet opias jubata*) engaged in stereotypic swimming from 7.2% to 0.5%. Laule (1993) documented reductions in coprophagy and self-biting in drills (*Papio leucophaeus*) as a consequence of a training programme designed to increase sensory stimulation and address socialization problems.

Opportunities for performing appetitive behaviours

Hughes and Duncan (1988) review the evidence that animals may need to perform some appetitive behaviours (such as nest building or foraging) even when the performance of those behaviours is not necessary in order to achieve a desired goal (such as a nest or food). Although it is clear that many behaviours are stimulus driven, in which case there may be no motivation to perform a behaviour in the absence of a specific stimulus, Hughes and Duncan point out that there are some behaviours that do seem to be internally driven. Preventing performance of these behaviours could result in frustration and stress. Certainly there are examples from studies in zoos that support this view.

Shepherdson *et al.* (1993) found that providing a fishing cat (*Felis viverrina*) with an opportunity to search for and hunt live fish resulted in profound and long-lasting changes in behaviours such as reduction in chronic inactivity and stereotypic behaviours. Bond and Lindburg (1990) found that carcass-fed cheetahs (*Acinonyx jubatus*) had greater appetites, fed for longer periods and displayed greater possessiveness of food than those fed a commercial diet. The carcass-fed animals spent more time smelling and handling food, and more time chewing and using their molars to slice food. Regurgitation and reingestion is an abnormal behaviour common in captive gorillas. Akers and Schildkraut (1985) showed that this behaviour could be reduced in some individuals by increasing the availability of browse, suggesting that the abnormal behaviour stemmed from a need to perform more foraging behaviour.

The preceding discussion of the behavioural benefits of environmental enrichment leads to a conclusion that is relevant to answering our original question about the stress-alleviating effects of enrichment in zoo animals. We conclude that, in zoos, *behavioural choices* for animals are increased by enhancing the physical and temporal complexity of animals' surroundings, providing cognitive challenges and supplying specific environmental features that satisfy appetitive motivation. An animal with a large repertoire of behavioural options is more likely to be able to exert control over its environment because of the increased likelihood that it has learned contingencies between its environment and its behaviour. Our conclusion is supported by empirical studies with laboratory and farm animals specifically designed to test the hypothesis that increased environmental complexity expands an animal's behavioural repertoire. Haskell *et al.* (1996) found that pigs in substrate-enriched pens exhibited a greater diversity of manipulative behaviours, as well as a more diverse behavioural repertoire, than pigs kept in substrate-impoverished pens. Likewise, Renner and Rosenzweig (1986), who compared object interaction behaviour of juvenile rats reared in either enriched or impoverished environments, found a greater diversity of manipulatory behaviour in the enriched rats. They concluded that object exploration strategies are experience dependent, and that the enriched rats acquire behavioural repertoires that are geared towards finding out about new elements of their environment.

Physiological evidence that environmental enrichment reduces stress in confined animals

'Coping' effectively with stress requires that an animal is able to adapt its neuroendocrine response to noxious stimulation by altering its behaviour or by learning about its situation and altering its expectations of the stimulation impinging on it (Levine, 1985). Indeed, the aversive effects of noxious stressors such as heat, cold and noise, for example, may be due to the lack of an opportunity to huddle, move away, flee or hide (Levine *et al.*, 1979). The behavioural evidence in zoo animals, therefore, suggests that stress-alleviating effects of increased behavioural choices are due to increased opportunities for performing appropriate coping responses to aversive situations.

There is little research directly examining the neuroendocrine consequences of increased environmental complexity. Laboratory studies with rodents and dogs have demonstrated that the pituitary–adrenal system is activated when the animal is unable to carry out appropriate behavioural responses to aversive stimulation (Weiss, 1968; Davis *et al.*, 1977; Dess *et al.*, 1983), but does facilitating this ability reduce pituitary–adrenal activity or produce other improvements such as improved immune function? In this section we review recent evidence that increasing the complexity of standard farm, laboratory and zoo enclosures has beneficial consequences on pituitary–adrenal activity and immune responses. Experiments that increase the complexity of standard farm, laboratory and zoo enclosures seem to have four general outcomes, discussed below.

No effect

A few studies report no effect of enrichment on baseline pituitary–adrenal function. This may be because the baseline conditions prior to the enrichment were not stress-inducing for the animals or, conversely, that the animals were already exhibiting a maximal pituitary–adrenal response. Byrne and Suomi (1991) found 'little discernible effect' of woodchip enrichment on urinary cortisol concentrations in rhesus monkeys, nor was there an effect on agonistic interactions or abnormal behaviour patterns. In juvenile rhesus monkeys, no effect on baseline cortisol or cortisol levels after adrenocorticotropic hormone (ACTH) challenge was found for a combined enrichment protocol of physical, feeding and sensory enhancements (Shapiro *et al.* 1993).

Individual differences in responses

In one study with mice, structural enrichment was found to induce changes in social organization, which caused increased cortisol for some individuals (Haemisch *et al.* 1994). Plasma corticosterone titres were significantly elevated in dominant males in the enriched groups compared with subordinate males, as well as dominant males in standard cage groups. These findings suggest that keeping adult male mice in structured cages can result in increased aggression towards intruders, a change in the social organization and altered endocrine states depending on the individuals' dominance position.

Decreased baseline level responses

A number of studies have demonstrated that enrichment decreases cortisol or improves immune function compared with baseline, unenriched conditions. Adding manipulable devices to the otherwise barren cages of laboratory-housed, adult rhesus monkeys lowered basal plasma cortisol concentrations and decreased abnormal behaviours (Line *et al.*, 1991). Using generation of fever as a measure of immune response to *Salmonella typhosa* challenge, Kuhnen (1998) found that the response was enhanced in golden hamsters (*Mesocricetus auratus*) housed in large enriched cages compared with small cages. The latter were thought to be associated with higher stress levels. In another experiment, juvenile rhesus monkeys were reared in cages enriched with a movable surrogate mother and movable objects (versus rearing in standard conditions with a non-movable surrogate mother). At age 8 months when infected with SIV/Delta virus the enriched monkeys demonstrated a slower progression of the disease (Clarke *et al.*, 1989). A zoo example of cortisol reduction by enrichment is provided by Carlstead *et al.* (1993). Captive leopard cats (*Felis bengalensis*), mentioned above, had chronically elevated cortisol levels when housed in the same building as lions and tigers. In a later phase of the experiment we changed their almost barren cages by providing a variety of hiding places and vegetation. This significantly reduced urinary cortisol concentrations to baseline levels and significantly reduced stereotypic pacing. Thus, providing enrichment in the form of increased cage complexity facilitated coping with a chronically present environmental stressor. Bettinger (Columbia, South Carolina, 1998, personal communication) presented preliminary evidence that salivary cortisol levels are lower in zoo elephants trained with positive reinforcement-based 'protected' contact compared with elephants trained with direct contact that involves some negative reinforcement.

Decreased responsiveness to some stressors

Most studies show that enrichment does not change baseline cortisol but does cause a lowered cortisol response to another, acutely stressful situation. Rats or pigs subjected to an arousing or frustrating situation are able to inhibit increases in pituitary–adrenal stress responses to the situation if provided with enrichment in the form of opportunities to perform behaviours such as chain-pulling, drinking or wheel-running (Heybach and Vernikos-Danellis, 1979; Levine *et al.*, 1979; Dantzer and Mormede, 1981). Morgan *et al.* (1998), in addition to finding that baseline cortisol was significantly lower in caged rhesus macaques when a music-feeder apparatus was available, also found that heart rate reactivity to brief restraint in the squeeze mechanism of the home cage was significantly less in animals with the music-feeder. Jarvis *et al.* (1997) reported that pregnant sows provided with nest-building material had lower increases in cortisol at the time of farrowing. When piglets are weaned, sows also exhibit increased cortisol secretion, but sows in large pens with straw show a significantly lower increase than sows in farrowing crates with no straw (Cronin *et al.*, 1991). Crockett *et al.* (1998) found a significant effect of Kong® toy presence on cortisol levels in laboratory-housed longtailed macaques, but only under two social contact conditions. Without the Kong®, cortisol levels were significantly higher in both social conditions (grooming contact with one opposite-sex partner or grooming contact plus mirror reflection of three other male–female pairs) than in no-contact conditions. With the Kong®, cortisol levels during no contact and social contact did not differ significantly. Thus, having a Kong® toy towards which the animals could direct behaviour seems to have lowered cortisol values in a social situation that was ordinarily stress-inducing.

The physiological evidence supports the hypothesis that enrichment facilitates coping with aversive, stress-inducing stimulation, while the behavioural evidence indicates that the mechanism for this effect is increased behavioural options. The differential effects of enrichment on individuals in social groups points out an additional factor that must always be considered when developing enrichment protocols. There are many anecdotal reports in zoos about enrichments, especially feeding enhancements, which increase competition and aggression when given to socially living groups of animals, particularly primates. These problems are usually solved by further increasing the number of enhancements so that animals are not forced to compete for access to them (Bloomsmith *et al.*, 1988).

Conclusions

The behavioural evidence we have reviewed indicates that inanimate enrichment increases the diversity of behaviours an animal uses to interact with its environment. The physiological evidence reviewed suggests that inanimate enrichment: (i) may decrease the physiological stress response directly elicited by aversive housing conditions; or (ii) may facilitate coping with sporadic, stress-inducing events or social situations that sometimes occur in captive environments. We conclude, therefore, that environmental enrichment can be an effective way of reducing captivity-induced stress by providing animals with increased behavioural options for responding to threatening or aversive stimulation in their surroundings. These options include the performance of appropriate coping and avoidance behaviours such as withdrawing or hiding, redirecting attention or activity, and performing active appetitive behaviours such as foraging and exploration, which increase control over the environment.

In zoos, we are often unable to identify the stressors acting on animals or which types of enrichment are most important for coping with them. Mostly we use purely behavioural means to assess the welfare status of zoo animals. Alert caretakers assess good welfare by the presence of a diverse behavioural repertoire, absence of abnormal behaviour and stereotypy, and low levels of aggression. Zookeepers often take a hit or miss approach to enrichment, but a consensus is emerging as to what types of enrichment are most effective in different species for maintaining behavioural diversity. In zoos, we assume, based on the limited physiological evidence, that providing animals with physically complex and temporally changing environments maximizes animals' choices for responding to the aversive characteristics of captive environments that may be chronically present. This will affect their development in captivity and their ability to reproduce normally as adults, and ultimately help to maintain genetic diversity in the captive population.

References

Akers, J.S. and Schildkraut, D.S. (1985) Regurgitation/reingestion and coprophagy in captive gorillas. *Zoo Biology* 4, 99–109.

Antelman, S.M. and Caggiula, A.R. (1980) Stress-induced behaviour: chemotherapy without drugs. In: Davidson, J.M. and Davidson, R.J. (eds) *The Psychobiology of Consciousness*. Plenum Press, New York, pp. 65–104.

Ballou, J.D. and Lacy, R.C. (1995) Identifying genetically important individuals for management of genetic diversity in pedigreed populations. In: Ballou, J.D., Gilpin, M. and Foose, T.J. (eds) *Population Management for Survival and Recovery*. Columbia Press, New York, pp. 76–11.

Beck, B.B. (1991) Managing zoo environments for reintroduction. In: *Proceedings of the American Association of Zoological Parks and Aquariums Annual*

Conference. American Association of Zoological Parks and Aquariums, San Diego, pp. 436–440.

Bloomsmith, M.A., Alford, P.L. and Maple, T.L. (1988) Successful feeding enrichment for captive chimpanzees. *American Journal of Primatology* 16, 155–164.

Bond, J. and Lindburg, D. (1990) Carcass feeding of captive cheetahs (*Acinonyx jubatus*): the effects of a naturalistic feeding program on oral health and psychological well-being. *Applied Animal Behaviour Science* 26, 373–382.

Boorer, M. (1972) Some aspects of stereotyped patterns of movement exhibited by zoo animals. *International Zoo Yearbook* 12, 164–166.

Bowen, R.A. (1980) The behaviour of three hand-reared lowland gorillas (*Gorilla g. gorilla*) with emphasis on the response to a change in accommodation. *Dodo Journal. Jersey Wildlife Preservation Trust* 17, 63–79.

Buchanan-Smith, H.M., Anderson, D.A. and Ryan, C.W. (1993) Responses of cotton-top tamarins (*Saguinus oedipus*) to faecal scents of predators and non-predators. *Animal Welfare* 2, 17–32.

Butler, R.A. (1957) The effect of deprivation of visual incentives on visual exploration motivation in monkeys. *Journal of Comparative and Physiological Psychology* 50, 177–179.

Byrne, G.D. and Suomi, S.J. (1991) Effects of woodchips and buried food on behaviour patterns and psychological well-being of captive rhesus monkeys. *American Journal of Primatology* 23, 141–151.

Carlstead, K. (1986) Predictability of feeding: effects on agonistic behaviour and growth in grower pigs. *Applied Animal Behaviour Science* 16, 25–38.

Carlstead, K. (1991) Fennec fox (*Fennecus zerda*) husbandry; environmental conditions influencing stereotypic behaviour. *International Zoo Yearbook* 30, 202–207.

Carlstead, K. and Shepherdson, D. (1994) The effects of environmental enrichment on reproduction of zoo animals. *Zoo Biology* 13, 447–458.

Carlstead, K., Seidensticker, J.C. and Baldwin, R. (1991) Environmental enrichment for zoo bears. *Zoo Biology* 10, 3–16.

Carlstead, K., Brown, J.L. and Seidensticker, J. (1993) Behavioural and adrenocortical responses to environmental change in leopard cats (*Felis bengalensis*). *Zoo Biology* 12, 321–331.

Carlstead, K., Fraser, J., Bennett, C. and Kleiman, D.G. (1999) Black rhinoceros (*Diceros bicornis*) in U.S. zoos: II. Behavior, breeding success and mortality in relation to housing facilities. *Zoo Biology* 18, 35–52.

Chamove, A.S., Anderson, J.R., Morgan-Jones, S.C. and Jones, S.P. (1982) Deep woodchip litter: hygiene, feeding and behavioural enhancement in eight primate species. *International Journal for the Study of Animal Problems* 3, 308–318.

Chamove, A.S., Hosey, G.R. and Schaetzel, P. (1988) Visitors excite primates in zoos. *Zoo Biology* 7, 359–369.

Clarke, M.R., Koritnik, D.R., Martin, L.N. and Baskin, G.B. (1989) Cage enrichment, physiology, and behaviour in nursery-reared rhesus monkeys. *American Journal of Primatology* 1 (Suppl.), 53–57.

Cociu, M., Wagner, G., Micu, N.E. and Mihaescu, G. (1974) Adaptational gastro-enteritis in Siberian tigers. *International Zoo Yearbook* 14, 171–174.

Crockett, C.M., Bellanca, R.U., Bowers, C.L. and Bowden, D.M. (1998) Cortisol responses of longtailed macaques to increased social stimulation. *American Journal of Primatology* 45, 175 (Abstract).

Cronin, G.M., Barnett, J.L., Hodge, F.M., Smith, J.A. and McCallum, T.H. (1991) The welfare of pigs in two farrowing/lactation environments: cortisol responses of sows. *Applied Animal Behavioural Science* 323, 117–127.

Dantzer, R. and Mormede, P. (1981) Pituitary-adrenal consequences of adjunctive activities in pigs. *Hormones and Behavior* 15, 386–395.

Davis, H., Porter, J.W., Livingstone, T., Hermann, T., MacFadden, L. and Levine, S. (1977) Pituitary-adrenal activity and lever-press shock escape behaviour. *Physiology and Behavior* 5, 280–284.

Denenberg, V.H. (1964) Critical periods, stimulus input, and emotional reactivity: a theory of infantile stimulation. *Psychological Review* 71, 335–351.

Denenberg, V.H. and Morton, J.R.C. (1962) Effects of environmental complexity on social groupings upon modification of emotional behaviour. *Journal of Comparative and Physiological Psychology* 55, 242–246.

Denenberg, V.H. and Morton, J.R.C. (1964) Infantile stimulation and emotional reactivity: a theory of infantile stimulation. *Psychological Review* 71, 335–351.

Dess, N.K., Linwick, D., Patterson, J. and Overmeier, J.B. (1983) Immediate and proactive effects of controllability and predictability on plasma cortisol responses to shock in dogs. *Behavioral Neuroscience* 97, 1005–1016.

Engel, G.L. (1967) A psychological setting of somatic disease: the giving up–given up complex. *Proceedings of the Royal Society of Medicine* 60, 553–555.

Epple, G. (1978) Lack of effects of castration on scent marking displays and aggression in a South American primate (*Saguinus fuscicollis*). *Hormones and Behavior* 11, 139–150.

Forthman, D.L., Elder, S.D., Bakeman, R., Kurkowski, T.W., Noble, C.C. and Winslow, S.W. (1992) Effects of feeding enrichment on behaviour of three species of captive bears. *Zoo Biology* 11, 187–196.

Gendreau, P.E., Freedman, N., Wilde, G.J.S. and Scott, G.D. (1968) Stimulation seeking after seven days of perceptual deprivation. *Perceptual and Motor Skills* 26, 547–550.

Gilloux, I., Gurnell, J. and Shepherdson, D.J. (1992) An enrichment device for great apes. *Animal Welfare* 1, 279–289.

Glatston, A.R., Geilvoet-Soeteman, E., Hora-Pecek, E. and van Hooff, J.A.R.A.M. (1984) The influence of the zoo environment on social behaviour of groups of cotton-topped tamarins, *Saquinus oedipus oedipus. Zoo Biology* 3, 241–253.

Haemisch, A., Voss, T. and Gartner, K. (1994) Effects of environmental enrichment on aggressive behaviour, dominance hierarchies, and endocrine states in male DBA/2J mice. *Physiology and Behavior* 56, 1041–1048.

Haskell, M., Wemelsfelder, F., Mendl, M.T., Clavert, S. and Lawrence, A.B. (1996) The effect of substrate-enriched and substrate-impoverished housing environments on the diversity of behaviour in pigs. *Behaviour* 138, 741–761.

Hediger, H. (1964) Wild *Animals in Captivity.* Dover Publications, New York.

Heybach, J.P. and Vernikos-Danellis, J. (1979) Inhibition of the pituitary-adrenal response to stress during deprivation-induced feeding. *Endocrinology* 104, 967–973.

Hosey, G.R. and Druck, P.L. (1987) The influence of zoo visitors on the behaviour of captive primates. *Applied Animal Behaviour Science* 18, 19–29.

Huck, U.W. and Price, E.O. (1975) Differential effects of environmental enrichment on the open field behaviour of wild and domestic Norway rats. *Journal of Comparative and Physiological Psychology* 89, 892–898.

Hughes, B.O. and Duncan, I.J.H. (1988) The notion of ethological 'need', models of motivation and animal welfare. *Animal Behaviour* 36, 1696–1707.

Inglis, I.R. (1975) Enriched sensory experience in adulthood increases subsequent exploratory behaviour in the rat. *Animal Behaviour* 23, 932–940.

Ings, R., Waran, N.K. and Young, R.J. (1997) Effect of wood-pile feeders on the behaviour of captive bush dogs (*Speothos venaticus*). *Animal Welfare* 6, 145–152.

Jarvis, S., Lawrence, A.B., McLean, K.A., Deans, L.A., Chirnside, J. and Calvert, S.K. (1997) The effect of environment on behavioural activity, ACTH, β-endorphin and cortisol in pre-farrowing gilts. *Animal Science* 65, 465–472.

Joffe, J., Rawson, R. and Mulick, J. (1973) Control of their environment reduces emotionality in rats. *Science* 180, 1383–1384.

Kastelein, R.A. and Wiepkema, P.R. (1988) The significance of training for the behaviour of Stellar sea lions (*Eumetopias jubata*) in human care. *Aquatic Mammals* 14, 39–41.

Konarska, M., Stewart, R.E. and McCarty, R. (1989) Sensitization of sympathetic-adrenal medullary responses to a novel stressor in chronically stressed laboratory rats. *Physiology and Behavior* 46, 129–135.

Kuhnen, G. (1998) Reduction of fever by housing in small cages. *Laboratory Animals* 32, 42–45.

Laule, G. (1993) The use of behavioural techniques to reduce or eliminate abnormal behaviour. *Animal Welfare Information Center Newsletter* 4, 1–11.

Laule, G. and Desmond, T. (1998) Positive reinforcement training as an enrichment strategy. In: Shepherdson, D.J., Mellen, J.D. and Hutchins, M. (eds) *Second Nature: Environmental Enrichment for Captive Animals.* Smithsonian Institution Press, Washington, DC, pp. 302–313.

Levine, S. (1985) A definition of stress? In: Moberg, G.P. (ed.) *Animal Stress.* American Physiological Society, Bethesda, Maryland, pp. 51–70.

Levine, S., Weinberg, J. and Brett, L.P. (1979) Inhibition of pituitary-adrenal activity as a consequence of consummatory behaviour. *Psychoneuroendocrinology* 4, 275–286.

Line, S.W.H., Markowitz, H., Morgan, K.N. and Strong, S. (1991) Effects of cage size and environmental enrichment on behavioural and physiological responses of rhesus macaques to the stress of daily events. In: Novak, M.A. and Petto, A.J. (eds) *Through the Looking Glass: Issues of Psychological Well-being in Captive Non-human Primates.* American Psychological Association, Washington, DC, pp. 160–179.

Maki, S., Alford, P.L., Bloomsmith, M.A. and Franklin, J. (1989) Food puzzle device simulating termite fishing for captive chimpanzees (*Pan troglodytes*). *American Journal of Primatology* 1 (Suppl.), 71–78.

Markowitz, H. (1982) *Behavioural Enrichment in the Zoo.* Van Nostrand Reinhold, New York.

Markowitz, H. (1997) The conservation of species typical behaviours. *Zoo Biology* 16, 1–3.

Markowitz, H., Schmidt, M.J. and Moody, A. (1978) Behavioral engineering and animal health in the zoo. *International Zoo Yearbook* 18, 190–194.

Mason, G.J. (1993) Forms of stereotypic behaviour. In: Lawrence, A.B. and Rushen, J. (eds) *Stereotypic Animal Behaviour.* CAB International, Wallingford, UK, pp. 7–40.

May, R. (1991) The role of ecological theory in planning the reintroduction of endangered species. *Symposia of the Zoological Society of London* 62, 145–163.

Mellen, J.D., Hayes, M. and Shepherdson, D. (1998) Captive environments for small felids. In: Shepherdson, D.J., Mellen, J.D. and Hutchins, M. (eds) *Second Nature: Environmental Enrichment for Captive Animals.* Smithsonian Institution, Washington, DC, pp. 184–204.

Meyer-Holzapfel, M. (1968) Abnormal behaviour in zoo animals. In: Fox, M.W. (ed.) *Abnormal Behavior in Animals.* W.B. Saunders, Philadelphia, pp. 476–503.

Miller-Schroeder, P. and Paterson, J.D. (1989) Environmental influences on reproductive and maternal behaviour in captive gorillas: results of a survey. In: Segal, E.F. (ed.) *Housing, Care and Psychological Well-being of Captive and Laboratory Primates.* Noyes Publications, Park Ridge, New Jersey, pp. 389–415.

Mineka, S. and Henderson, R.W. (1985) Controllability and predictability in acquired motivation. *Annual Review of Psychology* 36, 495–529.

Moberg, G. (1985) Influence of stress on reproduction: measure of well-being. In: Moberg, G.P. (ed.) *Animal Stress.* American Physiological Society, Bethesda, Maryland, pp. 245–267.

Moodie, E.M. and Chamove, A.S. (1990) Brief threatening events are beneficial for captive tamarins. *Zoo Biology* 9, 275–286.

Morgan, K.M., Line, S.W. and Markowitz, H. (1998) Zoos, enrichment and the skeptical observer. In: Shepherdson, D.J., Mellen, J.D. and Hutchins, M. (eds) *Second Nature: Environmental Enrichment for Captive Animals.* Smithsonian Institution, Washington, DC, pp. 153–171.

Pfaffenberger, C.J., Scott, J.P., Fuller, J.L., Ginsburg, B.E. and Bielfelt, S.W. (1976) Guide dogs for the blind: their selection, development and training. *Developments in Animal and Veterinary Sciences* 1. Elsevier, Amsterdam.

Rappaport, L. and King, N.E. (1987) The behavioural research program at the Washington Park Zoo. *Applied Animal Behaviour Science* 18, 57–66.

Renner, M.J. and Rosenzweig, M.R. (1986) Object interactions in juvenile rats (*Rattus norwegicus*): effects of different experiential histories. *Journal of Comparative Psychology* 100, 229–236.

Sambrook, T.D. and Buchanan-Smith, H.M. (1997) Control and complexity in novel object enrichment. *Animal Welfare* 6, 207–216.

Shapiro, S.J., Bloomsmith, M.A., Kessel, A.L. and Shively, C.A. (1993) Effects of enrichment and housing on cortisol response in juvenile rhesus monkeys. *Applied Animal Behaviour Science* 37, 251–263.

Shapiro, S.J., Bloomsmith, M.A., Porter, L.M. and Suarez, S.A. (1996) Enrichment effects on rhesus monkeys successively housed singly, in pairs, and in groups. *Applied Animal Behaviour Science* 48, 159–172.

Shapiro, S.J., Bloomsmith, M.A., Suarez, S.A. and Porter, L.M. (1998) Effects of social and inanimate enrichment on the behaviour of yearling rhesus monkeys. *American Journal of Primatology* 40, 247–260.

Shepherdson, D.J. (1998) Introduction: tracing the path of environmental enrichment in zoos. In: Shepherdson, D.J., Mellen, J.D. and Hutchins, M. (eds) *Second*

Nature: Environmental Enrichment for Captive Animals. Smithsonian Institution, Washington, DC, pp. 1–12.

Shepherdson, D.J., Carlstead, K., Mellen, J. and Seidensticker, J. (1993) Environmental enrichment through naturalistic feeding in small cats. *Zoo Biology* 12, 203–216.

Stein, D.G., Finger, S. and Hart, T. (1983) Brain damage and recovery: problems and perspectives. *Behavior and Neural Biology* 37, 185–222.

Stoskopf, M.K. (1983) The physiological effects of psychological stress. *Zoo Biology* 2, 179–190.

Suomi, S.J. (1987) Genetic and maternal contributions to individual differences in rhesus monkey biobehavioural development. In: Krasnegor, N., Blass, E., Hofer, M. and Smotherman, W. (eds) *Perinatal Development: a Psychobiological Perspective.* Academic Press, New York, pp. 397–420.

Thompson, V.D. (1989) Behavioral response of 12 ungulate species in captivity to the presence of humans. *Zoo Biology* 8, 275–297.

Tripp, J.K. (1985) Increasing activity in orangutans: provision of manipulable and edible materials. *Zoo Biology* 4, 225–234.

Uphouse, L. (1980) Reevaluation of mechanisms that mediate brain differences between enriched and impoverished animals. *Psychological Bulletin* 88, 215–232.

Vargas, A. and Anderson, S.H. (1999) Effects of experience and cage enrichment on predatory skills of black-footed ferrets (*Mustela nigripes*). *Journal of Mammalogy* 80, 263–269.

Warren, J.M., Zerweck, C. and Anthony, A. (1982) Effects of environmental enrichment on old mice. *Developmental Psychobiology* 15, 13–18.

Weiss, J.M. (1968) Effects of coping responses on stress. *Journal of Comparative Physiology and Psychology* 65, 251–260.

Weiss, J.M., Sundar, S.K. and Becker, K.J. (1989) Stress-induced immunosuppression and immunoenhancement; cellular immune changes and mechanisms. In: Goetzl, E.J. and Spector, N.H. (eds) *Neuroimmune Networks; Physiology and Diseases.* Wiley-Liss, New York, pp. 193–206.

Widman, D.R., Abrahamsen, G.C. and Rosellini, R.A. (1992) Environmental enrichment: the influences of restricted daily exposure and subsequent exposure to uncontrollable stress. *Physiology and Behavior* 51, 309–318.

Wilson, S.F. (1982) Environmental influences on the activity of captive apes. *Zoo Biology* 1, 201–209.

Wood, W. (1998) Interactions among environmental enrichment, viewing crowds, and zoo chimpanzees (*Pan Troglodytes*). *Zoo Biology* 17, 211–230.

Yanofski, R. and Markowitz, H. (1978) Changes in general behaviour of two mandrills (*Papio sphynx*) concomitant with behavioural testing in the zoo. *Psychological Record* 28, 369–373.

Understanding the Role of Stress in Animal Welfare: Practical Considerations[1]

17

T.L. Wolfle

215 Severn Avenue, Annapolis, Maryland, USA

Introduction

Stress is complex and individual-specific; not all agree on a single definition of stress nor would score individual animals similarly on a continuum of no stress to severe stress. Animal care personnel need clear recommendations regarding the many ramifications of stress as they are not normally experts in this complex subject but are often expected to understand and recognize stress in order to ensure the well-being of the animals in their care.

My goal is not to add to the complex pool of data on stress and distress, but to relate current understanding of stress to the day-to-day decisions made by animal care staff and Institutional Animal Care and Use Committees (IACUCs) as they strive to ensure the welfare of their charges. Although reference is made to US laws and policies, it is believed that the concepts discussed have wide applicability in other countries also. To simplify this discussion, I will refer to two types of stress: that caused by pain and that caused by factors other than pain.

Pain-induced stress

Pain-induced stress, as discussed in the National Research Council's report, *Recognition and Alleviation of Pain and Distress in Laboratory Animals*

[1] Information about the reports cited may be obtained from ILAR's web site www2/nas.edu/ilarhome

(NRC, 1992), is generally caused by requirements of a research protocol, by disease or by injury. Programmes of veterinary medical care and daily observations of each animal by trained individuals are normally adequate to minimize concerns about disease and injury, which are uncommon in well-run facilities and, when present, are attended to promptly.

Protocol-induced pain should be predictable. It should seldom be a surprise to the investigator or IACUC. When it is, it is cause for careful reconsideration of the research and the investigator's understanding of the consequences to the animal. This is not to say that all research must be pain-free. We know that cannot be the case, but research should be free of surprises that negatively affect the welfare of the animals. This is a point that deservedly receives careful consideration by IACUCs. Indeed, the Animal Welfare Regulations (CFR Title 9, Chapter 1, Subchapter A); the *Public Health Service Policy on Humane Care and Use of Laboratory Animals* (PHS, 1996); the *Guide for the Care and Use of Laboratory Animals* (NRC, 1996), the 'Agriculture Guide' (FASS, 1999), and several other key documents' major purpose is to avoid unintentionally induced pain and stress in research animals by investigators and care staff. Federal policy requires institutions to avoid imposing such pain and distress on animals and to achieve high degrees of animal welfare through programmes of adequate veterinary care, investigator and technician training, the composition and expertise of the IACUC, careful oversight and monitoring of animal use by the veterinary staff, diligence in the writing and review of protocols, and a constant search for suitable *in vitro* or alternative methods, including those that reduce the severity of interventions or improve the welfare of the animals.

One of the best checklists on this subject is the *US Government Principles for the Utilization and Care of Vertebrate Animals Used in Testing, Research, and Training* (IRAC, 1985). These are found in the 1996 edition of the *Guide for the Care and Use of Laboratory Animals* (NRC, 1996).[2] The preface to these 'Principles' states,

> The development of knowledge necessary for the improvement of the health and well-being of humans as well as other animals requires *in vivo* experimentation with a wide variety of animal species. Whenever US Government agencies develop requirements for testing, research, or training procedures involving the use of vertebrate animals, the following principles shall be considered; and whenever these agencies actually perform or sponsor

[2] The 'Principles' were developed by the Interagency Research Animal Committee, National Institutes of Health and published in the Federal Register on 20 May, 1985 by the White House Office of Science and Technology Policy. They are an integral part of the US Public Health Service's policy on animal care and have been endorsed by all government agencies that sponsor animal research. They were adapted from principles developed by the Committee on International Organizations of Medical Sciences (CIOMS), World Health Organization, which are recognized throughout the world. They may be found in Appendix D, pages 117–118 of the *Guide*.

such procedures, the responsible Institutional Official shall ensure that these principles are adhered to.

The nine government principles provide clear and succinct policies that institutions receiving federal funding for medical research must consider. It is unfortunate that they are not more often cited in lieu of the '3 R's' (Russell and Burch, 1959) for they are incorporated in the NRC Guide and the PHS Policy, adopted by all federal agencies sponsoring biomedical research, and include the Russell and Burch concepts of replacement, reduction and refinement in addition to other useful and basic principles.

Stress resulting from environmental causes

Situations other than pain that produce stress are grouped under the category of *environmental* causes of stress. Environmentally induced stress is often so subtle as not to be noticed. Its causes can include internal physiological and environmental events as well as experimental methods. There is considerable overlap between these categories of stress, as we shall see.

At one time or another, and to one degree or another, every research animal is adversely affected by its environment. Although avoidance of pain and pain-related stress and distress is imperative, its animal welfare importance pales in comparison with environmental causes of stress. Pain-induced stress in general, and protocol-induced pain in particular, is not the principal cause of stress in laboratory animals nor the greatest affront to the welfare of these animals. Pain is generally site-specific and produces visible signs that can and must be either avoided through careful protocol development and IACUC review or identified and ameliorated where possible through careful observation by well-trained investigators, technicians and veterinary staff.[3]

Perhaps some have not considered the causes of the other category of stress: that caused by non-painful environmental events, including conditions of routine husbandry, housing and social interactions. All research animals, whether on research protocols or not, are subject to conditions in their environment that provide for either comfort or discomfort. These conditions are so common, and include so many diverse situations, that they comprise a potentially vastly greater risk to the animals' welfare than does pain. Unlike pain-induced stress, environmentally induced stress is manifest in a myriad of ways. It lacks the specificity of pain with respect to the sensory system that mediates it and the physiological and behavioural reactions it elicits. Although pain tolerance can vary among individuals, the same novel

[3] These comments should not be construed to mean that research protocols in which pain to animals is an essential part should not be developed by investigators or approved by the IACUC. However, when pain is anticipated or required, ethical considerations and relevant federal and local laws and policies must be addressed.

painful stimulus generally produces similar reactions in all animals in a study. Such is typically not the case for novel environmentally induced stress where the response of each animal can be quite varied based on the individual histories of each animal. Whereas both pain and stress have adaptive qualities, their causes and manifestations are decidedly different and necessitate different treatment and avoidance approaches. Stress associated with pain abates with the relief of the pain, usually through the administration of analgesics or anaesthetics. Stress associated with environmental causes is often refractory to the administration of pharmacological agents (although anxiolytics are sometimes used in conjunction with changes in the environment) and instead requires environmental change.

Environmental events, such as entry into an animal room by an unfamiliar person or an experimental situation to which the animal is not adapted, can be stressful to the animal. This is commonly manifested as fear and anxiety, and is likely to be accompanied by rather profound physiological changes including alterations in the number of circulating blood cells and elevation in serum glucocorticoids. These environmentally induced physiological events may be the substrates for unanticipated manifestations of fear or anxiety at a later time.

However, let us look at another example. Let us say that a normally friendly dog had a difficult time in a research session. Some days later, when the investigator comes to look at the dog, the dog assumes an aggressive posture and sulks in the back of her cage. The attending animal care staff are aghast and fear they have done something to cause this behaviour. They review their exercise programme and give the dog extra positive human attention, which seems to have a positive effect on her. To them the dog seems fine, but when the investigator returns the same aggressive behaviour is manifested once again. Alas, it is not the cage environment and lack of socialization that has triggered this behaviour, and little improvement in the dog's behaviour towards the investigative staff is likely until the protocol is modified. A focus on environmental causes by the animal care staff ignores the underlying *internal* cause of this stress-induced behaviour. In this case, that behaviour, though evoked by the experience of the animal during the study, continues in the absence of that environmental stimulus and is sustained by the internal stressors, fear and anxiety in the presence of the investigator who, by association, has actually become the stimulus. The fear is linked to the conditions that initially caused it (the protocol) and is sustained in the absence of the experimental situation by the presence of persons attending the dog during the protocol (the investigator or technician). Lacking either of these antecedent stimuli, the dog is likely to show no signs of fear.

This example is similar to the fear shown by dogs as a result of improper early rearing practices, which might be far removed in time and location from the laboratory animal facility. It is thus important that consideration is given to the root cause of the fear and that it is not just assumed that it is caused by something in the immediate environment. The cause of such 'internal stressors' can sometimes be difficult to diagnose and even when diagnosed can have life-long consequences for the animal.

Again, our dog example. We all know that dogs that do not receive human attention early in life will probably never develop the bonds with people that socialized puppies do, even with significant attempts later in life to correct the early rearing deficit. This raises an issue related not only to the welfare of the animal but to the quality of the science itself, and it leads to an important point. Many dogs used in research are bred specifically for that purpose, either by the institution or by commercial breeders. Presumably, all dogs from a given source have received the same vaccinations as puppies, grew up in the same environment, and received the same quantity and quality of interaction with other dogs and with human care staff. Let us say that a group of these dogs is placed in a study in which a small blood sample is taken from the saphenous vein each day. Then, let us further suppose that one dog in particular seems difficult to restrain for this simple procedure. As time goes on, she becomes fractious to the point that the technicians tie her muzzle with gauze tubing before each procedure. Should it be surprising when this dog develops a fear of this test situation and of the people associated with it? Most probably something in this dog's past was missing which, had it been present, would have enabled her to not only cope with minimal restraint, but enjoy the human attention associated with it. Also, if the animal is distressed at this critical time of blood collection, the parameters of that blood are also likely to change. This, in turn, will add to the uncontrolled variability of the data and require more dogs to be used in the study. Lack of socializing of this one dog increased the stress level of that dog when being restrained by people and caused more dogs to be used. From a single lapse in the rearing protocol, the welfare of several other animals was potentially jeopardized.

The blame for this situation does not lie solely with the breeder of the puppy, but must be shared by the investigator, the veterinary staff and the institutional procurement department. The investigator should have noted the behavioural difference in this dog during the adaptation phases of the study, and the veterinary staff should have been sensitive to the fact that this dog was behaving differently from other dogs on this study or in the facility. The institution's procurement policy should have spelled out the required environment and rearing practices to be used with the animals. Some years ago I ran a large foxhound breeding colony for the National Institutes of Health. We developed a solid socialization programme that had dramatic results on the welfare of the dogs and their eagerness to be with people. The results were also well received by investigators who had long complained that the dogs were difficult to handle and were highly stressed during handling. At that time I also had the opportunity to visit a number of commercial breeders who were just beginning to develop protocols for the socialization of their dogs. The quality of these protocols varied greatly among the breeders. From that experience, I became aware of the importance of contract specifications that address not only breed, age, sex, size and vaccinations but also the type of environmental protocol under which the animals are reared.

Some argue that the cost of adding this behavioural quality assurance is too great since it requires social interaction both with other dogs and with people, especially during early developmental periods but also later in life in the research environment. The costs, however, are not excessive. Since most dogs are reared in litters, social interaction is seldom an issue during the period of time the dogs remain in the breeder's facility. Human interaction should not be an issue either. In rearing thousands of puppies at the NIH, we determined that remarkable and lasting effects could be achieved with as little as 5 min per week of focused human interaction. This interaction was conducted by caregivers entering the kennels and petting and feeding the dogs, or taking all members of a litter out on leashes at the same time. Since the process was accomplished in groups, the actual one-on-one time was probably less than 1 min per week. Note that the goal of these studies was not to determine how little attention could be given each dog, but to counter the view that socialization programmes were too expensive to be incorporated widely.

A positive result of this for the dogs was that they responded so dramatically to the attention that the carestaff and veterinarians could not avoid stopping at the kennel doors to pet the dogs as they went about their daily chores. This eagerness to be petted was a mixed blessing, however, for the dogs were so eager to greet *anyone* who came to the kennel door and were equally eager to bound out into the kennel aisle when the door was opened for any reason. These dogs romped down the long aisle to tantalize the dogs in adjacent kennels with their freedom. We were 'exercizing' dogs long before the USDA told us we should! I learned by this experience that the best adapted, least stressed dogs were those who remained at the kennel door whenever any human was in the vicinity. Those who are stressed in the presence of people remain in the back of the pack, or sulk at the back of the cage. These latter animals are the ones most likely to be labelled 'fear biters', and they have no place in a research facility. These simple observations make assessment of the animals' welfare (and the effectiveness of the performance standards used to achieve the desired goals) straightforward.

Similar observations have been made with singly caged macaques (Bayne *et al.*, 1993). Human interaction for only a few minutes a week, during which food treats were given to the animals, produced significant effects in calming the monkeys and reducing their stress as measured by reductions in agonistic and abnormal behaviour. The behavioural pathology of these animals was reduced even after the 'enrichment' was concluded. While some animals might not respond favourably to such human attention, it appears that, for animals that will be subject to human handling during their research career, very little time in association with people is required to assuage their fears. It would also seem that this time is money well spent for the comfort of the animals, the health and safety of the people who will handle them, and the quality of the research in which they are used.

Dogs and monkeys are mentioned here as examples, but many other species can similarly benefit by, or suffer from lack of, influences in early development including conspecific and human interaction of an appropriate quality and frequency. Some non-human primates, such as rhesus monkeys and chimpanzees, are extremely vulnerable to the quality of conspecific social interaction during early development and, lacking that, might never interact in a normal social manner as adults (Fritz and Fritz, 1979; Fritz and Howell, 1993; King and Mellon, 1994). Other species also are affected by seemingly benign human interaction, such as handling and weighing the animals. Chickens demonstrate greater feed efficiency and resistance to antigens (Gross and Siegel, 1982) and less fear (Jones, 1992) with selective handling.

These brief examples will suffice to make the point of the importance of environmental events on an animal's well-being. In order to understand the behaviour and state of well-being of an animal one must consider possible causal effects of antecedent conditions. The species-appropriate timing of these interactions varies greatly but the important point is that our influence on the welfare of the animal begins no later than yesterday! If you wait for tomorrow to begin, yesterday will never come for many animals. Though it may sound a little silly, this statement underscores the importance of considering each animal as an individual. In the examples above, it might be only one individual in the colony that manifests stress. In all likelihood, that animal's previous experience differed from the others in some significant way. There might be little we can do for that animal except accommodate her as much as possible. One should learn from this animal, however, and institute programmes to avoid similar situations for future animals.

Managing stress

The range of stress-related behaviours is large and all are not deserving of the same attention. One should remember Hans Selye's wise comment, 'The absence of stress is death'. Our goal is not to eliminate all stress, but to provide the animals with a suitable opportunity to adapt to new situations in order to avoid unnecessary or excessive stress. Pulling a juvenile dog on a leash as a method of leash training will have little success. Letting the dog explore while on the leash, however, will vastly improve the chances of success. That exploration seems to be an essential part of the dog's process of adaptation to the leash. Restraint of animals, such as the use of non-human primate chairs, is another situation in which adaptation is important. Introducing the animal to the chair in multiple sessions prior to a study, with a food reward and gentle handling, is commonly recommended. I have worked with many rhesus monkeys that would, of their own accord, run on pole and collar from their home cage to the restraint chair. They seemed rewarded by the opportunity to leave the home cage, even after a food reward was no longer offered.

During periods of adaptation, there is little question that the animals are under some degree of stress, just as you or I would be as we prepare for an examination. For many people – and presumably for some animals also – that stress is the essential stimulus for learning and adapting to a new situation. Without it, life would seem to be terribly boring. Just because learning can be stressful we most certainly would not want to eliminate that stress. However, some stress, such as that manifested by a fear-biter dog or a self-biting monkey, is of a different magnitude and does require our attention.

What is the difference between an animal that develops normal, species-typical behaviours and one which has destructive or maladaptive behaviours? We know that the process of adaptation to new environments – people, places or things – is important. However, the opportunity for the animal to adapt to a particular situation might have been days, weeks or even years ago. Perhaps out of sight, but clearly not out of mind! Animals that are maladapted to their environment are simply not coping with the situation because, in all likelihood, they were never given the chance to learn; or, as discussed above, abnormal behaviours were taught (reinforced) by virtue of the stress of an inadequate environment. Perhaps remedial adaptation is warranted, but it is also often the case that time and experience weigh too heavily on the animal and those behaviours will simply persist for life.

This discussion raises some interesting issues and the need for some definitions. Mild stress such as that accompanying learning and severe stress such as that accompanying self-biting deserve different consideration. While we want to use the former to motivate the animal to become adapted to a situation, we want to alleviate the latter. For the sake of distinction, we will consider that animals that are unable to adapt completely to stressors and show maladaptive behaviours are experiencing *distress*. It differs from stress by the manifestation of maladaptive behaviours or other pathological processes, such as gastric ulcers, immunosuppression, abnormal feeding and changes in social interaction with conspecifics or humans (NRC, 1992). Maladaptive responses that reduce the animal's distress can be reinforced and thus become a permanent part of the animal's behavioural repertoire. Coprophagy, hair-pulling and self-biting are common examples.

Focusing this discussion on the issues addressed by those responsible for animal care, I pose the following questions: what exactly constitutes distress and what intervention is required? Should all maladaptive behaviours be eliminated? Does the expression of stereotypic behaviour signal a state of poor well-being?

To answer these questions, we must again consider the individual animal and take a brief glimpse at the concept of time budgets. It is entirely possible that under some situations behaviour exhibited by an individual animal will be refractive to treatment, or that treatment options are unacceptable. Placing an animal in a small cage to inhibit pacing behaviour would not be a suitable option, but enriching that cage with climbing structures or deep bedding might be. An animal that practises excessive coprophagy could be placed on a barred floor where the faeces are out of reach, or the floor could be covered with deep bedding in which popcorn or other foraging objects are scattered. The former might alter the behaviour but the latter might increase the behaviour. One would seem to provide greater benefit to the animal than the other does, although the perception by humans of unwanted behaviour would continue. The selection of a proper environment should be based on risk–benefit analysis. In the above example, although many colony managers deem coprophagy undesirable, its risk to the animal is minimal while the benefit gained from foraging can be substantial (Maki *et al.*, 1989; Fritz *et al.*, 1992a; Fritz and Howell, 1993). Such may be the case for other behaviours commonly labelled as 'abnormal' like rocking and hair-pulling in non-human primates (Fritz *et al.*, 1992b).

Sometimes maladaptive behaviours just cannot be eliminated in an animal. What then? After providing the animal with reasonable accommodations, the next step is to apply preventive practices in order to avoid the antecedent conditions in future generations of animals. What led to these behaviours? Early rearing? Social deprivation? Restrictive housing practices? Analysis of all possible causes should lead to a limited set of most probable causes, around which remedial practices can be implemented in *other* animals. This strategy basically 'gives up' on the affected animal and places emphasis on avoidance of similar behaviours in future animals. One might ask, 'Is this humane?' Given the state of behavioural and medical science today, it would seem that it applies the concept of achieving the greatest good for the greatest number of individuals.

Time budgets were mentioned above. What do they have to do with stress? They relate to the manner in which one addresses alleviation of maladaptive or stereotypic behaviours. Why does an animal pace, or rock or become a self-biter? One argument would suggest that regardless of why the behaviour developed in the first place, it fills a void in terms of what the animal might have been accustomed to doing but can no longer do. Perhaps in a former life (or due to genetic predisposition) the animal spent major portions of its time foraging, or interacting socially with others, burrowing in the ground, or locomoting through a complex environment. With these things now lacking, new behaviours might emerge to fill the animal's time. Taking it one step further, management practices that seek to prevent one behaviour might trigger another less desirable one.

This raises an important question. How important are these stereotypic behaviours with regard to the welfare of the animal? Does stereotypy mean poor welfare? In a perfect world we would like all animals to exhibit a wide range of species-normal behaviours, and many would even like these behaviours to be socially acceptable to human observers too. However, as argued above, we cannot expect to 'cure' every stereotypy, nor should we seek success at all costs in this manner. Whether to fill a void left by a time budget out of balance or other causes, some stereotypic behaviours develop and seem refractive to intervention. Attempts to block them might lead to even less desirable behaviours. So, if they are adaptive, are they bad? These animals may be saying to us, 'in this environment, I'm doing pretty well. Don't change things too much or too quickly for I am adjusted to where I am'. If this animal's environment is impoverished, it should be corrected. However, we know that many of these behaviours are very pervasive and resist all attempts at environmental and social changes. Therefore, we come back to the notion that some behaviours deemed maladaptive and undesirable from a human's perspective should be left alone and energies placed on understanding and preventing recurrence in future individuals. This suggestion is also compatible with the report, *Psychological Well-being of Nonhuman Primates*, recently released by the National Research Council (NRC, 1998).

Ameliorating or avoiding stereotypic behaviours generally involves providing a suitable living environment. The living environment of research animals should be ecologically suited to the species in which a range of normal behaviours is possible. Unenriched cages in which normal postures, activities (including social interactions) and locomotion are not permitted are not conducive to well-being. A working definition for assessing well-being is that the animal should express 'a broad range' of species-typical activities, but not all activities expressed in nature are either necessary or appropriate in captivity. For example, breeding and parenting is often used to assess animal well-being but often breeding is not possible or desired; also, it is seldom possible to stimulate the lengthy locomotion used in searching for food that an animal must do in its natural environment. The assessment of well-being from such a limited set of behaviours is analogous to basing the adequacy of a cage environment on floor space alone (NRC, 1996, p. 25) and seldom satisfies the goal of ensuring the well-being of individual animals. Housing space should be provided that stimulates and supports the desired range of species activities. Well-being is based on how effectively individual animals utilize this space.

The bottom line

Some stress is good, some bad, some you seek to avoid, some you leave alone, some is essential for normal living, some is destructive. What is one to do? The bottom line for those entrusted with the care and overseeing of laboratory animals is to achieve a state of well-being in *individual* animals, through assessment of both physical and psychological health. With this as a goal, a focus *only* on reducing stress or distress is short-sighted, for well-being is more than the absence of stress. Consensus is emerging as to what constitutes well-being of individual animals and can be summarized as follows (modified from NRC, 1998):

- Sound physical health.
- Ability to cope effectively with day-to-day changes in its social and physical environment.
- Ability to engage in a broad range of species-typical activities.
- Absence of maladaptive or pathological behaviour.
- Presence of a balanced temperament.

When used together, these criteria provide a pretty good picture of the state of an animal's welfare.

Recognizing physically and emotionally healthy animals is one thing, but increasing the odds that animals will turn out this way is another issue. Again, from the NRC (1998) report, the five key strategies to pursue in order to optimize the likelihood of achieving desirable goals are to provide:

- appropriate social companionship;
- opportunities to engage in behaviour related to activities appropriate to the species, age, sex, and condition of the animals;
- housing that permits suitable postural, exploratory and locomotor expression;
- positive interaction with personnel; and
- freedom from unnecessary pain or distress.

These recommendations encompass the concepts of social interaction and environmental enrichment. Most species of animals used in research are social and should be given the opportunity to interact with compatible conspecifics. In fact, for some species such as dogs, non-human primates and most rodents, social companionship is often considered *the* most important element in reducing stress and achieving well-being. That is easy to say, but achieving compatible social groups is often difficult and requires great care and ongoing observations.

The Animal Welfare Regulations and the *Guide* provide minimum cage sizes for most laboratory animals. Many argue that focusing on the size of the cage, an 'engineering' criterion, is short-sighted and ignores the animal in the cage. A better criterion from the animal's perspective is the use of a 'performance' standard, the goal of which is to achieve housing that permits suitable postural, exploratory and locomotor expression, as noted above. Within such a framework, decisions as to what 'furniture' the animal needs to support its behavioural needs can be addressed. For non-human primates, this might include elevated perches and structures to permit locomotion above the ground, ability to forage, and opportunities to either engage in social interaction or have seclusion. These 'environmental enrichments' are essential features to consider when establishing goals for achieving well-being.

This done, there is one last step to provide the responsible *personnel* with relief from the distress that accompanies the fear that they are not doing all that might be done for the welfare of the animals. That step is to assess the success of the overall programme, and it involves only two decisions:

- Are individual animals in, or considered to be in, a state of well-being?
- Has the cause of unrelieved distress in any animal been shown to be derived from antecedent conditions? If so, is this documented in colony records and have practices been identified and implemented to benefit future generations of animals?

Summary

Stress in laboratory animals can be caused by many factors, but the most common cause is an inadequate physical and social environment. Selection of an environment that will support the normal behaviours of the animals requires knowledgeable personnel with a desire to implement changes and assess the value of those changes *to the animals*. Animal care and use programmes, such as veterinary care, husbandry and behavioural management, should be well documented at each research institution and understood by the personnel who apply them. These programmes, however, should retain flexibility to permit new innovations and professional judgement. Measures of their success should be based on the effect of their outcome *on the animals* (performance standards), not on the degree to which the specific protocols have been followed (engineering standards).

Experimental procedures, disease or injury can also cause stress. One of the primary functions of animal care and use review committees should be to seek to reduce the anticipated stress or distress, consistent with the goals of the protocol. Second only to a good preventive medicine and husbandry programme in which illness and injury is vigorously controlled, is a well-trained professional and technical staff. Such a staff is the first line of defence in identifying unanticipated adverse effects of a protocol, the early stages of disease, or injury. Upon recognition, a solid veterinary care programme and collaborative relationship between all those who have animal care responsibilities (investigators, veterinarians and technicians) can go a long way towards minimizing these causes of stress.

References

Animal Welfare Regulations Title 9, Animals and Animal Products, Subchapter A (Animal Welfare), Parts 1–4 (9CFR 1–4). See also public law 99-198, The Food Security Act, for the amendment to the Animal Welfare Act that established the requirement for 'a physical environment adequate to promote the phsychological well-being of nonhuman primates'.

Bayne, K., Dexter, S. and Strange, G. (1993). The effects of food treat provisioning and human interaction on the behavioral well-being of rhesus monkeys (*Macaca mulatta*). *Contemporary Topics* 32, 6–9.

FASS (Federation of Animal Science Societies) (1999) *Guide for the Care and Use of Agricultural Animals in Agricultural Research and Teaching.* FASS, Savoy, Illinois.

Fritz, P. and Fritz, J. (1979) Resocialization of chimpanzees: ten years of experience at the PFA. *Journal of Medical Primatology* 8, 202–221.

Fritz, J. and Howell, S.M. (1993) Psychological wellness for captive chimpanzees: an evaluative program. *Humane Innovations and Alternatives* 7, 426–434.

Fritz, J., Maki, S., Nash, L.T., Martin, T. and Matevia, M. (1992a) The relationship between forage material and levels of coprophagy in captive chimpanzees (*Pan troglodytes*). *Zoo Biology* 11, 313–318.

Fritz, J., Nash, L.T., Alford, P.L. and Bowen, J.A. (1992b) Abnormal behaviors with a special focus on rocking and reproductive competence in a large sample of captive chimpanzees (*Pan troglodytes*). *American Journal of Primatology* 27, 161–176.

Gross, W.B. and Siegel, P.B. (1982) Socialization as a factor in resistance to infection, feed efficiency, and response to antigens in chickens. *American Journal of Veterinary Research* 43, 2010–2012.

IRAC (Interagency Research Animal Committee) (1985) *U.S. Government Principles for Utilization and Care of Vertebrate Animals used in Testing, Research, and Training.* Federal Register, 20 May, 1985. Office of Science and Technology Policy, Washington, DC.

Jones, R.B. (1992) The nature of handling immediately prior to test affects tonic immobility fear reactions in laying hens and broilers. *Applied Animal Behaviour Science* 34, 247–254.

King, N.E. and Mellon, J.D. (1994) The effects of early experience on adult copulatory behavior in zoo-born chimpanzees (*Pan troglodytes*). *Zoo Biology* 13, 51–59.

Maki, S., Alford, P.L., Bloomsmith, M.A. and Franklin, J. (1989) Food puzzle device stimulating termite fishing for captive chimpanzees (*Pan troglodytes*). *American Journal of Primatology* 1 (Suppl.), 71–78.

NRC (National Research Council) (1992) *Recognition and Alleviation of Pain and Distress in Laboratory Animals.* Committee on Pain and Distress in Laboratory Animals. Institute of Laboratory Animal Resources, National Research Council, National Academy of Sciences. National Academy Press, Washington, DC.

NRC (National Research Council) (1996) *Guide for the Care and Use of Laboratory Animals.* Committee to revise the guide for the care and use of laboratory animals. Institute of Laboratory Animal Resources, National Research Council, National Academy of Sciences. National Academy Press, Washington, DC.

NRC (National Research Council) (1998) *The Psychological Well-Being of Nonhuman Primates.* Committee on well-being of nonhuman primates. Institute for Laboratory Animal Research, National Research Council, National Academy of Sciences. National Academy Press, Washington, DC.

PHS (US Public Health Service) (1996) *Public Health Service Policy on Humane Care and Use of Laboratory Animals.* US Department of Health and Human Services, Washington, DC. [PL 99–158, Health Research Extension Act, 1985]

Russell, W.M.S. and Burch, R.L. (1959) *The Principles of Humane Experimentation.* Methuen & Co., London. [Reprinted as a Special Edition in 1992 by the Universities Federation for Animal Welfare, Potters Bar, UK.]

Index